S0-BLI-468

V.I. Minkin, B.Ya. Simkin, R.M. Minyaev

Quantum Chemistry of Organic Compounds

Mechanisms of Reactions

With 66 Figures and 35 Tables

Springer-Verlag
Berlin Heidelberg New York London
Paris Tokyo Hong Kong Barcelona

Professor Vladimir I. Minkin
Professor Boris Ya. Simkin
Dr. Ruslan M. Minyaev

Institute of Physical and Organic Chemistry
Rostov University
344711 Rostov on Don/USSR

ISBN 3-540-52530-0 Springer-Verlag Berlin Heidelberg New York
ISBN 0-387-52530-0 Springer-Verlag New York Berlin Heidelberg

Library of Congress Cataloging-in-Publication Data

Minkin, V.I. (Vladimir Isaakovich). Quantum chemistry of organic compounds:
mechanisms of reactions /
V.I. Minkin, B.Y. Simkin, R.M. Minyaev. p. cm.
Includes bibliographical references and index.
ISBN 3-540-52530-0 (Berlin). —ISBN 0-387-52530-0 (New York)
1. Quantum chemistry. 2. Chemistry, Organic. 3. Chemical reaction,
Conditions and laws of. I. Simkin, B. IA. (Boris IAkovlevich)
II. Miniaev, R.M. (Ruslan Mikhailovich) III. Title.
QD462.M56 1990 547.1'28—dc20 90-10109 CIP

Printed in Germany

Typesetting: Thomson Press (India) Ltd, New Delhi
Offsetprinting: Colordruck Dorfi GmbH, Berlin; Bookbinding: Lüderitz & Bauer., Berlin
2151/3020-543210–Printed on acid-free paper

"Erst die Theorie entscheidet darüber,
was man beobachten kann"
Albert Einstein

Foreword

Chemistry is the science of substances (today we would say molecules) and their transformations. Central to this science is the complexity of shape and function of its typical representatives. There lies, no longer dependent on its vitalistic antecedents, the rich realm of molecular possibility called organic chemistry.

In this century we have learned how to determine the three-dimensional structure of molecules. Now chemistry as whole, and organic chemistry in particular, is poised to move to the exploration of its dynamic dimension, the busy business of transformations or reactions. Oh, it has been done all along, for what else is synthesis? What I mean is that the theoretical framework accompanying organic chemistry, long and fruitfully laboring on a quantum chemical understanding of structure, is now making the first tentative motions toward building an *organic* theory of reactivity.

The Minkin, Simkin, Minyaev book takes us in that direction. It incorporates the lessons of frontier orbital theory *and* of Hartree–Fock SCF calculations; what chemical physicists have learned about trajectory calculations of selected reactions, and a simplified treatment of all-important solvent effects. It is written by professional, accomplished organic chemists for other organic chemists; it is consistently even-toned in its presentation of contending approaches. And very much up to date. That this contemporary work should emerge from a regional university in a country in which science has been highly centralized and organic chemistry not very modern, invites reflection. It is testimony to the openness of the chemical literature, good people, and the irrepressible streaming of the human mind toward wisdom.

Cornell University Roald Hoffmann

Preface

The principal notions and conceptual systems of theoretical organic chemistry have been evolved from generalizations and rationalizations of the results of research into reaction mechanisms. In the sixties the data from quantum mechanical calculations began to be widely invoked to account for and predict the reactivity of organic compounds. In addition to and in place of the notions derived on the basis of the resonance and mesomerism theories that earlier had been treated semiquantitatively by means of correlation equations, novel research tools came to be employed such as reactivity indices, perturbation MO theory, or the Woodward–Hoffmann rules. It is very characteristic of these approaches, which have now taken firm root in the field of theoretical chemistry, that they, on the whole, imply an a priori assumption of the mechanism and probable structures of the transition states of reactions.

The current stage of theoretical research into reaction mechanisms associated with direct calculations of potential energy surfaces (PES) and reaction pathways, with precise identification of transition state structures, has developed in the wake of amazing progress in computing technology and the development of the methods of quantum chemistry. Analysis of the PES and reaction pathways furnishes unique information as to the detailed mechanism of chemical transformation which can, in principle, be obtained by none but the calculational methods. Such an analysis forms the basis for the so-called computer experiment providing for modelling of the reactions and structural situations which are hard or even impossible to realize empirically. Apart from being a correct explicative method, the quantum mechanical calculations of PES's are becoming a means by which mechanisms of chemical reactions may be predicted. This makes them all the more important for the broad community of organic and physical chemists engaged in studying reactivity and reaction mechanisms.

It is primarily these readers to whom the present book is addressed. Its first part (Chaps. 1–5) discusses the main characteristics of PES as well as the methods useful in its calculation and analysis. Particular attention is given to the question of adequacy of the calculational method to the character of a given problem determined by the specificity of structure and the type of transformation. Since most organic reactions are carried out in solution, up-to-date schemes are examined for taking the solvation effects into account. Any exact calculation starts from the selection of the most likely configurations of the reacting system. The theory of orbital interactions has proved very effectual for ascertaining these and determining the reaction paths. The notions of isolobal analogy rooted in this theory permit a unified conceptual approach to the reactions of both oganic and organometallic compounds.

The second part of the book (Chaps. 6–12) is devoted to the analysis of calculation data on the mechanisms of principal organic reactions. Chapters 7, 8, and 10 review the results of theoretical research into mechanisms of the hetero- and homolytic substitution and addition reactions in the gas phase as well as in solution. Owing to these results obtained over the last 5–10 years, some questions of stereo- and regiospecificity of the above reactions have been clarified and numerous structures calculated of transition states, ion pairs, unstable intermediate biradicals, and π-complexes as to which only vague conjectures had been possible before. In Chap. 9, it is shown how this information may be utilized for target-oriented structural modelling of the dynamic systems in which fast intramolecular rearrangements take place. Chapter 11 moves on to examine the reactions of electron and proton transfer as well as the role of steric and energy factors and the tunnelling mechanism in the realization of these important transformations. A brief analysis of pericyclic reactions concludes the book. The PES calculations have enabled a much more detailed insight to be gained into the workings of their mechanism.

The authors hope the present publication may invite a still greater attention to very interesting potentialities inherent in the quantum chemical analysis of reaction mechanisms and encourage more researchers to engage in further development of this promising field.

Rostov, June 1990

Vladimir I. Minkin
Boris Y. Simkin
Ruslan M. Minyaev

Table of Contents

Chapter 2
Quantum Chemical Methods for Calculating Potential Energy Surfaces . 61

Chapter 3
Effects of the Medium 88

Chapter 4
Orbital Interactions and the Pathway of a Chemical Reaction 106

Chapter 5
Substitution Reaction. 116

CHAPTER 1

Potential Energy Surfaces of Chemical Reactions

1.1 Introduction. Mechanism of Chemical Reaction and Quantum Chemistry

The mechanism of a chemical reaction is described by all elementary steps covering the transformation of starting reactants into products. Any full information on such a mechanism must include the sequence of these steps (stoichiometric mechanism) as well as the data on their nature. This concerns the data on the structure and energetics of all species involved in the reaction and the solvation shell for every point of the route which leads from the starting compounds, for a given elementary step, via transition states to its final products (intrinsic mechanism). These data may be rationalized to arrive at certain views as to stereochemistry, composition, structure, and relative energy of a transition states of the reaction.

One of the most important branches of theoretical organic chemistry deals specifically with the determination of these parameters. It should be noted that they cannot, with rare exceptions, be determined by experimental methods. Indeed, studying of reaction kinetics and isotopic effects, analysis of various correlational relationships of the steric structure of reaction products etc. give data which allow only indirect conclusions as to the overall reaction pathway since they all are invariably based on the studies of only the initial and the final state of every elementary step of the reaction. This situation may remind one of the "black box": direct access to the information therein is impossible, it can be deduced only through a comparison between the input and the output data.

Quantum chemistry has opened up basically new possibilities of studying chemical reactions. Indeed, the theoretical calculations enable the researcher to "peep" into the aforementioned "black-box", i.e., to calculate all critical parameters of the intrinsic mechanism of a reaction. Of course, the knowledge of the reaction mechanism is not the ultimate goal of the calculations, rather it is needed for revealing the general causes influencing the kinetics of a reaction and for establishing comprehensive concepts and theoretical models governing the reaction with an aim of finding effectual means that would make it possible to control the course of the chemical process. And it is precisely the quantum chemistry which plays a pioneering role in the development of such general approaches to understanding chemical reactions.

In order to describe in a most general manner the structural changes going on in a reacting system, in other words, the intrinsic mechanism of the reaction, it is necessary to solve the time-dependent Schrödinger equation. However, even in regard to the systems consisting of several atoms only, the task of obtaining even approximate solutions to this equation is extraordinarily complicated. Moreover, most chemical reactions exhibit at room temperature very insignificant quantum effects. Therefore, another approach, sufficiently rigorous, is employed to describe the dynamics of a chemical reaction, namely, the calculation of the potential energy surfaces (PES) and of the trajectories of interacting particles moving over these surfaces. In this approach, referred to as semiclassical, the problem is divided into two parts:

1) the quantum mechanical calculation of the potential energy for a system of interacting particles (statics), and
2) the solution of the classical equations of motion using the potential obtained for a description of the dynamics of the system.

1.2 Choice of a Coordinate System and the Representation of a PES

The PES of a system is given by the function describing its total energy (minus kinetical energy of the nuclei) of using the coordinates q of all nuclei the system contains. In case the system consists of N atomic nuclei, the number of the independent coordinates (degrees of freedom) that fully determine the PES is $(3N-6)$—for a linear system of N nuclei this number will be $(3N-5)$:

$$E(q) = E(q_1, q_2, q_3, \ldots, q_{3N-6}) \tag{1.1}$$

The form of Eq. (1.1) implies the possibility of separating the wave functions of the electrons and the nuclei. Such a differentiation, which corresponds to the Born–Oppenheimer approximation is feasible in virtue of a considerable difference between the masses of these particles. The electrons being lighter than the nuclei move much faster than these and instantaneously adapt themselves to the nuclear configuration which changes relatively slowly during a reaction, a conformational transition, or simply molecular virbrations. The Born–Oppenheimer approximation is satisfied fairly well for the vast majority of chemical reactions in the ground electron state of the molecules. When, however, the PES of an excited state approaches quite near the PES of the ground state or, indeed, intersects it in some parts of the configurational space, this approximation is no longer valid (see Sect. 1.6).

The concept of the potential energy surface forms the basis for all modern theoretical models describing the properties of individual molecules or their combinations that depend on the geometrical parameters of a system. There are among such properties kinetical, thermodynamic, spectral, and other characteristics. In order to represent a PES by means of Eq. (1.1), one has to calculate, using one of the current methods (see Chap. 2), the energy of the system for regular sets of the coordinates q, to construct the corresponding graphical patterns for the function $E(q)$ and to tabulate the data or, by making use of these, to try to find approximate analytical expressions of $E(q)$.

As coordinates, any set of parameters may be taken which do not depend on the absolute position and orientation of the nuclei in space. The internal coordinates, such as the bond lengths or other inter-nuclear distances, the valence and torsion angles, are the most convenient, but, in principle, the choice of coordinates is dictated by the specificity of the problem in hand. It is expedient that among the $(3N-6)$ independent coordinates there should be such structural parameters which are subject to the most drastic changes in the course of the transformation under study. Particularly successful systems of coordinates include the so-called reaction coordinate, i.e., the direction along which the wave function of the system changes especially fast during transition from the reactants to the products.

In the simplest case of a two-atom system ($N=2$, $3N-5=1$), the only coordinate describing the PES is the interatomic distance r. In this case, the dependence of the potential energy of the molecule [Eq. (1.1)] on r can be approximated quite well by the Morse curve:

$$E(r) = D_e\{1 - \exp[-a(r - r_e)]\}^2 \tag{1.2}$$

where D_e is the bond energy which consists of the dissociation energy D_o and the zero vibration energy (see Fig. 1.1).

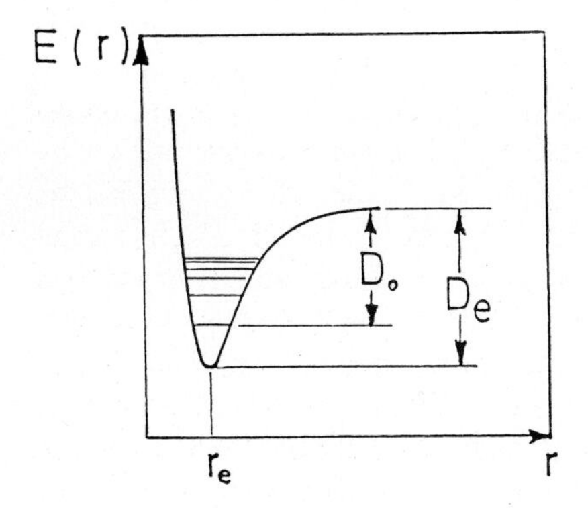

Fig. 1.1. Potential energy curve for a two-atom molecule (r_e is the equilibrium bond length, $D_o - D_e$ is the zero vibration energy)

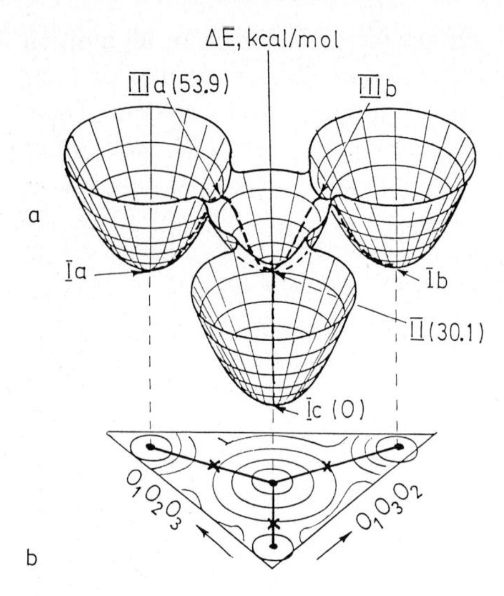

Fig. 1.2a, b. Potential energy surface (**a**) and its two-dimensional map (**b**) of an ozone molecule in the region of the stable isomers I (a–c) and II constructed from the data of ab initio calculations [4] depending on the angles $O_1O_2O_3$ and $O_1O_3O_2$. The data in parentheses near the numbers of structures are the relative energies of these structures (kcal/mol). The solid thick line on the two-dimensional PES map (**b**) shows the minimum-energy path of the topomerization reaction of the ozone molecule Ia ⇄ II ⇄ Ib (Ic). Dashed line on the PES (**a**) shows the total reaction path

For a three-atom ($N = 3$) nonlinear system of atomic centers, the function $E(q_1, q_2, q_3)$ is already a hypersurface in a four-dimensional space. Its graphic representation is not possible, so its visual perception is confined to the inspection of the sections at fixed values of one of the coordinates or is dependent on the assumption of a definite relationship between certain two coordinates. Thus, the PES of the ozone molecule O_3 featured by Fig. 1.2 has been constructed as a function of two internal coordinates, viz., the angles $O_1O_2O_3$ and $O_1O_3O_2$.

It is assumed, in virtue of the molecular symmetry, that the bond lengths O_1O_2 and O_2O_3 in Ia vary synchronously in the electrocyclic reaction. Assumptions based on similar grounds are made fairly frequently when calculating the PES of various reactions.

Ia ⇄ II ⇄ Ib, Ic   III

In such cases, a certain caution is called for regarding the conclusions as to the intrinsic mechanism since artificial restrictions of the degrees of freedom of a reacting system may sometimes involve serious deviations from the results of a more rigorous calculation (see Sect. 2.2).

The three deep minima on the PES of O_3 correspond to three degenerate isomers (topomers) Ia–Ic, while the central less deep minimum corresponds to the symmetrical intermediate II (Fig. 1.2a). A handy form of the representation of a three-dimensional PES is the two dimensional contour map containing curves obtained by projecting the function $E(q_1, q_2)$ onto the plane q_1, q_2 (Fig. 1.2b). As may be seen from Fig. 1.2a, the transition from funnels I to the less deep funnel of the metastable intermediate II requires the least energy when it takes place along the bottom of the valleys whose top points are the saddle points corresponding to the structures III, which represent the transition states of the reaction stages I$\rightleftarrows$II. The line going along the valley bottom of the three-dimensional (multi-dimensional in the general case) PES is called the total reaction path [5]—in Fig. 1.2a this path is indicated by a dashed line. A projection of the total reaction path onto the configurational space of nuclear variables, in this case onto the plane q_1q_2, is referred to as the minimal energy reaction path*, which is represented in the configurational space by a certain continuous curve $g(q_1, q_2) = 0$ the length of whose is usually regarded as the reaction coordinate. Figure 1.3 shows the energy profile of the interconversion reaction of two ozone topomers constructed as a function of the reaction coordinate.

This representation of the relationship between the energy and the structural changes occuring in the course of a reaction is the most frequently used. It clearly shows direct connection between the PES characteristics and the main theoretical concepts of chemical kinetics.

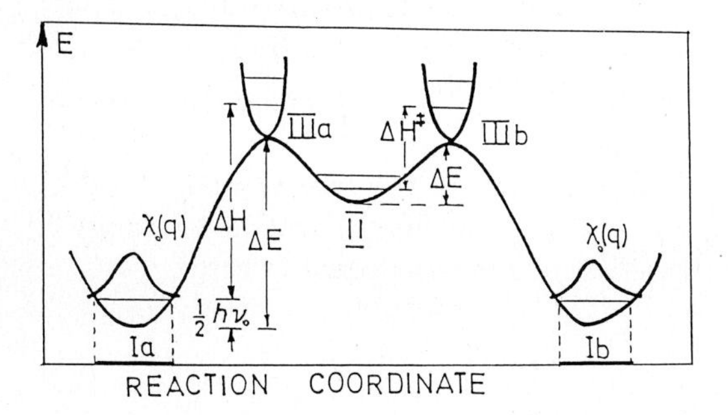

Fig. 1.3. Energy profile of the interconversion reaction of two ozone topomers Ia and Ib via symmetrical intermediate II. Maxima on the curve correspond to the transition state structures IIIa and IIIb (see Fig. 1.2). Nuclear wave functions $\chi_0(q)$ are given for zero vibration levels at the minima of Ia and Ib

---

*Below a stricter analysis of the notion of a reaction path will be given (Sect. 1.3.4)

1.3 Topography of the PES and Properties of a Reacting System

The PES function contains fairly full information on the mechanism of chemical interactions which can arise in a given system under different initial conditions (initial energy, relative orientation of reactants etc.). However, such functions can be obtained for a sufficiently wide spatial region only in the case of very simple atomic systems (2–4 centers). The reason for this lies primarily in the multidimensionality of the PES's. For a simplest reaction of the nucleophilic substitution:

$$CH_3F + Cl^- \rightleftarrows CH_3Cl + F^-$$

the PES is a hypersurface in the 13-dimensional space, i.e., the energy of the system depends on 12 degrees of freedom ($N = 6$). If a PES calculation were confined to only 10 fixed values of every variable, then the total number of separate geometrical configurations for which the energy should be calculated would amount to 10^{12}. Suppose we had a computer performing each calculation in one second, then the construction of this PES would require 10^{12} seconds, i.e., 10^5 years—the comment is superfluous.

However, in order to analyze the mechanism and the kinetics of a chemical reaction occurring under certain conditions, it is not necessary to know the full function of the PES. It would suffice to have information on certain portions of the PES, primarily those corresponding to the minima regions of Eq. (1.1) and to the saddle points. The search for these regions is directed at finding the so-called stationary or critical points of the PES.

1.3.1 Critical Points

The critical points of any given function $f(q)$ are those points of the configurational space in which the values of all first derivatives of the function relative to any independent variable q_i are equal to zero. In any critical point of a PES described by the coordinates $(q_1, q_2, \ldots, q_{3N-6})$ the first derivatives of energy E with respect to all coordinates (the gradients) become equal to zero.

$$\overrightarrow{\text{grad}}\, E \equiv \left(\frac{\partial E}{\partial q_1}, \frac{\partial E}{\partial q_2}, \frac{\partial E}{\partial q_3}, \ldots, \frac{\partial E}{\partial q_{3N-6}}\right) = (0, 0, 0, \ldots, 0) \equiv 0 \qquad (1.3)$$

In order to fully characterize the nature of such extremum points and thereby to establish the role they play on the PES of the system one needs to know the curvature of the PES in these points. To this end, second derivatives of the energy are calculated with respect to all coordinates in the extremum region. In its general form the set of second derivatives of the energy constitutes the matrix H referred to as the Hessian:

$$H = \begin{pmatrix} \frac{\partial^2 E}{\partial q_1^2} & \frac{\partial^2 E}{\partial q_1 \partial q_2} & \cdots & \frac{\partial^2 E}{\partial q_1 \partial q_{3N-6}} \\ \frac{\partial^2 E}{\partial q_2 \partial q_1} & \frac{\partial^2 E}{\partial q_2^2} & \cdots & \frac{\partial^2 E}{\partial q_2 \partial q_{3N-6}} \\ \cdots & \cdots & \cdots & \cdots \\ \cdots & \cdots & \cdots & \cdots \\ \frac{\partial^2 E}{\partial q_{3N-6} \partial q_1} & \frac{\partial^2 E}{\partial q_{3N-6} \partial q_2} & \cdots & \frac{\partial^2 E}{\partial q_{3N-6}^2} \end{pmatrix} \tag{1.4}$$

The critical point of a PES, in which the Hessian Eq. (1.4) has at least one zero eigenvalue, is called a *degenerate* critical point. *The degree of degeneracy* of a critical point is determined by the number of the zero eigenvalues of the matrix H. It is generally assumed that the degenerate critical points of a PES are of no great significance for the analysis of reaction mechanisms, we shall therefore consider only the *nondegenerate* critical points, i.e., such points in which the Hessian Eq. (1.4) has no zero eigenvalues.

The point of the PES in which the matrix of Eq. (1.4) is positively defined, i.e., all its eigenvalues are positive, is a minimum. Since the eigenvalues of the Hessian correspond to the force constants of normal vibrations of the system, all force constants are positive and any shift from this region of the PES increases the energy of the system[1]. Thus, the minima regions of the PES correspond to stable structures or intermediates.

In case the matrix H of Eq. (1.4) has in a given critical point only one negative eigenvalue, then this point is the first-order saddle point ($\lambda = 1$) (the order λ of critical points is determined by the number of negative eigenvalues of matrix H) and can characterize structurally the transition state of the reaction (see Sect. 1.3.3). The only negative value of the force constant corresponds to the imaginary frequency of the "normal" vibration of the system. Its vector, referred to as the *transition vector* [6], determines the direction and the symmetry of the reaction path in the highest energy point of the passage across the transition state region. Higher-order saddle points of the PES ($\lambda \geqslant 2$) with two or more negative force constants of normal vibrations are not as important for the theoretical analysis of reaction mechanisms and kinetics.

In the regions of the maxima on the PES, the matrix of Eq. (1.4) is negatively defined with all force constants having negative values. The system avoids getting into these regions which have no special physical importance such as

[1]Strictly speaking, the question is here of the matrix H of diagonal type obtained from Eq. (1.4) via orthogonal transformation of the coordinates to mass-weighted normal coordinates

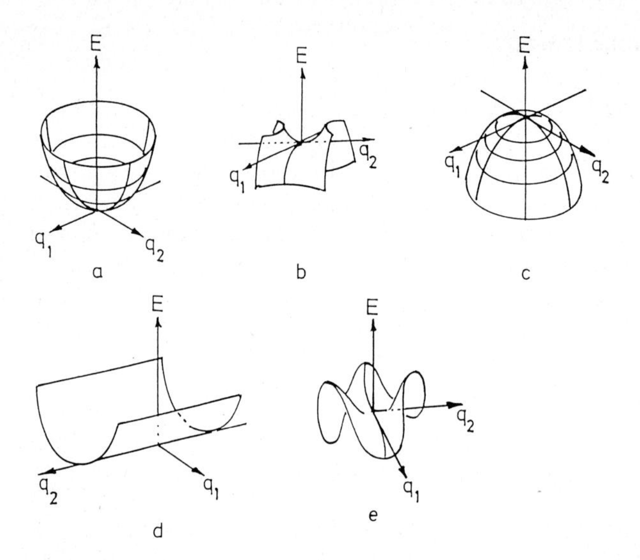

Fig. 1.4a–e. Assignment of nondegenerate (**a–c**) and degenerate (**d, e**) critical points of the three-dimensional PES $E\ (q_1, q_2)$—the coordinates q_1, q_2 are selected such as to correspond to the directions of the main curvature axes, **a** shows the point of a minimum with $\partial^2 E/\partial q_1^2 > 0$; $\partial^2 E/\partial q_2^2 > 0$; (the stability region); **b** denotes the saddle point with $\partial^2 E/\partial q_2^1 < 0$; $\partial^2 E/\partial q_2^2 > 0$; (the transition state structure); **c** the point of a maximum with $\partial^2 E/\partial q_2^1 < 0$; $\partial^2 E/\partial q_2^2 < 0$; **d** the degenerate point of a minimum with $\partial^2 E/\partial q_2^2 = 0$; $\partial^2 E/\partial q_1^2 > 0$; **e** the degenerate second-order saddle point (monkey saddle) with $\partial^2 E/\partial q_1^2 = \partial^2 E/\partial q_2^2 = 0$

is inherent in the regions of the minima and the saddle points. Figure 1.4 illustrates the above-given notions for the simplified case of a PES described by only two coordinates q_1 and q_2.

1.3.2 The Regions of the Minima on the PES

Only those geometrical configurations of the atoms which belong to the regions of the minima on the PES correspond to stable molecular forms, i.e., such species which are, in principle, amenable to experiment. It is assumed that the depth of the minimum on the PES will be such that it would contain at least one vibration level. Thus, a theoretical analysis of chemical transformations must start from ascertaining whether the structures of compounds involved in the reaction belong to the minima of the PES. For this purpose, it would be necessary to verify the validity of Eq. (1.3) and that the matrix H of Eq. (1.4) is positively defined in the point of configurational space related to a given structure. There exist quite effective methods for optimizing the molecular geometry based on the calculation of energy gradients. The overall procedure involves calculation of the total force field (complete set of the force constants) of a molecule and of the frequencies of molecular vibrations associated with these constants.

1.3.2.1 *Vibrational Spectrum of Molecules*

As was noted earlier, for a geometrical configuration that would correspond to a minimum on the PES to exist, the presence in this minimum of at least one vibration level is required. Therefore, after the verification of necessary conditions, the vibrational spectrum of the molecule can be found from the solution of the classical equations of motion with respect to the curvilinear internal coordinates q [7]. However, in practical calculations it is more convenient to make use of the cartesian coordinates $X_i Y_i Z_i$ for each atom $i (i = 1, 2, \ldots, N)$ in the molecule. In this case, the equations of motion have a simple form [8, 9]:

$$m_i \ddot{r}_i = -\sum_{j=1}^{3N} f_{ij} r_j \quad i = 1, 2, \ldots, 3N \tag{1.5}$$

where $r_i = (X_i - X_i^o)$ are the small shifts of the atom i from the equilibrium position X_i^o (point of a minimum on the PES) along the coordinate X_i; m_i is the mass of the atom i; $\ddot{r}_i$ is the second derivative of r_i with respect to time, and $f_{ij} = (\partial^2 E/\partial X_i \partial X_j)$ are the second derivatives of energy with respect to the cartesian coordinates X_i, X_j.

Quite effective analytical and numerical techniques have been developed in the framework of the Hartree–Fock method (SCF MO) for calculating the derivatives $f_{i,j}$ [8, 9]. Equation (1.5) is solved by standard methods [8]. The modes of normal vibrations have the form:

$$r_i = a_i \exp(2\pi v i t) \tag{1.6}$$

where the vibration amplitude a_i can be found from the equations:

$$\sum_{j=1}^{3N} f'_{ij} a_j = 4\pi^2 v^2 a_i \quad (i = 1, 2, \ldots, 3N) \tag{1.7}$$

in which f'_{ij} are the mass-weighted force constants related to f_{ij} by the following relationship:

$$f'_{ij} = f_{ij} m_i^{-1/2} m_j^{-1/2} \tag{1.8}$$

The eigenvalues of the matrix f'_{ij} equal $4\pi^2 v^2$, where v is the frequency of harmonic vibrations, and the eigenvectors of the matrix f'_{ij} indicate amplitudes of normal vibrations, which allows the type of vibration to be identified.

Six values of the calculated frequencies equal zero (are close to zero in real calculations), they correspond to six out of $3N$ variables which characterize translational and rotational motions of the molecule as a whole.

The importance of a careful verification by means of the described scheme for calculating the vibration spectrum where the structure under theoretical investigation actually belongs to a minimum point on the PES has been convincingly demonstrated by a recently conducted refinement of the structure of tetralithiotetrahedrane.

In a number of nonempirical calculations, starting from Ref. [10], this molecule appeared to prefer (65 kcal/mol, STO-3G basis set) non-classical structure IV to the classical form in which the lithium atoms are linked with carbon by normal two-center two-electron bonds. However, a recent calculation on IV (STO-3G, 4–31G basis sets) supplemented with a calculation of its vibration spectrum has shown that this structure does not correspond to the true minimum on the PES of C_4Li_4. The vibration spectrum of IV contains two imaginary vibration frequencies. The shifts determined by the eigenvectors related to these frequencies bring IV into the minimum on the PES associated with the structure IVa. The results of the calculations in Ref. [11] are quite important in view of considerable interest in experimental synthesis and reactivity of tetralithiotetrahedrane:

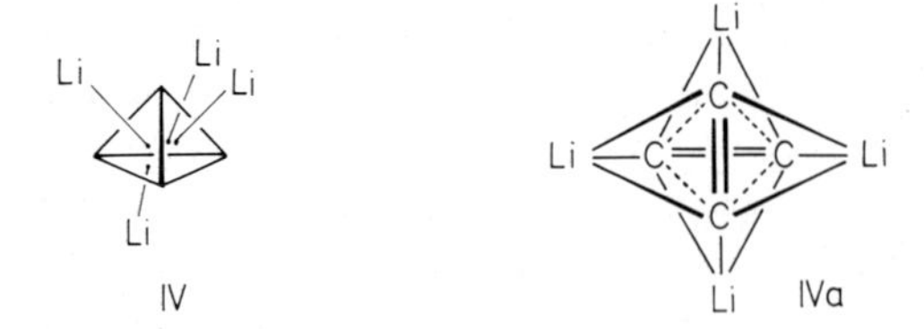

Computer time expenditure for calculating a force field employing the basis sets of orbitals usual in nonempirical calculations (Chap. 2) is comparable with that spent on obtaining the ground state wave function. The accuracy of computing harmonic vibration frequencies depends strongly on the method used for quantum chemical calculations. For example, for nonempirical methods using a basis set of moderate size, the calculation of this quantity is subject to an error (usually overestimation) of 10–15%. In semiempirical methods of the MINDO/3 [12] and MNDO [13] type, the accuracy is somewhat higher. It should be noted that part of the error in the calculation of vibration frequencies is due to harmonic approximation and cannot be blamed on any shortcomings of the quantum mechanical methods; when anharmonicity is taken into account, the accuracy is improved. The data of Table 1.1 may give an idea of the

Table 1.1. Ab initio (DZ-type basis set) [9], semiempirical (MINDO/3) [12] and experimental (gas phase) frequencies (cm^{-1}) of normal vibrations of hydrogen cyanide

Symmetry and type of vibration	3–21G	6–31G*	MINDO/3	Experiment
$\Sigma^+ CH(\nu_{CH})$	3691	3679	3536	3311
$\Pi CH(\delta)$	990	889	769	712
$\Sigma^+ C \equiv N(\nu_{CN})$	2395	2438	2268	2097

possibilities inherent in a theoretical calculation of the vibration spectrum of organic molecules.

Knowledge of the frequencies of normal vibrations permits a stricter evaluation of the relative energies of various structures which arise in the course of a chemical reaction, namely, it allows the energy contributions of their zero vibrations $\Sigma(1/2h\nu_i)$ (see Fig. 1.3) to be taken into account, which should then be included in the total energy of every such structure.

1.3.2.2 Calculation of Thermodynamic Functions of Molecules

Thermodynamic and activation parameters—such as the constants of equilibria and rates, the free energies of equilibria and activation—associated with a certain temperature are employed when conducting experimental studies of equilibria, thermochemistry, and kinetics of reactions. The energy quantities calculated by the methods of quantum chemistry are related, even when the energies of zero vibrations are allowed for, to the temperature of 0 K. A transition from the differences between the internal energies to thermodynamic quantities, which provides for direct comparison of calculational and experimental data, can be made when structural characteristics of reaction participants and their vibration spectra are known. These parameters, in their turn, can be calculated if the form of the PES function of Eq. (1.1) is known. Making use of the data obtained, one may be means of the well-known relationships of statistical mechanics and thermodynamics—see Ref. [14] and Refs. [15, 16]—derive partition functions and calculate absolute entropy and heat capacity of all individual molecules, ions, and intermediate structures taking part in a given reaction.

In a rigid molecule approximation (internal rotation and inversion barriers appreciably exceed kT), one may single out contributions from separate degrees of freedom of the translational, rotational, and vibrational motions to the entropy S and the heat capacity, with anharmonicity of vibrations and some other effects neglected:

$$S = S_{\text{trans}} + S_{\text{rot}} + S_{\text{vibr}} \tag{1.9}$$

$$C_P = C_{P(\text{trans})} + C_{P(\text{rot})} + C_{P(\text{vibr})} \tag{1.10}$$

The contributions from the translational degrees of freedom may be calculated without the data of quantum chemical computations as they depend only on the external conditions (T, P) and the molecular mass m:

$$S_{\text{trans}} = \tfrac{5}{2}R\ln T + \tfrac{3}{2}R\ln m + R\ln[(2\pi)^{3/2}k^{5/2}h^3] + \tfrac{5}{2}R - R\ln P \tag{1.11}$$

$$C_{P(\text{trans})} = \tfrac{5}{2}R \tag{1.12}$$

The rotational contributions to the entropy and the heat capacity are:

$$S_{\text{rot}} = R\left[\tfrac{3}{2}\ln T + \tfrac{3}{2} + \ln\frac{2(k^3 I_A I_B I_C)^{1/2}(2\pi)^{7/2}}{\sigma h^3}\right] \quad (1.13)$$

$$C_{P(\text{rot})} = \tfrac{3}{2}R \quad (1.14)$$

where σ is the symmetry number; I_A, I_B, I_C are the principal moments of inertia calculated from the data on internuclear distances in a PES minimum.

The contributions to the entropy from the vibrational degrees of freedom are given in harmonic approximation by:

$$S_{\text{vibr}} = R\sum_i g_i \ln[1 - \exp(-hc\nu_i/kT)] + \frac{Rhc}{kT}\sum_i \frac{g_i \nu_i \exp(-hc\nu_i/kT)}{1 - \exp(-hc\nu_i/kT)} \quad (1.15)$$

$$C_{P(\text{vibr})} = R\left(\frac{hc}{kT}\right)^2 \sum_i \frac{g_i \nu_i^2 \exp(-hc\nu_i/kT)}{[1 - \exp(-hc\nu_i/kT)]^2} \quad (1.16)$$

where ν_i and g_i are, respectively, the frequency and the degree of degeneracy of the i-th vibration.

The greatest contributions to the vibrational components of Eqs. (1.15) and (1.16) are made by the terms defined by low frequency deformation vibrations. Table 1.2 lists some calculation data on the entropy and the heat capacity in comparison with the experimental results.

Having calculated energy contributions from the zero vibrations and individual entropy terms of the reactants A and the products B, one may calculate the free energy of reaction at an arbitrary temperature T:

$$\Delta G^T = H_A^T - H_B^T + \frac{1}{2}\sum_{\nu \in A} h\nu_i - \frac{1}{2}\sum_{j \in B} h\nu_j - T(S_{\text{vibr}}^A - S_{\text{vibr}}^B + S_{\text{rot}}^A - S_{\text{rot}}^B + S_{\text{trans}}^A - S_{\text{trans}}^B) \quad (1.17)$$

Table 1.2. Values of entropy and heat capacity calculated by the MINDO/3 [16] method and obtained experimentally at 298 K

	S, kcal/mol		C_p, kcal/(mol K)	
Compound	calc.	obs.	calc.	obs.
CH_3Cl	56.1	55.9	10.4	9.7
$CH_2{=}CH_2$	52.6	52.4	11.0	10.2
$CH_2{=}C{=}O$	58.3	57.8	12.9	12.4
Thiophene	68.3	67.9	19.3	17.4
Benzene	65.2	64.3	20.9	19.6

The dependence of the heat of reaction on temperature is defined by Kirchoff's law:

$$\Delta H^T = \Delta H^{T_0} + \int_0^T (\Delta C_P) dT \tag{1.18}$$

Table 1.3 lists free energies for six industry-important reactions at 300–1500 K calculated by the MINDO/3 method and found experimentally. The calculation of ΔG is subject to an error of a mere 0.5–1 kcal/mol, with the exception of the reaction of formation of hydrogen cyanide due to unsatisfactory description of triple bonds by the MINDO/3 method (see Chap. 2).

A more detailed analysis of the determination of macroscopic properties of a substance from the data of theoretical quantum chemical calculations may be found in the review of Ref. [15] devoted to this problem. The difference between the energies of two minima (generally, of two critical points on the PES) calculated theoretically and that between the free energies [Eq. (1.17)] found experimentally may disagree by as much as 100 kcal/mol, which underlines the importance of correct comparison of the theoretical results with the experimental data. Within this disagreement, any conclusion as to the energetics of a reaction, which disregards statistics, may prove erroneous.

1.3.2.3 Topological Definition of Molecular Structure

Geometrical configurations of a set of atomic nuclei related to the points of minima on a PES represent molecular structures. Theoretical chemistry describes the mechanisms of all chemical transformations by means of the molecular structures whose chief characteristic are the chemical bonds linking various atoms within a molecule. It is the changes in the molecular structure which determine the essence of the chemical reaction. Therefore in order to correctly describe the structure, it is important to define rigorously this notion on the basis of the first principles of quantum mechanics.

A point in the configurational space in the minimum of a PES is, within the framework of the Born–Oppenheimer approximation, a quantum-mechanically well-defined characteristic. However, a geometrical definition of molecular structure directly related to this point can hardly be fully satisfactory. The fact is that in the energy ground state a molecular system is located not on the PES itself but on the hypersurface of the zero vibration level (see Fig. 1.3, the one-dimensional case) assuming with the probability $|\chi_0(q)|^2$ (where $\chi_0(q)$ is the nuclear wave function in the corresponding minimum of the zero vibration state) any geometrical configuration within this hypersurface (see Fig. 1.2b, hatched areas). These changes (deformations) in the geometrical configuration regarding the molecules in the ground state, which correspond to narrow and deep minima on the PES, are, as a rule, much smaller than the corresponding interatomic distances in the molecule. Such deformations may be quite considerable

Table 1.3. Free energy changes for several industry-important reactions obtained from a MINDO/3 calculation and experimental data, according to Ref. [16]*

Reaction	ΔG kcal/mol					
	300 K		900 K		1500 K	
	MINDO/3	exper	MINDO/3	exper	MINDO/3	exper
$N_2 + 3H_2 \rightarrow 2NH_3$	−7.8	−7.7	24.1	24.2	57.5	58.4
$2CH_4 \rightarrow CH{\equiv}CH + 3H_2$	68.0	68.0	32.0	31.6	−7.0	−7.4
$CH_2{=}CH_2 + 0.5O_2 \rightarrow H_2\overset{O}{C{-}C}H_2$	−19.7	19.4	−8.1	−7.7	5.0	3.9
$CH_4 + 2H_2O \rightarrow 4H_2 + CO_2$	27.0	27.1	−1.8	−2.0	−33.7	−34.3
$CH_4 + NH_3 + 3/2O_2 \rightarrow HCN + 3H_2O$	−119.0	−118.0	−133.8	−129.4	−143.4	−139.8
$CH_4 + Cl_2 \rightarrow CH_3Cl + HCl$	−25.6	−25.6	−27.5	−27.5	−29.9	−29.5

*When parametrizing the MINDO/3 method, the contributions from zero vibration energies are taken into account, therefore, when using Eq. (1.17), these contributions must be left out for stereochemically nonrigid molecules and ions and also at the start of reaction when the initial structure is still intact. Thus, the geometrical definition of molecular structure does not correspond to the content of this concept commonly adopted in chemistry. This discord can easily be explained: the definition of molecular structure, as formulated in terms of the classical structural theory, rests on topological rather than geometrical principles, in other words, on the determination of bond sequences and stereochemical types of coordination centers, but not on a quantitative description of all structural parameters. Such a definition cannot be translated unambiguously into the language of quantum mechanics for lack of appropriate operators that would correspond to both the chemical bonding and the molecular structure as a whole. As has been pointed out in this connection [17, 18], a quantum mechanical definition of structure would require introduction of additional postulates into the completely formulated quantum theory, which, evidently, may not be regarded as its further development.

A real prospect for a correct definition of molecular structure on the basis of quantum mechanics emerged thanks to a topological approach to the analysis of the electron density $\rho(q)$ distribution in a system characterized by this density. This approach developed by Bader [19], based on the topological properties of the charge distribution in molecular systems, provides an invariant pattern of the distribution of chemical bonds between the atoms in a molecule and determines unambiguously the type of a molecular structure for both the classical (Butlerov-type) and the nonclassical (see Ref. [20]) molecular systems.

The topological characteristics of the distribution of electron density $\rho(\mathbf{r})$ in a molecular system are determined by the properties of the gradient field $\nabla\rho(\mathbf{r}, q)$ where $\mathbf{r}$ is the coordinate of the point in the real space where ρ is calculated within the given nuclear configuration q. This field is represented by the trajectories $\nabla\rho(\mathbf{r})$ as shown in Fig. 1.5.

The gradient trajectories $\nabla\rho$ cover the whole space, they can begin and end only at the critical points or at infinity. Critical points of a gradient field are the points in which

$$\nabla\rho = 0 \tag{1.19}$$

They are classified by their rank n and signature m and are designated as (n, m). *The rank* of a critical point is the number of nonzero eigenvalues for the matrix of second derivatives of ρ with respect to cartesian coordinates $(\partial^2\rho/\partial x_i \partial x_j)$ $(x_i = x, y, z)$ and *the signature* is the difference between the positive and the negative eigenvalues of this matrix. *A maximum* can be described as the point $(3, -3)$ (all three eigenvalues of the matrix $\|(\partial^2\rho/\partial x_i \partial x_j)\|$ are negative). The maxima coincide with the positions of nuclei in the molecule (see Fig. 1.5). A critical point, which is the local maximum in two directions and the local minimum in the third, is designated as $(3, -1)$, its signature is $1 - 2 = -1$. This point is referred to as *the bond critical point* or simply the bond point.

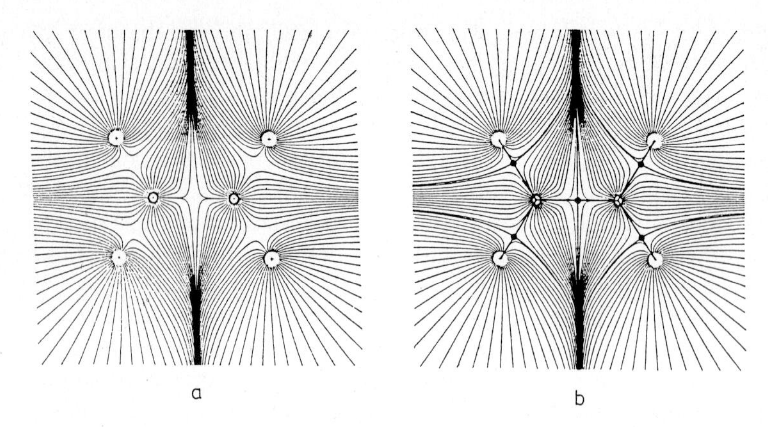

Fig. 1.5a, b. Two-dimensional representation of the gradient paths $\nabla\rho(r)$ for the ethylene in its equilibrium geometrical configuration as calculated by the ab initio (STO-3G) method [19]. The black dots denote the bond critical points. *Trajectories* linking the neighboring nuclei through the separating critical point are the bond paths (**b**). *Solid lines* (**b**) traced through the critical points perpendicular to the bond paths separate atomic basins. (Reproduced with permission from the American Chemical Society)

Furthermore, there exist a ring point (3, + 1) and a cage point (3, + 3) which describes, respectively, the presence of a ring or a three-dimensional cage in the molecule. The ring point has two ascending and one descending direction, while the cage point represents the true minimum of ρ.

All gradient pathways are divided into two types. The first type comprises the paths passing through the local maxima of the function $\rho(\mathbf{r}, q)$, which correspond to the positions of atomic nuclei. A region of the space bounded by the gradient paths coming from infinity to the given molecule constitutes the basin of the corresponding atom in the molecule.

Each two neighboring atomic basins are separated by a nodal surface on whose intersection line with the molecular plane lies the bond point (3, − 1). Also a combination of several neighboring atomic basins produces isolated basins, e.g., those of the CH_2, CH_3 groups, whose properties depend on the electron density distribution in the common basin.

The gradient pathways of the second type determine the bond distribution in a molecular system. This type includes two gradient paths, which link the bond point located between each two neighboring atomic basins with the nuclei corresponding to these basins. Addition of these two paths produces a line in space along which the electron density retains its maximal value. This line is referred to as the bond path. A complete network of the bond paths determines the molecular graph for a given nuclear configuration.

Figure 1.6 shows molecular graphs for various molecules in their equilibrium geometrical configurations. As a rule, such graphs resemble ordinary structural formulas. However, in some cases they indicate certain important features.

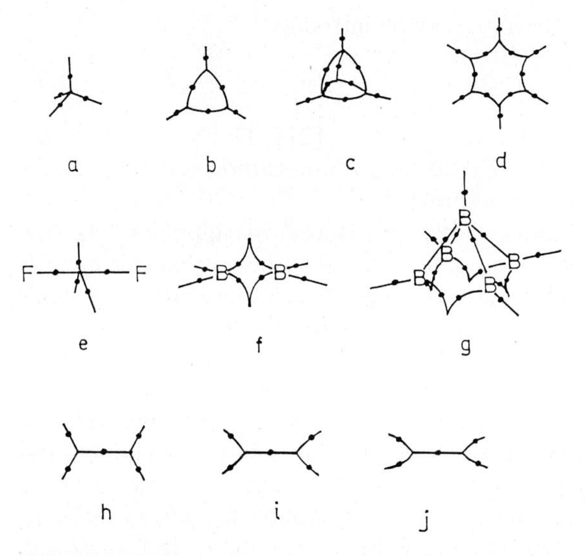

Fig. 1.6a–j. Molecular graphs defined by topological properties of the electron density distribution, according to ab initio (STO-3G) calculations of the gradient field. **a** methane; **b** cyclopropenium cation; **c** tetrahedrane; **d** benzene; **e** $CH_3F_2^-$; **f** diborane B_2H_6; **g** pentaborane B_5H_9; **h** ethylene in equilibrium geometry configuration, HCH = 115.6°; **i** ethylene; HCH = 85.6°, the energy relative to the ground state structure is 45 kcal/mol; **j** ethylene, HCH = 58.8°, the energy relative to the ground state structure is 189 kcal/mol. (Adapted from Refs. [21–23])

Thus, in the case of the structures with small cycles (Fig. 1.6b and c) the bond paths are bent, which is an evidence in favor of the concept of bent (banana) bonds in these compounds. For pyramidal pentaborane B_5H_9 the molecular graph does not contain any bond paths between the basal boron atoms, which means that there is no concentration of the electron density in the basal plane (Fig. 1.6g).

Bader's calculations have shown that small and, in some cases, even large structural variations do not affect the character of a molecular graph. Indeed, as may be seen from Fig. 1.6h–j even very large deformations of the angle HCH in ethylene associated with huge energetic excitations do not lead to any alterations in the molecular graph, i.e., to a distortion of the initial structure. This makes it possible to single out on the PES of Eq. (1.1) a nuclear space region $Q(q)$, which may be associated with a permanent molecular structure. The molecular structure itself is in this case defined as an equivalent class of molecular graphs. On such a conceptual basis, a given molecular structure is described by a set of "configurations" with a given number of the bond paths connecting the same nuclei in each molecular graph.

The magnitude of the electron density ρ_b at the bond point characterizes the strength and the character of a bond. Bader has shown that in the case of the

CC bonds an empirical bond order n may be introduced

$$n = \exp\{6.458(\rho_b - 0.2520) \qquad (1.20)$$

where ρ_b is calculated with the 6–31G* basis set [21]. This yields the values of n equal to 1.00, 1.62, 2.05, and 2.92 for the carbon–carbon bonds in ethane, benzene, ethylene, and acethylene, respectively.

To characterize a bond type, Bader introduced the quantity $\nabla^2\rho$, the Laplacian of ρ [19]. If at the bond point $\nabla^2\rho_b < 0$, it means that this bond is covalent or polar. The case of $\nabla^2\rho_b > 0$ characterizes ionic and hydrogen bonds as well as the noncovalent van der Waals-type molecular interactions.

The ab initio calculations [22, 23] of the Laplacian at the bond points show that $\nabla^2\rho_b < 0$ for the C—C bonds in hydrocarbons; it grows but still remains negative for the N—N and O—O bonds in H_2NNH_2 and HOOH. But with the bonds F—F, S—N, and S—S in, respectively, the molecules F_2, S_4N_4 and S_8^{2+} [24] $\nabla^2\rho_b > 0$. Such behavior of $\nabla^2\rho_b$ indicates that the character of the local distribution of electron density at the bond itself and in the regions adjacent to it depends strongly on the type of bond. When $\nabla^2\rho_b < 0$, the electron density concentration is greater at the bond than in the neighboring regions, while with $\nabla^2\rho_b > 0$ the situation is reverse: the electron density at the bond is depleted as compared with the neighborhood. The Laplacian $\nabla^2\rho$ can be represented graphically in the form of two-dimensional maps for various spatial sections of a molecule. Such representation is convenient for determining the optimal positions for an attack upon the molecule by electrophilic or nucleophilic species [22, 23].

Currently, the analysis of the molecular structure based on topological examination of the electron density has gained wide acceptance [19, 21–26]. This approach has led to some unexpected conclusions as to the character of bonds in certain compounds. For example, a study of the C—Li bonds in organolithium compounds [27] has shown that in compound V, notwithstanding the large distance between the carbon atoms C_1 and C_4 equal to 3.121 Å, there exists a covalent chemical bond indicated by the molecular graph Va.

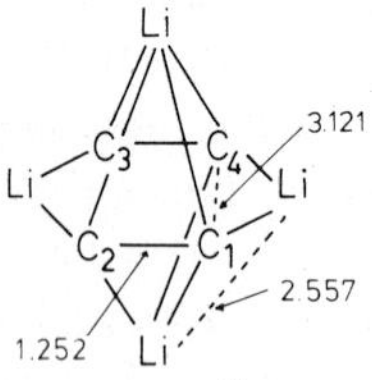

Li
C_3
C_4
3.121
Li
Li
C_2
C_1
2.557
1.252
Li
V

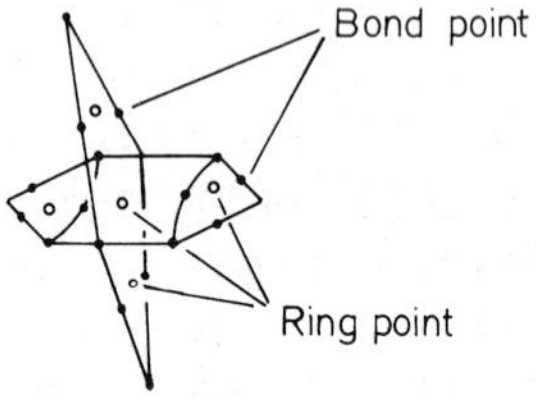

Bond point
Ring point
Va

1.3.2.4 *Structural Diagrams*

Based on the above-described concept of molecular structure, all configurational space of the nuclear coordinates $Q(q)$ may be divided into a finite number of

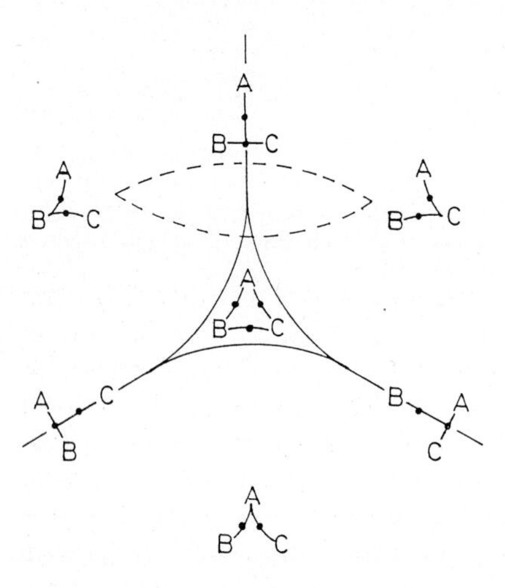

Fig. 1.7. Section of the structural diagram of a type ABC molecular system [25]. Coordinates of the plane are defined by two out of three independent parameters of the molecular system

structural regions, each of which corresponds to a certain molecular structure. At the boundaries between such regions a jumpwise alteration of the structural type takes place [25]. Such a procedure parallels the division, suggested by Mezey [5], of the PES and, accordingly, of the configurational space $Q(q)$ into non-intersecting domains. As Mezey has shown, this division may be done proceeding from the properties of the PES function $E(q)$ of Eq. (1.1) or its gradient $-\nabla E(q)$.

Within every domain, all geometrical configurations, however different they may be, belong to one and the same structure. This structure is unambiguously defined by the molecular graph, which fully retains its type within a given domain. At the domain boundary, the molecular structure changes abruptly. Thus, there is a one-to-one correspondence between Bader's structural regions [25] and Mezey's domains of configurational space [5].

The separation of the configurational space into structural regions may conveniently be represented in the form of a structural diagram [25, 26].

Figure 1.7 shows a structural diagram for the three-atomic (three-fragment) molecular system. This diagram corresponds to the PES of ozone topomerization (see Fig. 1.2). On the periphery, three structurally stable regions are located inside which geometrical alterations do not change the sequence of bond paths and, consequently, of the molecular graph of the system. Such a change does occur only when the boundary surfaces are reached which separate the structural stability regions. At these surfaces, a sudden change takes place of the type of the molecular graph. For example, upon the transformation A—B—C → B—C—A, the transition structure has a molecular graph different from the graphs of the initial and the final structure. In it, the bond path of the migrant atom A ends not at another atomic center but at the critical point of the saddle type, as is exemplified in Fig. 1.8 by the isomerization reaction HCN → CNH [25].

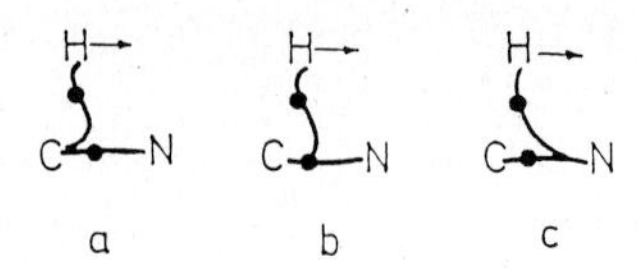

Fig. 1.8a–c. Molecular graphs in the neighborhood of the conflict structure for the HCN→CNH isomerization. Graphs **a** and **c** represent the stable structures, graph **b** shows the conflict structure. (Reproduced with permission from the American Chemical Society)

The structures of this type are unstable—any geometrical deformations, however small, which cause deviation from a boundary surface, destroy them.

One of the points of the configurational space $Q(q)$ lying on the boundary surface and possessing minimal energy in this region corresponds to the transition state of reaction (III in Fig. 1.2). Inter-section of three boundary surfaces singles out one more structural stability region in the center of the structural diagram. As may be seen from Fig. 1.2, in the case of the ozone molecule there really exists a corresponding cyclic isomer. In Fig. 1.7 dashed lines show two structurally distinct paths for interconversion reaction of three acyclic isomers. Either path is associated with passing over the region of unstable structures with one path involved in the formation of an intermediate cyclic structure. Which particular path is preferred by a real molecular system, is dictated by energy considerations depending on the PES of a given reaction. Whereas for the isomerization HCN→CNH a cyclic structure is unfavored and serves as a transition state, in isomerization reactions of cyclopropane (see Sect. 1.4) and in topomerization reactions of ozone (Fig. 1.2) the cyclic structure is included in all transformations. The 1, 2-sigmatropic shifts of the methyl group in the 1-propyl cation occur in a similar manner [26].

The topological approach has widened the concept of molecular structure (new definition of unstable structures), but especially important is that it has laid solid physical foundation under this central concept of chemistry by showing the possibility of rigorously defining the structure on the basis of the fundamental principles of quantum mechanics. This has put an end to uncertainty which hitherto prevailed in regard to this question.

The authors of Refs. [19, 21–26] hold the view that the electron-topological approach to the analysis of molecular structure should, in principle, not be restricted to the Born–Oppenheimer approximation, however, this question is still unresolved (see Ref. [18] and other authors cited there).

1.3.3 Saddle Points on the PES. Transition States

The determination of a transition state on the PES of a reacting system allows the kinetic parameters of a reaction to be calculated, using to this end standard relationships from the theory of absolute reaction rates. The geometry of the transition state structure pre-determines the stereochemical outcome of reaction, the curvature of the PES in the transition state region determines the magnitude of isotopic effects while the position of the transition state structure on the PES influences energy distribution in the products of chemical transformation.

The quantum chemical calculations are the only source of direct information on the structure and the energetics of transition states. Although these cannot in principle be observed experimentally since the Schrödinger equations has no stationary solutions in the points not corresponding to minima on the PES, one may, according to the Bersuker theorem [28], still envisage an indirect experimental approach to a transition state structure. The fact is that over the saddle point of a PES, which corresponds to a transition state, there always exists such an excited state energy surface which at this point exhibits a minimum. Hence one may attempt to study this transition state by appropriate spectral methods.

A points on the PES of a reacting system corresponds to a transition state if the following conditions are satisfied [29]; 1) it is a critical point, i.e., Eq. (1.3) is satisfied; 2) the force constant matrix has at this point a single negative value (the theorem of Murrell–Laidler [30]); 3) it possess the highest energy along the total reaction path represented by a continuous line in the configurational space linking the initial reactants and the products; 4) it has the minimal energy among all the points on the PES which satisfy the first three conditions.

1.3.3.1 Localization of the Transition States on the PES

A search for the transition state points on a PES is a much more complex problem than the determination of energy minima, moreover, direct experimental verification is in this case impossible. Several approaches to this problem have been worked out differing among one another in strictness and methodology.

The conventional approach to finding the transition state structures consists in the direct visualization of the saddle point on the PES in the form shown in Fig. 1.2. Evidently, this approach is confined to the analysis of those PES's which represent the functions of not more than two variables. Moreover, a point by point computation of a PES would require an enormous computer time consumption.

The reaction coordinate method [31, 32] often employed in practical calculations, represented an important improvement in this area of research. Its essence is the following: The coordinate q_i of the system, which undergoes the most drastic changes during reaction, is singled out. It is gradually varied while all or part of other independent variables are optimized at each step, i.e., the total energy of the system is minimized with respect to these coordinates. Then having constructed the E vs q_i function and determined its highest point, one finds the expected transition state of the reaction.

Unfortunately, in the general case such an approach has turned out to be incorrect and may lead to errors. As an example, let us consider the reaction of hydrogen addition to singlet methylene when the reactants draw together along the C_{2v}-symmetry route.

The genuine transition state of the reaction was found by analyzing the energy contour diagrams constructed from the ab initio calculation data [33] on the dependences of the energy on all three independent geometrical parameters R, θ, and r (Fig. 1.9). The transition state denoted in the Fig. 1.9 by a cross possesses a fairly high energy of 26.7 kcal/mol because the reaction, given the geometry of approach with C_{2v} symmetry kept intact, is forbidden by the orbital symmetry conservation rules.

Recent ab initio calculations [33], with the electron correlation energy taken into account by means of the MP2 perturbation theory and the stationary point identified by exact calculation of the Hessian matrix, have given two negative force constants for the structure considered earlier as a transition state. This indicates that the least-motion path on the PES is not an actual reaction path. Figure 1.9 shows that when the use is made of the reaction coordinate regime and the reactants are gradually brought closer to each other while reducing R, optimizing the angle θ, and maintaining the distance r at a magnitude characteristic of the transition state, we, moving along the PES, do not reach this state. At $R = 2.0$ Å the calculated reaction path loses the needed continuity and avoids passing the transition state point.

The reason for such behavior of the reaction coordinate lies in the fact that the reaction path direction cannot be rigorously determined in its starting or end points corresponding to either the reactants or the products (see Sect. 1.3.4).

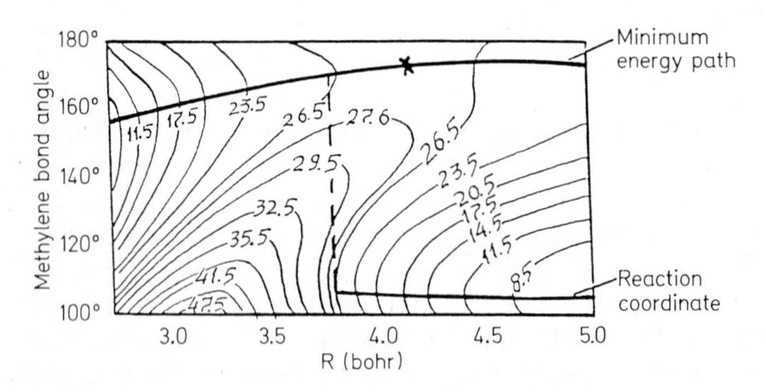

Fig. 1.9. Contour map of the PES of addition reaction of a hydrogen molecule to singlet methylene when reactants draw together along the C_{2v} symmetry path [with r(H—H) = 1.4 bohrs] constructed as a function of the distance R and the angle θ. The transition state structure is denoted by a cross. Numbers in the breaks of the curves are the energy levels in kcal/mol [33]. 1 bohr = 0.529 A. (Reproduced with permission from the American Chemical Society)

It depends in this case on the choice of both the reference point and the coordinate system [34]. Only in case the internal coordinate being varied is really sufficiently close to the genuine reaction coordinate, the reaction coordinate method yields correct results regarding the localization of the transition states.

Several effective methods for direct localization of the transition state points have been evolved which do not require a calculation of the whole PES [35–37]. The Newton–Raphson scheme [38] is a standard method for determining any critical points, however, it converges toward the saddle point only in a region sufficiently close to it. One of the best known methods is the McIver and Komornicki [39] minimization of the norm of the gradient S_g:

$$S_g = \sum_i \left(\frac{\partial E}{\partial q_i} \right)^2 = (\nabla E)^2 \tag{1.21}$$

The function $S_g = f(q)$ has minima at all critical points of the PES of Eq. (1.1) where $\nabla E = 0$, which makes it possible to use the quite convenient technique of nonlinear optimization. In order to correctly identify the nature of every critical point found when the value of S_g becomes zero, it is also necessary to calculate the curvature of the PES at these points, i.e., the matrix of Eq. (1.4) since Eq. (1.21) has redundant minima. Also this method converges to the needed stationary point only if the initial geometry is located in a region close to this point.

There is another group of methods useful in the search for the saddle points. The search starts from a minimum and proceeds uphill toward a saddle point [40–44]. An evaluation of the energy gradient as well as the Hessian of Eq. (1.4) is necessary in this case. Clearly, the computations of the first and the second derivatives with respect to all independent coordinates require a great deal of additonal computer time, but ultimately the information obtained proves to be quite useful (see Sect. 1.3.2.2).

In addition to the above-mentioned methods, there exist some other schemes for locating a transition state based on original ideas, such as the X-method [45], the Halgren–Lipscomb or synchronous transit method [46], the method of a hypersphere (or a circle in the two-dimensional case) [47] and others. These are considered in detail in the review articles [35–47].

1.3.3.2 Symmetry Selection Rules for Transition State Structures

In those cases when the expected transition states of reactions may retain certain symmetry elements of the initial reactants and the products, invoking of the group-theoretical arguments may help formulate the rules of selection of the transition state structures by their symmetry [6, 48]. These rules, which are considered here in a simplified form, allow certain structures to be discarded straight away without checking them by calculating the matrices of force

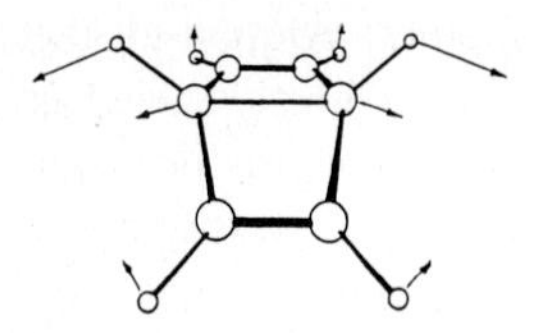

Fig. 1.10. Transition vector corresponding to the normal vibration mode of a_1 symmetry ($\nu = 1995\,cm^{-1}$) with negative force constant for the C_{2v}-transition state structure of the V–VI reaction. Arrows indicate direction of atomic shifts. Calculation [49] was done by the MNDO method with configuration interaction taken into account. (Reproduced with permission of the Royal Society of Chemistry)

constants. Equally important is that a consistent application of these rules reveals a number of regularities inherent in the intrinsic mechanism of reactants.

The key concept for the formulation of the rules in question is the concept of the transition vector (see Sect. 1.3.3.1). This vector may be regarded as a quite short, albeit finite, portion of the part of the reaction path whose beginning lies at the transition state point. A displacement of the system in the direction defined by the transition vector lowers its potential energy. Figure 1.10 shows as an example the form of the transition vector for the electrocyclic reaction of disrotatory rearrangement of the *cis*-Dewar benzene into benzene [49].

VI

The transition vector necessarily possesses all the symmetry elements of the nuclear configuration of a transition state, i.e., corresponds to one of the irreducible representations of the symmetry point group of this state. Figure 1.10 shows, for example, that the transition vector for the transformation of Dewar benzene into benzene corresponds to the fully symmetrical a_1-type vibration of the C_{2v} transition state structure, i.e., the transition vector is symmetrical with respect to the reflection in each of the two symmetry planes of the molecule and to the rotation about the C_2 axis.

Stanton and McIver [6] have formulated the following theorems intended to regulate the rules of selection of the energetically lowest transition states of reactions by the symmetry properties of the corresponding transition vectors:

1) The transition vector cannot belong to a degenerate representation of the point group of symmetry of the transition state.
2) The transition vector is antisymmetrical with respect to those symmetry operations of the transition state which transform the reactants into the products.
3) The transition vector is symmetrical with respect to those symmetry operations which do not transform the reactants into the products.

So as to use these properties of the transition vector for selecting various possible structures of transition states of a reaction, it is necessary to examine the effect of each individual operation of the transition state point group. If such operation does not bring about a transformation of the reactants into the products, it has the character +1, in the reverse case the operation has the character −1. After comparing the results obtained with the known tables of the characters, the structures of the transition states whose transition vectors do not meet the demands of Theorems 1–3 can be discarded. Let us consider briefly some important consequences that follow from these theorems.

Theorem 1 is simply a group-theoretical reformulation of the Murrell–Laidler theorem, it implies that a PES cannot have more than one negative curvature direction at the transition state point.

The most important corollary of Theorem 2 is this: no nuclear configuration can represent a transition state of a given reaction if the rotation about the third or a higher odd-order axes transforms the reactants into the products. Without attending to the proof of this statement, we illustrate it with some examples.

The calculation data on the PES of the ozone pseudorotation reaction (Fig. 1.2) show that the symmetrical structure II, which has a third-order symmetry axis, is not a transition state of the reaction but rather an energy-rich intermediate, i.e., interconversions of the topomers proceed nonconcertedly in two steps.

Another form of the PES is realized in a series of topomerizations of the *T*-shaped structures VIIa ⇄ VIIb ⇄ VIIc. In this case, the symmetrical structure IX with a third-order symmetry axis is not a transition state either, however, it represents not a local minimum, but a maximum on the PES.

The mechanism of such topomerization, first calculated on the anion SH_3^- (X = S, R = H), consists in successive unbending of each of the axial bonds with all three ligands moving in a plane common with that of the central atom [50]. The basic pattern of the PES shown by Fig. 1.11 remains unchanged for various compounds of type VII, for example, for an interesting class of the d^8 complexes (X = Ni, Pd; R = PR'_3) [51]. As may be seen from Fig. 1.11, the reaction path

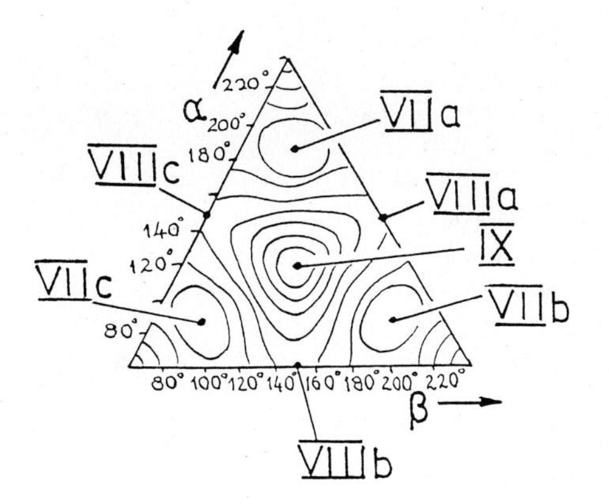

Fig. 1.11. General pattern of the PES for topomerization reactions of the type XR_3 structures [50, 51] VIIb ⇄ VIIb ⇄ VIIc

circumvents the structure IX, which occupies the top of a high hill in the center of the PES. Analogous is the PES of the valence isomerization reaction of the cyclopropenide anion which fluctuates between the structures X and XI [52] while the symmetrical structure XII is not a transition state. Likewise, in the more complex five-fold degenerate valence isomerization reaction of the cyclopentadienyl cation XIII in the singlet electronic state the symmetrical structure XV with a fifth-order symmetry axis is not a transition state [52, 53]. The reaction path of interconversion of the structures XIII (pseudo-rotation) runs along the PES valley circumambulating the hill of structure XV.

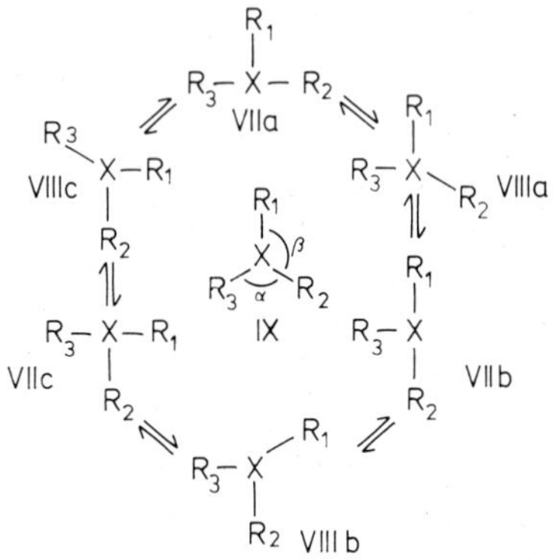

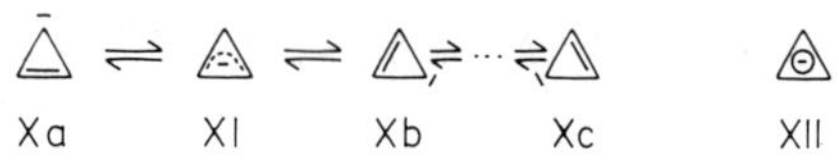

It should be emphasized that these examples are not intended to show that the structures of transition states can never have the symmetry axes of the third, fifth and the higher odd orders. Thus, in the reaction of nucleophilic substitution at the tetrahedral carbon atom (Sect. 5.1) a transition state structure possessing C_3-symmetry cannot be ruled out since the rotation about the C_3-axis Y—C—X does not affect transformation of the reactants into the products and so corollary of Theorem 2 is not violated.

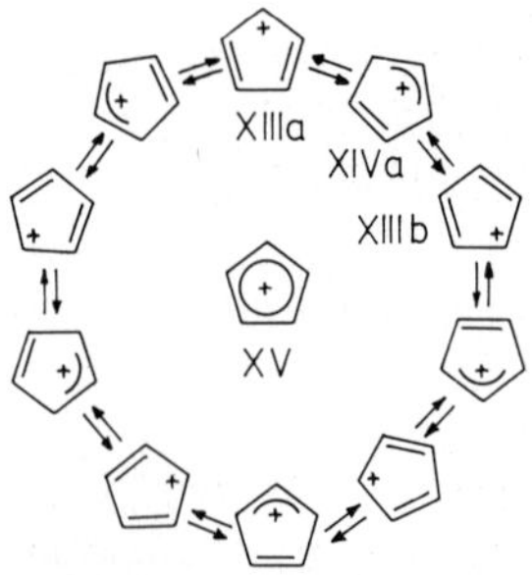

For possible transition state structures which belong to higher symmetry groups with the symmetry axes of the fourth and the higher even orders, the selection rules are not as rigid as in the case of the structures with the odd-order axes. For instance, the topomerization of the rectangular D_{2h} structures of cyclobutadiene XVI occurs according to a number of ab initio and semiempirical calculations, via a transition state of D_{4h} symmetry XVII which has only one negative force constant. This transition state corresponds to a b_{1g} distortion XVII → XVI [54, 55].

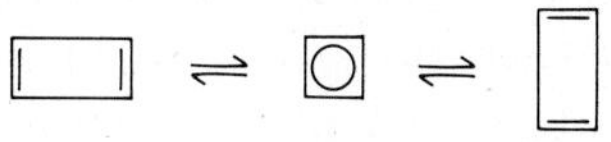

XVI a ⇌ XVII ⇌ XVI b

Also in the case of the reaction of hydrogen isotope redistribution the structure XVIII with D_{4h} symmetry of the nuclear skeleton cannot, unlike the D_{3h} and Td structures XIX, XX, be rejected as a possible candidate for a description of the transition state [56]. On the other hand, the calculated activation energy for the process with the transition state XVIII exceeds more than by a factor of three the experimental value of 44 kcal/mol and is higher than the hydrogen molecule dissociation energy of 108.8 kcal/mol. So the structure XVIII must be rejected on purely energetic grounds and the elucidation of the true structure of the transition state requires a detailed study of the PES by means of the techniques described in the previous Section. Such a calculation points to a more complex mechanism admitting of a trapezoid transition state XXI and an involvement of the stage $H_2 + 2D$ [57]:

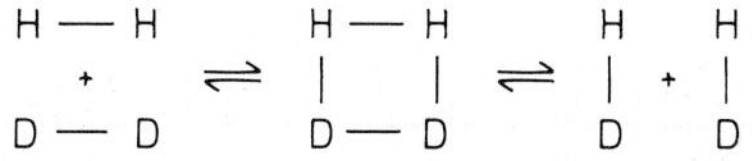

H—H + D—D ⇌ XVIII ⇌ H—D + H—D

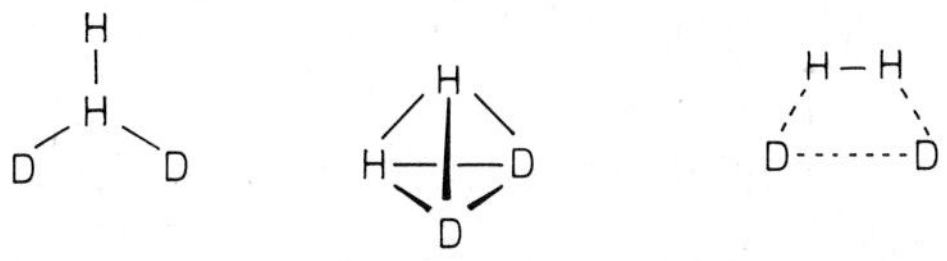

XIX XX XXI

A number of additional restrictions placed upon the transition state structures by symmetry demands are closely connected with the general features of the pathways of chemical reactions, they will be discussed below in Sect. 1.3.4.3.

1.3.3.3 Calculation of Activation Parameters of Reactions and of Kinetic Isotopic Effects

Once the structure of the transition state of a reaction is determined, its vibration spectrum and the energy ΔE in regard to the reactants and the products are known, one may proceed to calculating the activation parameters of the reaction directly comparable with the experimental values obtained as a result of kinetics studies.

From the diagram in Fig. 1.3 the relationship follows between E and the enthalpy of activation at the temperature 0 K:

$$\Delta H^{\neq} = \Delta E + \sum_{j}^{3N-7} \frac{1}{2} h\nu_j - \sum_{i}^{3N-6} \frac{1}{2} h\nu_i \tag{1.22}$$

where ν_j and ν_i are the normal vibration frequencies of the structures of the transition state and the starting reactants (or the products)—for the case of the transition state structures a contribution from imaginary frequency vibrations is ruled out. Allowance for the arbitrary temperature can be made using Eqs. (1.16), (1.18).

For calculating the entropies of activation the same relationships are used as in the case of reaction entropies (Sect. 1.3.2.2), while the free activation energy $\Delta G^{\neq}$ is calculated by means of Eq. (1.17).

A calculation of the vibration frequencies for the transition and the ground state structures of a reacting system provides for a correct solution of one more important problem of chemical kinetics, namely, a theoretical assessment of the change in reaction rates during isotopic substitution, i.e., the calculation of the kinetic isotopic effect. This effect arises owing to changes in the entropies of activation and in the zero vibration energy of the reactants as a consequence of the difference in masses of individual atoms in isotopomers which affects the values of the force constants and the vibration frequencies [see Eq. (1.8)].

Disregarding the tunnel effects (see Sect. 1.5) and staying within the approximation rigid rotator-harmonic oscillator, one may, for the biomolecular reaction ($A + B$), calculate the kinetic isotopic effect (ratio between the reaction rate constant K_1 of the compound with the light isotope and the rate constant K_2 of the compound containing the heavy isotope) from the Bigeleisen equation:

$$\frac{k_1}{k_2} \cdot \frac{S_2^A S_2^B S_2^C}{S_1^A S_1^B S_1^C} = \frac{\nu_1^{\neq}}{\nu_2^{\neq}} (VP)(EXC)(ZPE) \tag{1.23}$$

$$VP = \prod_{i}^{3N-6} {}^{A} \frac{u_{2i}}{u_{1i}} \prod_{i}^{3N-6} {}^{B} \frac{u_{2i}}{u_{1i}} \Bigg/ \prod_{i}^{3N-7} {}^{C} \frac{u_{2i}}{u_{1i}} \tag{1.24}$$

$$EXC = \prod_{i}^{3N-6} {}^{A} \frac{1 - e^{-u_{1i}}}{1 - e^{-u_{2i}}} \prod_{i}^{3N-6} {}^{B} \frac{1 - e^{-u_{1i}}}{1 - e^{-u_{2i}}} \Bigg/ \prod_{i}^{3N-7} {}^{C} \frac{1 - e^{-u_{1i}}}{1 - e^{-u_{2i}}} \tag{1.25}$$

$$ZPE = \frac{\exp\left[\sum_i^{3N-6} {}^A (u_{1i} - u_{2i})/2\right] \exp\left[\sum_i^{3N-6} {}^B (u_{1i} - u_{2i})/2\right]}{\exp\left[\sum_i^{3N-7} {}^C (u_{1i} - u_{2i})/2\right]} \quad (1.26)$$

The symbols A, B, and C refer to the reactants and the transition state, respectively, S denotes the symmetry numbers, $v^{\neq}$ is the imaginary vibration frequency of the transition state, $u_i = h_i/kT$. Eqs. (1.24) and (1.25) describe the change in the rate constants defined by the difference between the entropy terms, Eq. (1.26) characterizes the contribution from the difference between the zero vibration energies of molecules with the light and the heavy isotopes. As regards the monomolecular reactions, only the products and the exponents referring to reactants A remain in the numerators of Eqs. (1.24)–(1.26).

Table 1.4 lists calculation data on activation and thermodynamic parameters as well as kinetic isotopic effects for three reactions of the retroene type. In the first stage, a search for the transition state structure was conducted and its compatibility with the demands of the Murrell–Laidler theorem verified. Afterwards the vibration frequencies of the reactants and the transition state structure were calculated whose values were used in the corresponding equations. Underestimation of the kinetic isotopic effect in the last two reactions is related to underestimation of the role of the tunnel mechanism (see Sect. 1.5). An exact reproduction of the values of kinetic isotopic effects is a more reliable check on the accuracy of the calculated transition state structures than that of the values of activation entropies. This is explained by the fact that the calculated values of normal vibration frequencies, corresponding to the negative force constants, are directly included into Eqs. (1.24)–(1.26) that determine the magnitude of the kinetic isotopic effect.

Table 1.4. Calculated (MINDO/3) and experimental (in parentheses) values of thermodynamic activation parameters and the kinetic isotopic effect of retroene decompositions at 650 K [59]

Quantity	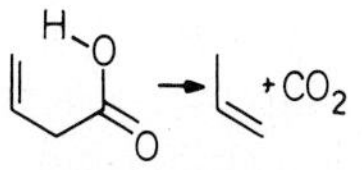 [structure: H–O ... O] → + CO_2	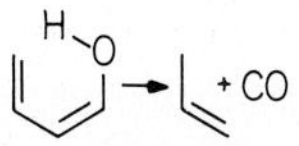 [structure: H–O] → + CO	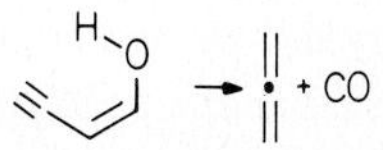 [structure: H–O] → + CO
$\Delta H^{\neq}$, kcal/mol	46.3 (39.3 ± 1.6)	58.6 (39.3 ± 1.3)	63.4 (38.2 ± 5.1)
$\Delta S^{\neq}$, cal/mol/K	−13.9 (−10.2 ± 2.5)	−9 (−11.2 ± 1.9)	−9.1 (−11.3 ± 5.1)
ΔH, kcal/mol	−6.8 (−6.8)	25.4 (15.3)	31.5 (15.9)
ΔS, cal/mol/K	28.3 (30.5)	39.8 (33.6)	31.3
$v_H^{\neq}$, cm^{-1}	1395	808	654
$v_D^{\neq}$, cm^{-1}	1118	676	542
K_H/K_D	2.47 (2.7)	1.78 (2.7)	1.72 (2.7)
$K(^{12}C)/K(^{14}C)$*	1.035 (1.031)		

*At 550 K

1.3.4 Pathway of a Chemical Reaction

1.3.4.1 Ambiguity of the Definition

Five years after Born and Oppenheimer had developed the milestone concept of potential energy surface, Pelzer and Wigner suggested that chemical reaction crossed on its way a saddle point on the PES. In 1935 Eyring postulated that it is the motion along the unique line passing through a saddle point on the PES which determines the course of a chemical reaction. This line was identified with the minimal energy reaction path (MERP). Surprisingly, although discussions around the concept of the MERP have been continuing up to the present day, the Eyring notion of it has in fact remained unchanged [14].

It would be useful to recall here some notions introduced in Sect. 1.1.1. In accordance with Mezey [5], we shall refer to the line of the steepest descent connecting two adjacent minima and passing through the saddle point on the PES in the $(3N-5)$-dimensional space as *the total reaction path.* This curve is analogous to the three-dimensional path on the PES in Fig. 1.2a. The arc length of the total reaction path denotes *the total reaction coordinate* [5]. The projection of the total reaction path onto the configurational space represents the MERP—see Fig. 1.2b. Accordingly, the arc length along the MERP is *the reaction coordinate.* The MERP depicts the continuous transformation of the nuclear geometric configuration reflecting the motion of the reactant to the product in the course of chemical reaction. In the general case, the reaction coordinates is a complex function of the overall nuclear coordinates $(3N-6)$ and its functional dependence may sharply change at certain portions of the MERP. It is usually not possible to derive an analytical function of the reaction coordinate for a multidimensional PES.

Disregard of the complex character of the reaction coordinate and the practice of singling out one structural parameter whose monotonical variation is intended to describe the course of the reaction may often lead to serious errors. The example of the reaction $CH_2 + H_2$ in Sect. 1.3.3.1 has already shown that a calculation of the reaction path in the reaction coordinate regime of this type may fail to detect the true transition state on a PES. A particularly frequent miscalculation stemming from the use of a simplified reaction coordinate is the alleged "detection" of differing pathways for the direct and the reverse reactions. A clear idea of how such an artifact may arise is given by Fig. 1.12 which shows a split of the reaction path on the model PES that depends on two parameters. By moving from the minimum at point A gradually increasing the coordinate X (through minimizing the energy at each point with respect to the coordinate Y), we obtain the reaction path shown in Fig. 1.12 by a thick line [1] which does not run through the saddle point. The reverse motion from B to A, with the reaction coordinate X being diminished, gives a different reaction path [2]. Moving along it while continuously reducing the coordinate X, we shall not be able to reach the point A at all, unless we diminish still further the parameter

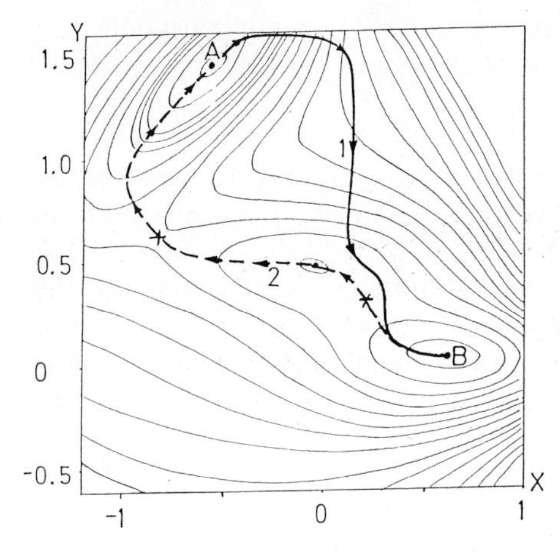

Fig. 1.12. Distinct reaction paths for the direct $A \rightarrow B$ (*solid line*) and the reverse $B \rightarrow A$ (*dashed line*) reactions on the model two-parameter surface. $E = \sum_{i=1}^{4} A_i \exp(a_i(x_i - x_i^{(o)})^2 + b_i(y - y_i^{(o)})(x - x_i^{(o)}) + c_i(y - y_i^{(o)})^2$ [35, 60]. The transition states are marked by X

X. The true reaction path represented in Fig. 1.12 by a dashed line is a function of two variables $f(X, Y)$.

Such splits of the paths of the forward and the backward reaction were first found in the calculation of the PES of pericyclic reactions in the reaction coordinate regime. They were called the "chemical hysteresis" [61]. Of course, this hysteresis has no physical sense, although such a possibility was not originally rejected out of hand, it simply follows from the shortcoming of the procedure shown in Fig. 1.12.

Another inconvenience that emerges in the conventional approach to determining the MERP and the reaction coordinate results from an ambiguity of these concepts caused by their invariance relative to the transformation of coordinates. The property of invariance is possessed only by the critical point of the PES, while the form of the MERP may change drastically depending on the choice of a coordinate system [15, 35]. At the same time, it is clear that there must also exist a physically meaningful interpretation of the MERP since the energetically most favored motion of a reacting system over the PES leading to the transformation of the reactants into the product has to correspond to a quite definite sequence of the nuclear configurations whose character cannot depend upon a coordinate representation.

1.3.4.2 A More Accurate Definition of the MERP and the Reaction Coordinate

A solution to the problem of constructing a strict theoretical model that could harmoniously accomodate the above-considered notion concerning the pathway of chemical reaction was suggested by Fukui [62]. He introduced the concept of the intrinsic reaction coordinate (IRC) defined as a classical trajectory of a system's motion over the PES crossing the saddle point of a transition state and representing infinitesimally slow slipping down of the reacting system from this point into the valleys of the reactants and the products. Such a treatment satisfies the classical equations of motion:

$$\dot{q}_i = \frac{\partial H}{\partial p_i}; \qquad \dot{p}_i = -\frac{\partial H}{\partial q_i} = -\frac{\partial E}{\partial q_i} \tag{1.27}$$

where H is the classical Hamiltonian, p_i is the impulse corresponding to the coordinate q_i and $\dot{q}_i$, $\dot{p}_i$ are the derivatives with respect to time.

If the coordinate system is chosen in such a manner that the kinetic energy T may be expressed as:

$$T = \frac{1}{2}\sum_i p_i^2 \tag{1.28}$$

then the trajectory in question will coincide with the line of the steepest descent from the point of a transition state [34]. Equation (1.28) is satisfied for the mass-weighted coordinates $\xi_i = m_i^{1/2} q_i$ with the cartesian coordinates $q_i(x, y, z)$. Therefrom follows the dimensionality of the IRC (atomic mass)$^{1/2}$ · Å and the necessity to define it as the MERP on the mass-weighted PES. By integrating the Eqs. (1.27) on condition that $p = O$ at every point (the condition of infinitesimal velocity of motion along the MERP defined by the IRC) or using some other computer-time saving techniques [34, 62], the IRC can be calculated.

The starting point of the IRC lies at the saddle point of the transition state, which has to be found by some independent method (see Sect. 1.3.3.1). Since the energy gradient equals at this point zero, the initial direction of the IRC cannot be determined through solving the Eqs. (1.27), however, at this point it exactly coincides with the direction of the normal vibration with the negative force constant, i.e., with the direction of the transition vector. Calculations of the IRC for some reactions, such as the isomerization HC≡N→C=N—H [34], S_N2 nucleophilic substitution [34], elimination of hydrogen fluoride from ethyl fluoride [63], pyrolysis of HCOOEt [64] and others [65, 66] have confirmed that the MERP's obtained by means of the IRC formalism can be fairly well reproduced using less sophisticated approaches based on chemical intuition in the selection of a reaction coordinate and on reliable localization of the transition state. These calculations have helped evolve a new approach to studying the chemical reactions—the so-called *reaction ergodography* which consists in a procedure of plotting the IRC for a given reaction [62]. This procedure includes an analysis of all dynamic aspects of important physical and chemical properties of the compounds exhibited along the IRC. The analysis of the reaction path of the methanethiol dehydrogenation process depicted in Fig. 1.13 [66] may serve as an example.

Other authors have also contributed a great deal to the development of the present-day understanding of the reaction path [34, 67, 68]. In 1984, Quapp and Heidrich pointed out [69] that the true PES does not necessarily depend

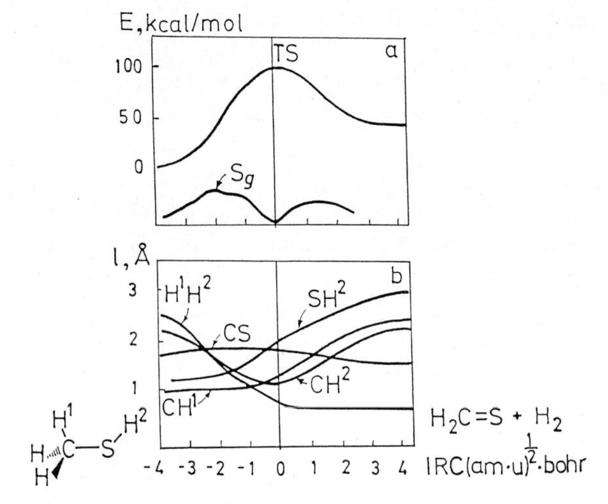

Fig. 1.13a, b. Changes occurring along the reaction path of **a** the potential energy and gradient norm (S_g), and **b** the bond lengths. (Adapted from Ref. [66])

on the atomic masses and, consequently, the use of mass-weighted coordinates gives the same path of the steepest descent as the one obtained without them. Moreover, it was shown on the basis of Riemannian geometry [70, 71] that any line belonging to the PES may be described in an invariant tensoric form, so the steepest descent line beginning from the saddle point on the PES (transition state), given in this form, does not depend on the choice of the coordinate system. Based on the foregoing, the total reaction path and the MERP may both be defined as the trajectory, orthogonal to the equipotential contours of the PES, which connects the energy minima through a common saddle point from which it slopes downward along the two steepest descent lines in, respectively, the full $3N-5$ space and the configurational $(3N-6)$ space [35, 69]. The lengths of the arcs along the total reaction path and the MERP are the total reaction coordinate and the reaction coordinate, respectively. Any point on these curves may be taken as the starting point, although a critical point (the minimum or saddle point) would be the most convenient choice. The configurational space is a hyperplane in the full $(3N-5)$ space, in Ref. [71] it is referred to as the dynamic plane. Physical motion along the MERP from the reactants to the products takes place over this hyperplane. Figuratively this motion may be represented as a bird's flight over the earth surface which turns into hovering over the minima. The minima of the PES are the "nests" of a chemical system where it is born acquiring the molecular structure and converting into the "fully-fledged" molecule[1].

[1] We can't help taking the risk of altering a little the rhyme which Ken Kesey used as an epigraph to his well-known novel: "One flew east, one flew west, One flew over the molecule's nest"

1.3.4.3 Symmetry Demands on the Reaction Path

It was shown in Sect. 1.3.3.2 that the symmetry properties that characterize the transition state of an elementary reaction are predetermined by the symmetry of its initial and final states. The transition vector being a small section of the reaction path passing through the transition state point retains all symmetry elements of the latter. Defining, as was done above, the reaction path as the steepest descent line from the saddle point to the minima on the PES which correspond to the reactants and the products, one may show that all along this path the nuclear configuration symmetry of the interconverting structures of reactants and products must stay unchanged [62, 72–74]. Instantaneous nuclear configurations that emerge on the MERP as the reaction goes on cannot acquire new symmetry elements, since their symmetry cannot be higher than that of the stable structures of the initial and the final state. This principle implies a number of useful corollaries which supplement the demands placed upon the symmetry properties of transition states:

a) The linear transition states are necessarily related to the linear structures of reactants and products since only the linear configurations possess the C_∞ axis of symmetry.
b) The planar transition states must lead to the planar structures of reactants and products for only the planar configurations do not change when reflected in the plane.
c) The achiral transition states cannot lead to optically active reactants and products.

It is important to note that these rules are valid only on condition that the MERP is a continuous line without bifurcation points and all critical points lying on this line are nondegenerate [75, 76]. Next we give some examples to illustrate how the rules a), b), c) operate in chemical reactions.

In the isomerization reaction of the T-shaped structures VIIa ⇄ VIIb ⇄ VIIc, all nuclear motions, both with the identical and the different ligands R, occur in a common plane. No pyramidal structures are present on the MERP.

All nondegenerate reactions must obey the requirements a)–c). For the degenerate transformations (in which a simple exchange of positions of equivalent nuclei occurs), the symmetry considerations admit of supplementary, in comparison with the structures of the reactants and the products, symmetry elements in transition state structures provided that the symmetry operations corresponding to these elements lead to the mutual exchange between nuclear configurations of the reactants and the products [6].

This statement may be illustrated by the following example. The transition state XVIII of the isotopic exchange reaction (Sect. 1.3.3.2) has, over against the symmetry elements of nuclear configurations of the reactants and the products, only one more, viz., the fourth-order axis C_4. But it is precisely a 90° rotation of the nuclear configuration XVIII about this axis which causes the transformation of the reactants into the products. Hence, the appearance on this reaction path of the structure XVIII is allowed by the symmetry rules.

Different is the case of the degenerate intramolecular isomerization of sulfuranes SR_4 (each ligand is given its own index). The CNDO/2 calculations of the PES of this reaction [77] have shown that C_{2v} structures of the type XXII are the stable forms of sulfuranes SH_4 and SF_4, which is in accord with the experimental data on SF_4. The transition state XXIII in the topomerization XXII ⇄ XXIIa characterized by mutual exchange between the pairs of the equatorial and the axial ligands (respectively, Nos. 3, 4 and 1, 2 in XXII) has C_{4v} symmetry (Fig. 1.14). The calculated activation barrier of ~ 15 kcal/mol for this topomerization (Berry pseudorotation) reproduces quite well the experimental value.

XXII(R), C_{2v} ⇌ XXIII(R), C_{4v} ⇌ XXIIa(R), C_{2v}

XXIV, D_{4h}

XXII(S), C_{2v} ⇌ XXIII(S), C_{4v} ⇌ XXIIa(S), C_{2v}

Although the C_{4v} structure has an additional, as compared to C_{2v}, symmetry element, i.e., the fourth-order axis, it is easy to ascertain that the rotation about this axis does indeed lead to a transformation of XXII into XXIIa. Another possible mechanism of topomerization includes a planar structure XXIV with D_{4h} symmetry. This structure would link not only the topomers XXII(R), XXIIa(R), but also the enantiomers of XXII (*R*, *S* forms). However, the XXIV structure with D_{4h} symmetry cannot serve as a transition state since the operations of the additional symmetry elements, i.e., the rotation about the fourth-order symmetry axis and the reflection in the inversion center, do not result in the transitions XXII(*R*) → XXIIa(*R*) or XXII(*R*) → XXIIa(*S*). Indeed, although the configuration XXIV corresponds to a first-order saddle point, it is not the lowest transition state in the topomerizations XXII ⇄ XXIIa. The calculations show the relative energy of XXIV to be very high exceeding the barrier of the dissociation reaction $SR_4 \rightarrow SR_2 + R_2$. Thanks to the above-noted special features of the achiral planar structure of sulfuranes XXIV, their derivatives can be obtained in an optically active form. This form represents a

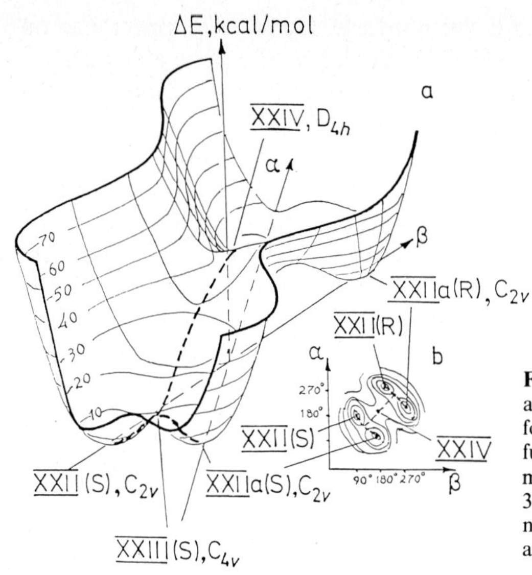

Fig. 1.14a, b. Potential energy surface (**a**) and its schematic two-dimensional map (**b**) for the enantiomerization reaction of sulfurane SH_4 as calculated by the CNDO/2 method. α is the angle 1-S-2, β is the angle 3-S-4 in the structures XXII–XXIV. The numbers in breaks of the lines on the PES are the isoenergy levels in kcal/mol [77]

new type of chirality of tetracoordinate structures, called the cuneal chirality; it has recently been realized in cyclic sulfuranes [78]:

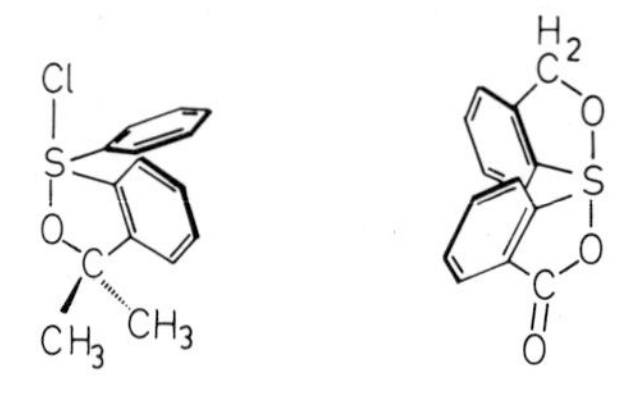

1.3.4.4 *Chiral and Achiral Pathways of Degenerate Reactions*

The possibility of an additional symmetry element, absent in the reactants and the products, appearing in the transition state structures of degenerate reactions widens the choice of theoretically admissible pathways of these reactions. A reaction may develop both along a path that includes a transition state of higher symmetry and along the path on which the general restrictions are valid imposed by the symmetry demands. We are confronted here with an interesting situation of an apparent violation of the microscopic reversibility principle since there may exist on the PES two or more energetically equivalent reaction paths and transition states which are interconvertible through an operation associated with the additional symmetry element allowed for degenerate reactions [79, 80].

The most common and important case are the narcissistic reactions. Under this type the degenerate transformations are classified [81] in which the structure of the products is viewed as a reflection of the structure of the reactants in a mirror plane, which is a symmetry element absent in both the reactants and the products. These reactions include the pyramidal inversion XXV and the cyclic inversion XXVI, the valence isomerizations, such as XXVII, and, most importantly, numerous reactions of enantiotopomerization:

XXV XXVI XXVII

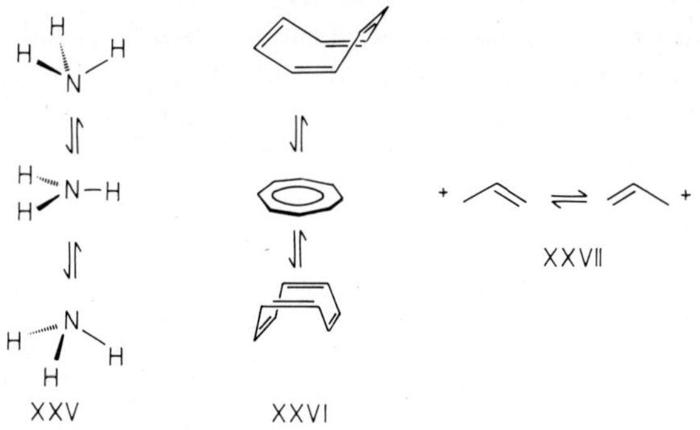

An example of special significance for stereochemistry is given by a conceivable enantiotopomerization of methane, which is accompanied by an inversion of the bond configuration at the tetrahedral carbon atom [82, 83]:

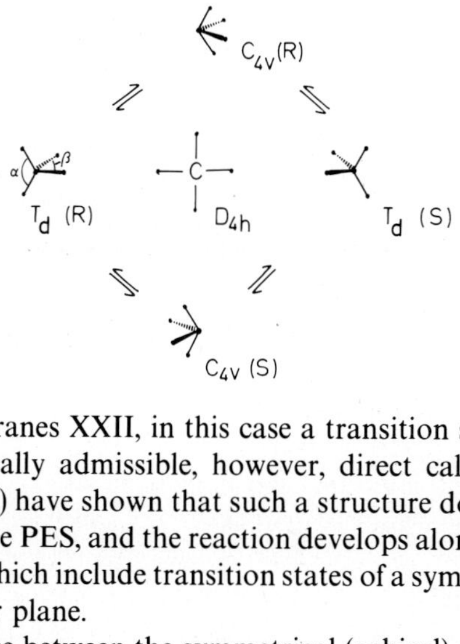

$C_{4v}(R)$ T_d (R) D_{4h} T_d (S) C_{4v} (S)

Similarly to sulfuranes XXII, in this case a transition state with planar D_{4h} structure is theoretically admissible, however, direct calculations (see review listed under Ref. [83]) have shown that such a structure does not correspond to the saddle point on the PES, and the reaction develops along one of two possible enantiomeric paths which include transition states of a symmetry lower than D_{4h} and without a mirror plane.

Generally, a choice between the symmetrical (achiral) and the asymmetrical (chiral) routes of the reaction in question is dictated by the number of independent geometry parameters, antisymmetrical relative to the mirror plane and adequately defining the reaction coordinate, as well as by the degree of their interrelation [81]. In case the reaction coordinate is described by the change of a single parameter, for example, of the pyramidalization angle α in the ammonia inversion XXV, the reaction path takes its course via the achiral transition state lying in the mirror plane. The same type of the transition state and reaction route is retained when the reaction coordinate is described by the function of two antisymmetrical variables whose variation is strictly synchronized. If, for example, the angles α and β in the methane Td-structures increased simultaneously from 109.5 to 180° (i.e., the reaction coordinate corresponded in its initial portion to symmetrical deformational vibration), then the reaction would follow the achiral route through a transition state with D_{4h} symmetry. In actual fact, the asymmetrical vibration, with the variation of these anyles being asynchronous, has a lower frequency, so the PES of the methane enantiotopomerization contains two routes through a transition state with chiral C_{4v}-bond configuration (Fig. 1.15).

The calculation of the surface featured in Fig. 1.15 has been done by the simplest EHMO method. But even a more rigorous treatment [20, 83] gives a similar energy sequence in the Td, C_{4v}, and D_{4h} structures. The situation changes when all symmetry constraints (two symmetry planes and equivalence of all C—H bond lengths) are removed and a complete geometry optimization is carried out. In such a case, the structure C_{4v} reduces its symmetry to the C_{2v} form (see Fig. 1.16) which is associated with the true saddle point on the PES. (For a more detailed analysis of this problem see Ref. [20].)

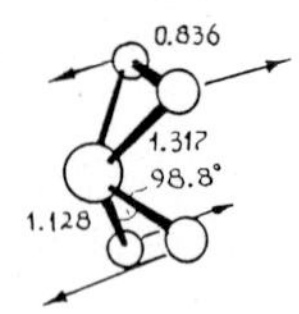

Fig. 1.15. Potential energy surface for the enantiotopomerization reaction (degenerate enantiomerization) of methane as a function of the angles α and β (tetrahedral compression mode). The calculations have been performed by the EHMO method [20,82] with the imposed condition of equality of all C—H bond lengths

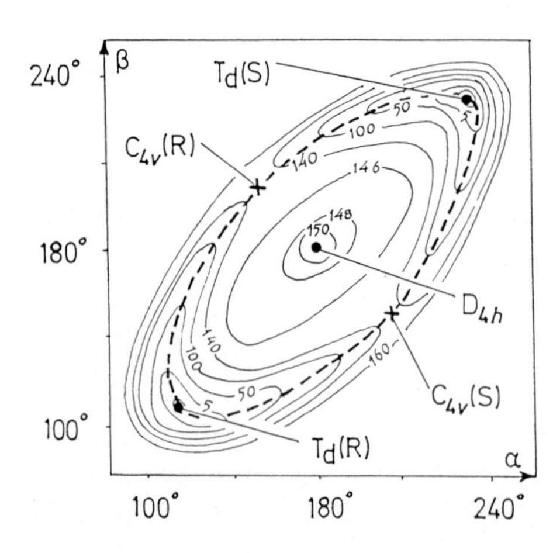

Fig. 1.16. Geometries of transition state structures in the reaction of enantiotopomerization of tetrahedral methane as calculated by the MINDO/3 method [20]. Bond lengths are in Å, *arrows* indicate the transition vector components

1.3.5 Empirical Correlations of the Reaction Pathways

The duration of an elementary act of reaction is the order reciprocal to kT/h, i.e., it amounts at 300 K to approximately 10^{-13} s. This time span embracing all reaction phases from its start to the formation, upon overcoming the activation barrier, of the products is too small to be able to register experimentally the sequence of the restructuring which the nuclear configuration of a reacting system undergoes. Moreover, since the system while moving along the reaction path from one PES minimum towards another one is not in its stationary state, such an experiment is impossible in principle.

There are, however, possibilities of making approximate assessments of the character of structural changes occurring along the reaction path. They are based on the experimental data that characterize properties of the ground states of the reactants and the products. In view of the profound significance for theoretical chemistry of the concept of reaction path and bearing in mind some ambiguities in its definition, it is highly important to establish in regard to this concept certain correlations with physically meaningful properties amenable to precise characterization. Several approaches are known useful in the search for such correlations.

1.3.5.1 Molecular Vibrations and the Reaction Coordinate

When choosing between possible deformation types of the molecular framework in the course of a reaction, one may compare the corresponding force constants or vibration frequencies. The most likely type of deformation will correspond to a minimal vibration frequency.

It is, for example, known that the dehydrohalogenation of vinyl halogenides XXVIII proceeds preferably by the mechanism of *trans*-elimination. The stereoselectivity can be explained as follows [84]. The structure of bent acetylene emerging upon *trans*-elimination of the hydrohalogen elements transforms much more readily into an equilibrium linear form than the structure forming in the case of a *cis*-elimination. The frequency of the symmetrical normal vibration π_g, which corresponds to the deformations of XXIX, equals 612 cm^{-1} which is much less than the value of 729 cm^{-1} for the antisymmetrical vibration π_u of the structure XXX:

$H^1H^2C{=}C(X)H^3$ (XXVIII) $\xrightarrow{trans\text{-}E}$ $H^1C{\equiv}CH^3$ (XXIX) $\xrightarrow{\pi_g}$ $HC{\equiv}CH$

$H^1H^2C{=}C(X)H^3$ (XXVIII) $\xrightarrow{cis\text{-}E}$ $H^2C{\equiv}CH^3$ (XXX) $\xrightarrow{\pi_u}$ $HC{\equiv}CH$

However, only in the initial reaction phase the form of normal vibration can approximate the reaction coordinate whose typical feature is its correspondence with the imaginary vibration frequency (Sect. 1.3.3). A more rigorous approach based on the data on the force constants of a reacting molecule describes the reaction coordinate in terms of the so-called interaction displacement coordinates $(j)_i$ [85].

These coordinates represent changes of $(n-1)$ internal coordinates R_j, the changes which minimize the potential energy of the system for a given displacement of R_i (compare with the method of reaction coordinate). The method uses the force constants of molecular vibrations (interaction compliance) calculated somewhat differently than in Eq. (1.4), namely as the forces that need to be applied to achieve measurable distortions in R_i with the potential energy minimized with respect to other coordinates:

$$\gamma_{ii} = \left(\frac{\partial^2 E}{\partial R_i^2}\right)_{F_j=0} \tag{1.29}$$

The reaction coordinate $\tilde{R}_i$ corresponding to the MERP is calculated for some fixed displacements of each of n internal coordinates R_i as:

$$\tilde{R}_i = R_i + \sum_{j\neq i}^{n} (j)_i R_j = R_i + \sum_{j\neq i}^{n} \left(\frac{\gamma_{ii}}{\gamma_{ij}}\right) R_j \tag{1.30}$$

The use of Eq. (1.30) for calculating a reaction coordinate admits, in fact, the possibility of describing the properties of a reacting system (large molecular distortions) by means of the parameters of a static system since the force constants γ are calculated for extremely small nuclear displacements. As calculations show [85], this admission proves justified in the case of intramolecular rearrangements.

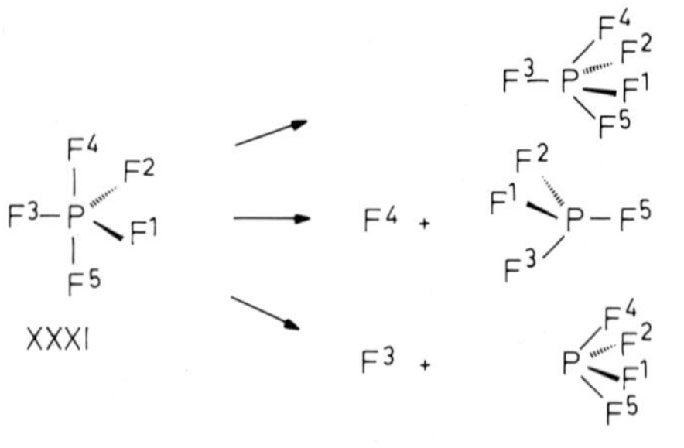

Consider, for example, the molecule PF_5 XXXI for which detailed data on the force field are available. Deformations of the valence angle of the equatorial bonds F_1—P—F_2 distort this structure to a square-pyramidal structure XXXII, which corresponds to a reaction path calculated [86] for the well-known polytopal rearrangement, the Berry pseudorotation. A stretching of the axial bond P—F_4 leads, when calculating by means of Eq. (1.30), to a tetrahedral structure PF_4 XXXIII, which parallels the scheme of a S_N2-type reaction, while the stretching of the equatorial bond P—F_3 gives rise to a C_{2v} structure of PF_4 XXXIV.

In order to calculate a reaction coordinate from Eq. (1.30), data are required on the total force field for a given molecule. Comprehensive experimental information on all force constants is, however, known for very few small molecules. One way out of this impasse consists in combining the experimental force constants γ available for some characteristic vibrations with the constants calculated in harmonic approximation [87] by the scheme described above (Sect. 1.3.2.1).

1.3.5.2 The Principle of Least-Motion

Back in 1938 Rice and Teller [88] formulated the general principle which stated that those elementary reactions are the most favored which exhibit the fewest possible alterations in the positions of atomic nuclei and in electronic configuration. The part referring to the electron configuration was later developed into the Woodward–Hoffmann rules, while that concerning the nuclear shifts became known as the principle of least motion of nuclei or simply the principle of least-motion (PLM) [89, 90][1].

Implicitely, as the principle of minimal structural changes, the PLM gained long ago currency among organic chemists as one of their basic concepts. The mathematical formulation of the PLM rests on a mechanical model of the molecule in which the energy of structural deformation, when initial reactants (r) turn into products (p), is assumed to be proportional to the sum of the squares of the changes in the positions of the nuclei common for r and p:

$$E = \sum f_i(q_i^p - q_i^r) \tag{1.31}$$

where f_i is the force constant often set equal to unity.

The direction of reaction associated with the minimal deformation energy is considered preferable. The changes in the coordiates q_i corresponding to the values of $E_{\min}$ permit conclusions to be drawn as to the reaction path and reaction coordinate. For example, in the case of the dehydrohalogenation of vinyl halogenides XXVIII considered earlier the calculated value of $E_{\min}$ for *trans*-elimination of hydrohalogenide amounts to 0.32 Å ($f_i = 1$), while for the *cis*-elimination it is 1.24 Å whence a conclusion may be made in favor of the former direction.

The principal equation of the PLM [Eq. (1.31)] coincides with the standard relationships for the potential energy of small vibrations, hence it is valid only in the earliest stage of a reaction. There are suggestions [84, 89] that a promising

[1] Salem called attention to an earlier formulation of the "principle de la moindre déformation moléculaire" advanced in 1924 by the French researchers Muller and Peytral, see Salem L (1982) Electrons in Chemical Reactions: First Principles. Wiley and Sons, New York

development of the PLM might consist in replacing in Eq. (1.31) the coordinates q_i^p of a product by the coordinates of a transition state structure. This innovation depriving the PLM of its chief attraction, the utmost simplicity, is hardly acceptable seeing that the transition state structure cannot be determined experimentally.

Numerous reactions violate the PLM requirements. Thus, as may be seen from Fig. 1.9, the addition of hydrogen to methylene does not take place along the least motion path, i.e., with retention of C_{2v} symmetry of the reacting system (Sect. 1.3.3.1).

The reason for this violation of the PLM requirements lies, apparently, in the fact that the driving force of chemical reactions has a much more complicated nature than that given by a condition of the Eq. (1.31) type. Indeed, this condition may be used in calculating a reaction coordinate only as a crude approximation.

1.3.5.3 *Structural Correlations of the Pathways of Chemical Reactions*

Consider the simplest energy profile of the reaction $A \rightarrow B$ shown in Fig. 1.17. As has repeatedly been emphasized, only those geometrical configurations can be observed experimentally which correspond to the stable structures A and B and occupy minima regions on a PES. The rest of the potential energy curve can be obtained only by calculation. Assume now that through some little structural alterations in A and B the positions of initial minima can be shifted. If these changes are systematic and correlated with the reaction coordinate of the transformation $A \rightarrow B$ (Fig. 1.17), then the procedure of making structural determinations in a series of the compounds A', A'',... B', B'' (representing monotonical perturbations in the parent A and B structures) may be regarded as a method for experimental investigation of structural changes along the reaction path [92–94].

Let us consider the essence of this approach using as an example the reactions of nucleophilic addition to the carbonyl group [95]:

$$R_1R_2R_3N + RR'C{=}O \rightarrow RR'C\begin{matrix} \diagup O^- \\ \diagdown \overset{+}{N}R_1R_2R_3 \end{matrix}$$

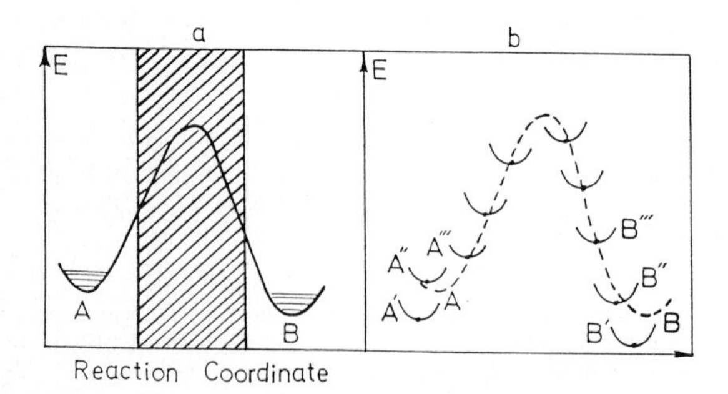

Fig. 1.17a. Energy profile of the concerted reaction $A \rightarrow B$. The shape of the energy curve can be studied experimentally only in the region of the minima of A and B. **b** Variation of the structure of A and B displaces the equilibrium structures of the derivatives $A', A'', A''', \ldots B', B'', B''' \ldots$ along the reaction path

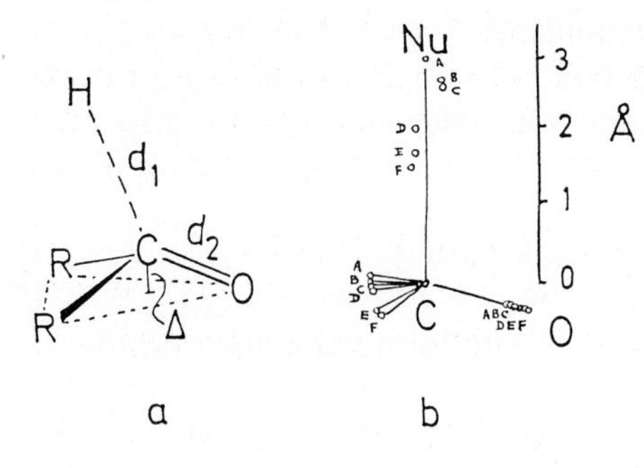

Fig. 1.18a. Structural parameters of the reactive site of nucleophilic addition to the carbonyl group according to Refs. [92,95]. At $d_1 \rightarrow \alpha$, $\Delta = 0$ and d_2 corresponds to the length of the double bond C=O. **b** Projection of the reaction coordinate onto the plane NCO. The points *A,B,C,D,E,F* correspond to the structures XXXV–XL. The arrows show the direction of the nucleophile lone-pair orbital. (Reproduced with permission from the American Chemical Society)

The route of this reaction may be traced by analyzing structural changes occuring in the amine and carbonyl fragments in a family of compounds such as natural aminoketones.

In these structures, the distances N...CO decrease progressively in the order XXXV – XI with the value of d_1 being, even for XXXV, considerably smaller than the sum of the van der Waals radii N and C (3.21 Å) thus indicating attractive interaction between the amine and the carbonyl fragments. Other geometrically independent structural parameters of the reactive site are the bond length d_2 of C=O, which gradually grows with the decrease in d_1, and the angle of pyramidalization (Fig. 1.18a). A good correlation has been found between these parameters.

$$d_1 = -1.701 \log \frac{\Delta}{\Delta_{\max}} + 1.479 \text{ Å}$$

$$d_2 = -0.71 \log\left(2 - \frac{\Delta}{\Delta_{\max}}\right) + 1.426 \text{ Å} \tag{1.32}$$

XXXV, d_1 = 2.91 Å   XXXVI, d_1 = 2.76 Å   XXXVII, d_1 = 2.581 Å

XXXVIII, d_1 = 2.457 Å   XXXIX, d_1 = 1.993 Å   XL, d_1 = 1.64 Å where Δ_{max} is the maximal angle of pyramidalization when the distance d_1 is shortened to the length of the ordinary bond C—N. The form of Eqs. (1.32) coincides with the well-known Pauling equation which relates the length of the bond with its order:

$$d = d_0 - c \log n \tag{1.33}$$

Ab initio calculations [95] of the reaction path for nucleophilic additon of the hydride ion to the carbonyl group of formaldehyde ($H^- + CH_2 = O \rightarrow CH_3O^-$) have shown that for the calculated quantities d_1, d_2, there indeed exist the relationships of the type of Eqs. (1.32)–(1.33). This verifies the correctness of the empirical correlation of the reaction path presented in Fig. 1.18 in the snap-shot form. The calculated angle of approach of the nucleophile to the carbonyl plane equals 170°.

Another illustrative example of a fairly exact reproduction of the typical features of a reaction coordinate by means of X-ray structural data is given by an analysis of distortions of bond configurations of the pentacoordinate phosphorus atom in cyclic phosphoranes [96].

The pentacoordinate structures with trigonal-bipyramidal configuration belong to the type of the stereochemically nonrigid structures. The most typical for these structures polytopal rearrangement XLI → XLIa (the Berry pseudorotation), which results, through passing via the square-pyramidal structure XLII, in the pair-wise exchange of the ligand positions, requires that, e.g., PF_5 (XXXI → XXXII) overcome an activation barrier estimated from various data to amount to a mere 3–4 kcal/mol.

XLI $\theta_{15} = 180°$, $\theta_{24} = 120°$ ⇌ XLII $\theta_{15} = 150°$, $\theta_{24} = 150°$ ⇌ XLIa $\theta_{15} = 120°$, $\theta_{24} = 180°$

As a result of the transition from the trigonal-bipyramidal XLI to the square-pyramidal form, the most drastic interrelated changes are experienced by the angles θ_{15} and θ_{24}. Holmes [96] analyzed X-ray structural data on 34 phosphoranes and placed them in a descending order with regard to the magnitude of the angle θ_{15}, which conversely, was an ascending order with respect to the angle θ_{24}. The experimental data agree very well with the theoretical expectations for the variation of these angles along the coordinate of the Berry pseudorotation. This trend may be expressed in percent of the transition XLI → XLIa (0% corresponds to an ideal trigonal bipyramid XLI while 100%—to the square pyramid). Some representative structures of phosphoranes given below show that the sequence of 34 studied compounds practically models the total reaction path.

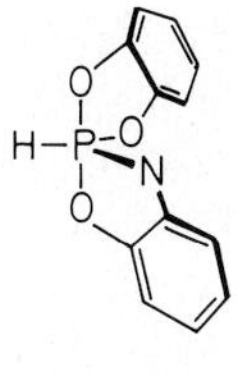

XLIII (5%) 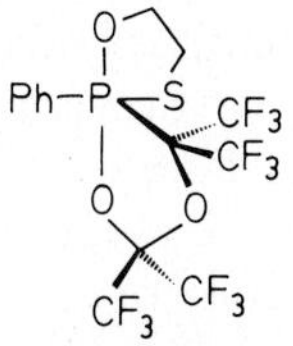 XLIV (29%) 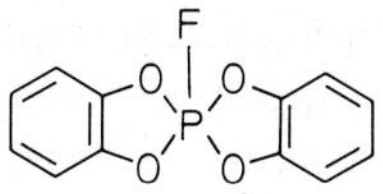 XLV (65%) 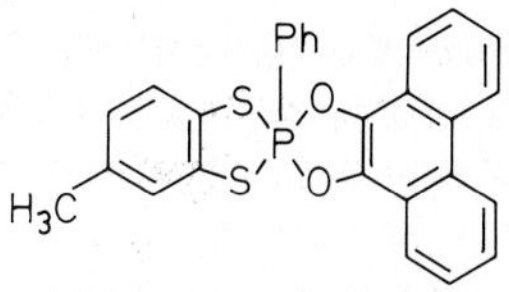 XLVI (97%)

The form of the structures XLIII–XLVI indicates that in the series of the compounds studied the surrounding of the central atom, let alone the periphery of the molecule, changes dramatically. Sometimes, in order to perform structural correlation even compounds with differing central atoms are taken, as is the case with the Berry pseudorotation for transition metal complexes [97]. So as to make sure that the noted structural trends do indeed reflect characteristic changes, while reaction is in progress, rather than the side effect of the crystal structure packing, one has to ascertain whether the independent structural parameters are correlated and change systematically. If a correlation actually exists among two or more independendent parameters that describe the structure of a given fragment with various surroundings, then the correlation functions derived quite probably define the MERP on the PES in the corresponding parameter space. This statement formulated by Dunitz [92, 94, 98] is known as the principle of structural correlation of the pathways of chemical reactions.

1.4 Dynamic Approach

So far we have considered the static approach to understanding chemical reaction, defined, in terms of this approach, as infinitely slow motion of nuclei along the MERP belonging to a given PES. In this approach, the following factors are left out of consideration: the kinetic energy of the system; the vibrational excitation of the reacting molecule arising upon collisions with other molecules; redistribution of the excitation energy and its localization in certain types of vibration; rotation of activated molecules—i.e., all those dynamic effects which adapt the reacting molecule to the deformation defined by the reaction coordinate. Owing to the presence of kinetic energy, the total energy of a system always exceeds the level given by the PES and the dynamic trajectories may greatly differ from the MERP depending on the initial conditions (the coordinates and impulses).

The currently most important technique of the dynamic approach is based on the calculation of classical trajectories. In such an approximation, nuclei of a chemical system in question are treated as classical particles moving under forces defined by the PES. The trajectories represent the solutions to the Hamiltonian (or Lagrangian) of Eq. (1.27).

To solve the dynamic problem, an overall PES system (including the whole space of changing internal coordinates) rather than the critical points only has to be constructed. The PES is approximated by analytical functions the methods of whose determination have been fairly well elaborated mathematically [99, 100]. The parameters that determine the initial state of the system (coordinates and impulses) may be defined as changing gradually or selected randomly, for example using the Monte Carlo method. One more point is of particular importance when calculating the trajectories of the particles over the PES. The equations of motion will reproduce them exactly only on condition that the operator of kinetical energy does not contain cross-terms, which may be achieved through selection of appropriate orthogonal linear combinations of internal coordinates [56].

Next, the forces need to be calculated that have an effect on the atoms of the reacting system. To this end, the first derivatives of the potential energy with respect to coordinates should have to be calculated in a great number of points and the values obtained approximated by analytical expressions. However, so as to reduce calculation work, a different technique is commonly applied: the forces along a trajectory are calculated from the quantum mechanical expressions for the potential surface [100, 101]. This procedure takes about 80% of the total calculation time, the rest is spent on integrating the equations of motion.

The merit of the dynamic calculations consists in that they broaden substantially the notion of the internal mechanism of reactions linking it with the actual conditions of chemical transformations. Thus, the calculations [101] of dynamics of the reaction $CH_2 + H_2$ have shown that the initial conditions (original orientation, impulses) are crucial in determining the form of the dynamic trajectory which is close to the MERP only for a narrow interval of the redundant energy (Fig. 1.9). Bearing in mind that carbenes, in whatever manner they are generated, are always vibrationally excited particles, one may appreciate the importance of this conclusion for selecting a method for obtaining carbene.

The monomolecular reactions of topomerization of cyclopropane represent the most thoroughly studied example of a theoretical calculation of dynamics of an organic reaction [100, 102]. (1) is the route of diastereotopomerization (in the nondegenerate case—of optical isomerization). Both reactions include

(1) XLVII (2) XLVIII (FF) $(EF)^{\ddagger}$

(EE) $(EF)^{\ddagger}$ (FF) XLYIIa

the formation, upon stretching of the C—C bond, of the trimethylene biradical XLVIII (FF)—FF stands for "face-to-face".

The PES of stereomutation of cyclopropane was derived by varying all 21 independent geometry parameters in an ab initio calculation (STO-3G basis set, CI 3×3). For the optical isomerization, a calculation in the static approximation revealed a MERP associated with nonsynchronous conrotatory motion of two terminal methylene groups through a transition state (EF)—EF stands for "edge-to-face" (Fig. 19a). The transition state energy is 58 kcal/mol.

For calculating the trajectories, six initial parameters have to be set: the three angles α, θ_1, θ_2 (see XLVIII), the total energy of the molecule E_{tot}, the initial part of the total energy required for rotation E_{rot} and its distribution over two methylene groups.

Figure 1.19b shows the trajectory of the motion of cyclopropane XLVII towards its enantiotopomer XLVIIa for the case when the total energy equals 61 kcal/mol exceeding the transition state energy by a mere 3 kcal/mol. Calculations show that the reaction XLVII → XLVIII → XLVIIa proceeds only when about half that energy is the kinetical energy of rotation E_{rot} of the methylene groups while the rest is localized at the vibrations of the C—C bond. As may be seen from Fig. 1.19b, the dynamic trajectory, which links on the PES the enantiotopomers XLVII and XLVIIa, roughly follows the MERP (Fig. 1.19a) though by no means coinciding with it.

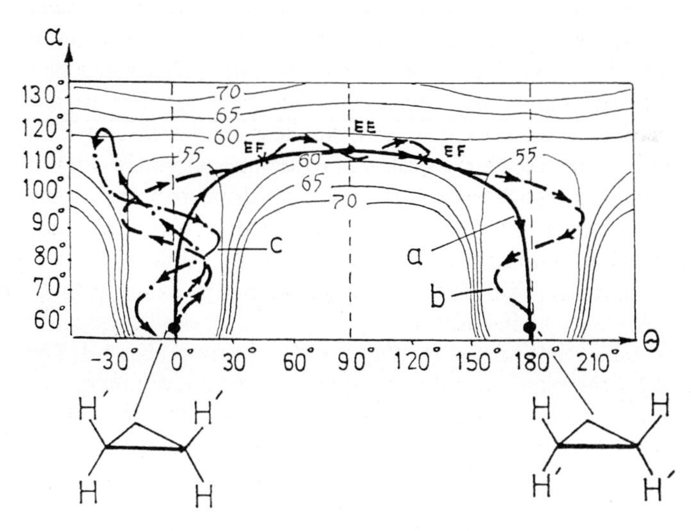

Fig. 1.19. Section of the PES of the XLVII–XLIIa reaction along the MERP for synchronous conrotatory motion of the methylene groups. Assignments of the angles α and $\theta(\theta = \theta_1 = \theta_2)$ are given in the structure XLVIII. Numbers in the line breaks are the relative energies in kcal/mol. *a* Static trajectory (MERP); *b* reactive trajectory at $E_{tot} = 61$ kcal/mol, $E_{rot} = 31.6$ kcal/mol; *c* nonreactive trajectory at $E_{tot} = 61$ kcal/mol, $E_{rot} = 24$ kcal/mol. (Adapted from Ref. [100])

A different situation arises when $E_{tot} \geqslant 65$ kcal/mol. In this case quite a few variants are possible depending on the initial magnitude of the rotation energy. For $E_{rot} \leqslant 10$ kcal/mol there are no trajectories along which the reaction might proceed (Fig. 1.19c). Trajectories capable of reaction fall first within the range of $12 \leqslant E_{rot} \leqslant 20$ kcal/mol, but then in the range of $25 \leqslant E_{rot} \leqslant 30$ kcal/mol they become anew reactionless. As E_{rot} rises still higher reaching 33 kcal/mol, the reaction once again can occur, while at $E_{rot} > 35$ kcal/mol it is unrealizable. Thus, there are certain alternating energy zones where the reaction in question (XLVII → XLVIIa) is prohibited.

There are cases when dynamic calculations change altogether our notions about mechanisms of certain reactions. A well-known example is given by the reaction $H_2 + I_2 \rightarrow 2HI$ which was earlier thought to be a simple bimolecular transformation. Semiempirical calculations of the barrier of the bimolecular process would yield the value of 66 kcal/mol in good agreement with the experimental activation energy of 65 kcal/mol. However, the trajectory calculations have shown the true mechanism of this reaction to be much more complex. Its most likely channel is $H_2 + I_2 \rightarrow H_2 + 2I \rightarrow I + H - I - H \rightarrow 2HI$.

Such calculations of reaction dynamics are still few being hampered by high demands on the quality of the PES and considerable computer-time expenditure. But the examples adduced show that the study of dynamics is, as an essential complement to the static analysis, undeniably important for understanding reaction mechanisms.

There is, apparently, no need for large-scale molecular-dynamic calculations in every concrete case. Representative reactions should, however, be analyzed from this angle so as to understand possible deviations of the true trajectories of motion over the PES from the MERP and to reveal optimal energy zones of a reaction defined by the character of the vibration excitation. Furthermore, the importance of such an analysis consists in the fact that here, unlike the case with the MERP, there exists in principle a possibility of experimental verification of correctness of the trajectories calculated. In this connection, one may refer in the first place to such experiments as the collisions in crossing molecular beams and the selective laser excitation of vibration-rotational states [3, 103].

An important simplification of the theoretical investigation into dynamics has been suggested by Miller, Handy and Adams [104] who proposed taking into account only the most important part of the PES, i.e. the reaction channel. The PES is approximated by the reaction path and its quadratic environment. All motions of nuclei are divided into the motion along the reaction path and the $(3N - 7)$ harmonic vibration motions transverse to it. In this case, *the classical reaction path Hamiltonian* is given by:

$$H(P_s, S, P_k, Q_k) = \tfrac{1}{2}P_s^2 + U_o(s) + \sum_{k=1}^{3N-7} [\tfrac{1}{2}P_k^2 + \tfrac{1}{2}\omega_k^2(S) \cdot Q_k^2] + A \quad (1.34)$$

where s and p_s are the reaction coordinate and the impulse corresponding to it, resp.; $U_o(s)$ is the potential along the reaction path; P_k, ω_k, and Q_k are the impulse, the frequency and the coordinate of the normal vibrations transverse to the reaction path; A are the terms that describe nonadiabatic interactions. The reaction path Hamiltonian has been applied to calculate the unimolecular dissociation of formaldehyde $H_2CO \rightarrow H_2 + CO$ [104, 105], the unimolecular isomerization $HNC \rightarrow HCN$ [106] and some other reactions—see Ref. [107].

The reaction of monomolecular dissociation of the H_2CO molecule in the singlet ground state proceeds via a planar transition state of C_s symmetry—see Ref. [107] and references therein—with retention of the symmetry plane all along the reaction path (see Sect. 1.3.4.3). The latest ab initio calculations (MP4 SDTQ, see Sect. 2.2) have produced the value of activation barrier of dissociation equal to 85.9 kcal/mol [108], while the experimental data are by 5–6 kcal/mol lower. Calculations of the rate of this reaction by means of the reaction path Hamiltonian have shown that it indeed proceeds at an energy 5–10 kcal/mol lower than the classical limit, which can be explained by the effect of tunnelling. A similar result has been obtained also in the case of the $HNC \rightarrow HCN$ isomerization [106].

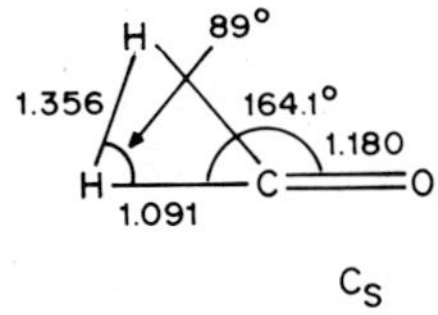

On the whole, the procedure of comparison of the experimental data on reaction kinetics with the results of dynamics calculations is, admittedly, more laborious than that with the results of static approximation based on the theory of the transition state. To arrive at the theoretical value of the reaction rate in dynamic approximation, one has to calculate the probabilities (reaction sections), which depend on the distribution of initial conditions, of transitions from the region of the reactants into that of the products for every trajectory as well as to derive a rate versus relative initial energy function [109]. The total rate constant to be compared with the experimental value is obtained through averaging over all individual constants.

1.5 Tunnelling Effects in Chemical Reactions

Theoretical analysis of kinetics of a chemical reactions, whether it is done with the aid of the theory of transition states or by chemical dynamics methods, rests on the classical notion of the necessity to overcome an activation barrier, i.e. the saddle point on the PES of a molecular system. Meanwhile, a sizeable contribution to the total rate of some reactions is made by underbarrier trajectories—this is a purely quantum mechanical effect of the system oozing through the energy barrier so that there exists a certain probability of a transition from the reactants to the products even when the internal energy of the system is lower than the energy of the transition state. The condition for the predominance of the tunnelling effect is given by the inequality [110, 111]:

$$h/2\pi d\sqrt{2mE^{\neq}} > k_BT/E^{\neq} \quad (1.35)$$

where $E^{\neq}$ and d are the height and the half-width of the barrier on the PES, and m is the mass of the tunnelling particle.

It is evident from Eq. (1.35) that the relative probability of tunnelling strongly depends on the form of the barrier, rising when it is narrow; it also rises for the particles of small mass and with falling temperature T. The magnitude of T at which the tunnelling effects prevail over the classical overbarrier transitions is defined by

$$T_t = \frac{h}{2\pi^2 kd}\sqrt{\frac{E^{\neq}}{2m}} \quad (1.36)$$

At $E^{\neq} \approx 1$ eV (23 kcal/mol) and d ≈ 3 Å, $T_t \approx 50$ K for protons and 1600 K for electrons. In electron transfer reactions, the tunnelling effect may show up even for the distances of several tens of Ångström units (for d ≈ 30 Å and $E^{\neq} \approx 1$ eV, $T_t \approx 160$ K).

In typical organic reactions developing at ambient temperature, the role of the tunnelling effects is usually insignificant. When, however, the reaction coordinate is determined predominantly by the shifts of light nuclei, particularly protons, the contributions from tunnelling may become appreciable and, in some cases, even decisive, as will be shown below.

An overall scheme for quantitative assessment of the influence of tunnelling effects upon the reaction rate has been developed by Miller [113, 114]. The simplest method for calculating the tunnelling rate constant is based on the theory of transition state with correction for tunnelling. This correction consists in formal replacement of the classical motion along the reaction coordinate with the quantum motion. This approach was first formulated in the works by Bell [115].

A rigorous theory of transition state with consistent inclusion of the results obtained by solving the dynamic problem was evolved by Kupperman [116]. In more exact terms, the tunnelling correction amounts to a replacement of the probability of tunnelling $P(E)$ by the quantum mechanical probability of the passage of a particle across the one-dimensional barrier $V(s)$ with $V(s)$ being the potential along the minimal energy path. The dependence of the transverse vibration frequency on the reaction coordinate may be taken into account by adding to $V(s)$ the adiabatic vibration energy of transverse oscillators: $(n + \frac{1}{2})h\nu(s)$. To simplify the calculations of the $P(E)$, the one-dimensional barrier $V(s)$ is commonly approximated by a parabola or by the Eckart function.

For a monomolecular reaction, in terms of the transition state theory (the RRKM approximation—Ref. [14]), the total reaction rate obtained by averaging over all vibration states is:

$$k(E) = A\left(\frac{E - E^{\neq}}{E}\right)^{m-1} \tag{1.37}$$

where E is the total energy of the molecule, m is the number of the vibrational degrees of freedom of the ground state and A is the frequency factor (in s^{-1} units) calculated from the data on frequencies of the ground and the transition states:

$$A = \left(\prod_{i=1}^{m} \nu_i\right) \Big/ 2\pi \left(\prod_{i=1}^{m-1} \nu_i^{\neq}\right) \tag{1.38}$$

So as to take into account the contribution from tunnelling, the assumption is made that the motion along the reaction coordinate is separated from all other degrees of freedom. In this case, one may obtained for the energy barrier described by a parabola a new expression for the rate constant of the monomolecular reaction which includes the possibility of one-dimensional tunnelling:

$$k(E) = \frac{(m-1)! \prod_{i=1}^{m} h\nu_i}{2hE^{m-1}} \sum_{n} P(E)\left[E - E^{\neq} - h\sum_{i=1}^{m-1} \nu_i^{\neq}\left(n_i + \tfrac{1}{2}\right)\right] \tag{1.39}$$

where n_i are the quantum vibration numbers and $P(E)$ is the probability of one-dimensional tunnelling along the reaction coordinate versus energy calculated from the relationship:

$$P(E) = e^{\varepsilon}/(1 + e^{\varepsilon}) \tag{1.40}$$

where $\varepsilon = 2\pi E/h\nu_{im}$ with ν_{im} standing for the imaginary frequency of the vibration of the transition state which predetermines the direction of its decomposition.

As is evident from Eqs. (1.37)–(1.40), for calculating the rate constant that would take account of tunnelling through the potential barrier one needs to know the matrix of the force constants both for the initial structure and for the transition state structure (see Sect. 1.3.3). The application of the scheme under discussion to calculating the rates of some simple organic reactions has permitted the role of the tunnelling mechanism in the determination of these rates to be assessed from a different angle. One of the reactions best studied in this respect, for which detailed non-empirical calculations were carried out with precise localization of the transition state by the gradient method [114, 117], is the monomolecular decomposition of formaldehyde:

$$H_2C{=}O \longrightarrow H_2 + C{\equiv}O$$

The standard way for this reaction to be realized is photodissociation. It develops from a highly excited vibrational level of the electron ground state, which is reached upon radiationless transition from the first excited singlet electron state. The total energy of the molecule E consists of the energy of the $S_0 \rightarrow S_1$ photoexcitation (80 kcal/mol) and the energy of the ground state zero vibrations (16.7 kcal/mol). This value of 97 kcal/mol is lower than the activation barrier of 103.6 kcal/mol calculated from Eq. (1.22), which comprises the energy of zero vibrations of the transition state (11.7 kcal/mol) and the energy level of the transition state saddle point (92 kcal/mol). Under these conditions, the classical rate constant should equal zero. At the same time, the calculation of this constant from Eq. (1.39), with tunnelling allowed for, yields the value of $k \approx 10^6 \, s^{-1}$.

One more reaction, for which the crucial rôle of the tunnelling mechanism was demonstrated, is the 1,2-sigmatropic hydrogen shift in unsaturated carbene vinylidene XLIX—a prototype of unsaturated carbenes:

$H_2C{=}C: \longrightarrow$ [L: transition state, 1.428, 52, 1.185, 1.074] $\longrightarrow HC{\equiv}CH$

XLIX L La

Detailed nonempirical calculations—extended basis set, inclusion of configurational interaction [118]—permitted one to look afresh at the problem of experimental fixation of XLIX, which is now actively attended to [119].

Table 1.5 lists the frequencies of normal vibrations of vinylidene and the transition state L of its rearrangement into acetylene La which has, in acordance with the symmetry requirements (Sect. 1.3.3.2), planar structure ($E^{\neq} = 6.4$ kcal/mol). The calculation of the rearrangement rate using the data of Table 1.5 leads to the conclusion in favor of extremely fast tunnelling across the barrier even when the energy of XLIX does not exceed the energy of its zero vibrations, i.e., in the absence of any vibrational excitation whatsoever. The calculated life-time (reciprocal value of the rate constant) of vinylidene is a mere

Table 1.5. Frequencies of normal vibrations (cm^{-1}) calculated for vinylidene XLIX and the transition state L (ab initio, *DZ* basis set, CI [118])

XLIX		L	
ν^{as}_{CH}	3244	ν_{CH}	3454
ν^{s}_{CH}	3239	ν_{CH}	2699
ν_{CC}	1710	ν_{CC}	1844
δ_{CCH}	1288	δ_{H^2C}	937
δ (off plane)	787	δ (off plane)	573
δ_{CH_2} (pyramid.)	444	reaction coordinate	1029 im

Table 1.6. Calculated classical and tunnelling rates of cyclobutane radical cation isomerization at various temperatures

T(K)	k (class.) (S^{-1})	k (tunnell.) (S^{-1})
100	1.034×10^{8}	1.67×10^{10}
198	3.25×10^{10}	1.67×10^{10}
298	5.96×10^{11}	1.67×10^{10}
398	1.17×10^{12}	1.67×10^{10}

10^{-11} s, dropping to 10^{-12} s with the redundant energy of 2 kcal/mol. Clearly, the preparative isolation of vinylidene is out of the question despite the fact that a shallow minimum on the PES corresponds to the structure of XLIX. On the other hand, the life-time of carbene is, at temperatures close to 0 K, sufficient for its spectrum to be observed.

Until recently, the tunnelling of heavy atoms was thought to be unlikely. This at first sight selfevident conclusion has, however, proved to be not always in conformity with truth. Carpenter was the first to have taken notice of it [120]. He made an estimate of the rate constant of tunnelling in the automerization of cyclobutadiene reaction, according to which this process could constitute $> 97\%$ of the total rate constant below 0° C. A series of works [121–123] that followed this study warranted the conclusion that the bond shift reactions of [4*n*]-annulenes and the interconversion of Jahn–Teller isomers, i.e., the species derived from a common symmetric precursor by operation of the Jahn–Teller effect, could proceed by a mechanism that is primarily heavy-atom tunnelling. This process is important in situations in which the distance the heavy atom has to move so as to rearrange is on the order of the de Broglie wavelength of that atom. In the case of cyclobutadiene, the distance (< 0.2 Å) the carbon atoms move during the reaction is comparable with the de Broglie wavelength of carbon. Generally, a tunnelling of "organic" elements can occur only if the geometries of the interconverting isomers closely resemble one another. Table 1.6 lists the classical and the tunnelling rate constants of the cyclobutadiene radical cation isomerization. The tunnelling rate is predicted to predominate at temperatures below 196 K.

1.6 Description of Nonadiabatic Reactions

The Born–Oppenheimer approximation (adiabatic approximation) becomes unsatisfactory when the potential energy surfaces draw closer or intersect so that the energy difference between them turns comparable with the vibrational quantum $h\nu$. In the region of the mixing of electron states, a strong interaction between the electron and the nuclear motion arises, which was termed the vibronic interaction. The narrow energy gap between the ground and the excited

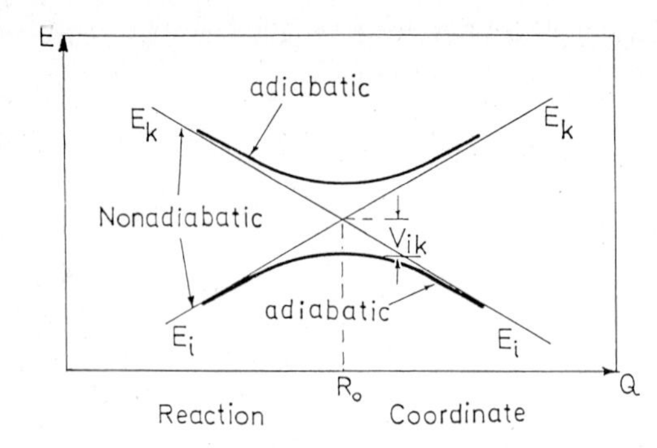

Fig. 1.20. Intersection of the potential surfaces. V_{ik} is the energy of interaction between the electron states i and k in the region of their drawing together

states means, in terms of classical mechanics, that the velocity of nuclear motion gets close to that of the motion of electrons.

From the quantum mechanical point of view, this suggests that in the region of intersection or quasiintersection of the terms the operator of the kinetic energy of nuclei cannot be neglected and that a solution has to be sought to the general electron-nuclear equation of Schrödinger [124].

The intersection or drawing together of the PES's ordinarily occurs in a quite small and localized region of internal coordinates so that one may retain the notion of adiabatic surfaces using a special approach only for the intersection region.

Figure 1.20 shows intersection of the potential surfaces and the region of the adiabatic (solid curve) and nonadiabatic (diabatic) (dotted line) potentials. During the reaction, the system moving along the reaction coordinate usually stays on the lower potential surface E_i with the probability of staying there not depending on time. However, at low values of V_{ik} and high rates of passage across the region where the PES's draw closer (R_0), a finite probability arises of a jump over to another adiabatic surface with the energy of E_k. In terms of the dynamic approach (Sect. 1.3), this means that after passing the nonadiabatic region, two reaction channels are open to the system: one over the PES E_i (adiabatic) and another over the PES E_k (nonadiabatic).

The probability of the P_{ik} transition depends essentially on the form of the potential surfaces E_i and E_k. Its correct calculation for real multidimensional PES's constitutes a complex mathematical problem, which is why it is a common practice to perform in this case one-dimensional approximation (the reaction coordinate is approximated by one parameter) and make use of the classical expression for the probability of a transition between the PES's E_i and E_k known as the Landau–Zener formula.

In the nonadiabatic region, a PES can satisfactorily be described by hyperbolas with asymptotes:

$$\begin{aligned} E_i' &= -l_i(R - R_0) \\ E_k' &= -l_k(R - R_0) \end{aligned} \tag{1.41}$$

where l_i, l_k are the tangents of the angles of inclination against the axis of the reaction coordinate, and R_0 is the point of maximal proximity of the terms on the reaction coordinate.

The probability of a transition between the PES's E_i and E_k in the case of uniform motion along the coordinate R at a constant velocity v is:

$$P_{ik} = \exp[-4\pi^2 V_{ik}^2/hv(l_i - l_k)] \tag{1.42}$$

where $V_{ik} = \langle \psi_i | V(t) | \psi_k \rangle$ with $V(t)$ being the term of the potential interaction energy in the Hamiltonian operator depending on the time of motion along the reaction coordinate $R = vt$.

A correct calculation of nonadiabatic reaction rates requires the construction of the PES's of the ground state and of at least one more, i.e., the lowest excited state over the whole region of variation of coordinates. Being quite laborious, these calculations have been performed only for several simplest reactions [125]. By way of example, Fig. 1.21 shows those regions of the configurational space where there occurs intersection of the PES of the ground electron state of the parent carbene CH_2 with the lower excited states.

The general condition for the intersection of the PES's of different electron states is given by the equalities.

$$E_i = E_k; \qquad V_{ik} = 0 \tag{1.43}$$

which may be satisfied for multiatomic systems in some regions of the configurational space through certain variations of the system's geometry even if the intersecting terms belong to identical symmetry types [125]. In many cases, a refinement of the Hamiltonian of the system through inclusion of additional terms into it or the refinement of the initial wave function may help avoid crossing of the nonadiabatic surfaces or even pushes them apart (Fig. 1.20). As a result, the possibility is created of returning to adiabatic description of the interacting PES's, and in the area of their close proximity and osculation the reaction course may be analyzed by means of the relationship (1.42). A

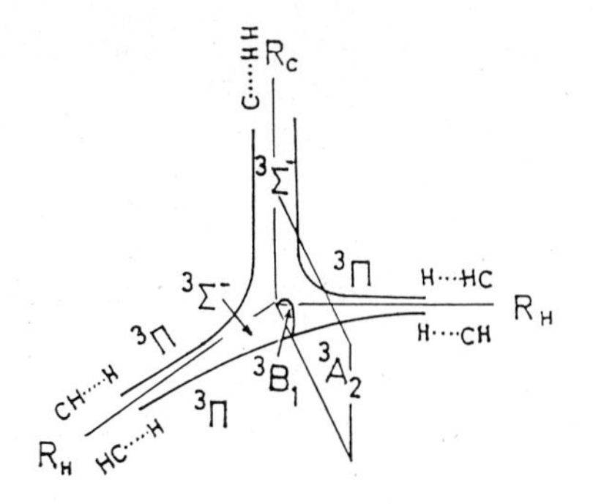

Fig. 1.21. Intersection of the ground state $^3A''$ PES of CH_2 with those of excited electron states [124]. The Pekeris coordinates are used: $R_A = 1/2(R_{AB} + R_{AC} - R_{BC})$; $R_B = 1/2(R_{AB} + R_{BC} - R_{AC})$; $R_C = 1/2(R_{AC} + R_{BC} - R_{AB})$. (Reproduced with permission from the American Chemical Society)

consistent classification by the reaction types of the variants of avoided crossings of potential surfaces has been suggested by Salem and his coworkers [126–128].

The first type comprises the reactions in which intersection of the surfaces occurs with retention of a certain symmetry element thus garanteeing the validity of the second equality of Eq. (1.43). In this connection, one may point to the important case of retention of the symmetry plane which provides for separation and orthogonality of the σ and π orbitals and for the absence of interaction between the electrons in these orbitals. An example is given by the photochemical reaction of splitting off of hydrogen by carbonyl compounds leading to the biradical state and the transition of one of the electrons from the σ to the π orbital:

$$\underset{2\pi\sigma}{H_2C{=}O} + \underset{2\sigma}{H{-}CH_3} \longrightarrow \underset{3\pi\ 2\sigma}{H_2\dot{C}{-}\ddot{O}{-}H} \quad \underset{1\sigma}{\cdot CH_3}$$

As may be seen from Fig. 1.22, even an insignificant distortion of the nuclear configuration (acoplanarization of the formaldehyde molecule by about 0.1 Å) results in the surfaces of the ground and the excited states being pushed apart as a result of the mixing of σ and π which gives rise to additional terms of the electron-nuclear interactions in the Hamiltonian.

The second type of the reactions, in which the pushing apart of the potential surfaces and the avoiding of their crossing are achieved through improvement on the form of the wave function, are those associated with the electron transfer (see Sect. 9.1) which occurs in a certain area of the configurational space. An example, known from inorganc chemistry, is the dissociation of NaCl into ions from the parent covalent form. The reactions of geometric isomerization of olefines through rotation about a double bond belong to the same category. The unstable structure on the PES of the ground term has in the transition state

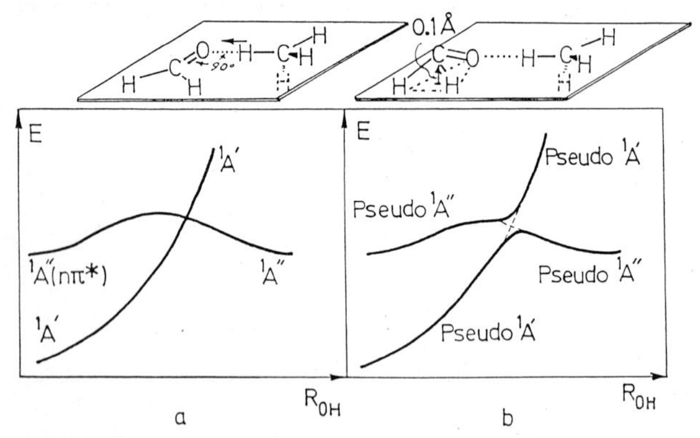

Fig. 1.22a. Crossing of energy surfaces of electron states of different symmetry for the formaldehyde + methane reaction with the symmetry plane retained; **b** avoided crossing when nuclear symmetry is distorted [127]. (Reproduced with permission from the American Chemical Society)

region (90° rotation) a biradical nature LI, while the excited ($\pi\pi^*$) state corresponding to this structure is bipolar (LII):

LI LII

Upon introduction of the acceptor substituents, the above-mentioned terms lie in polar solvents in close proximity and can even cross if the wave function employed is either purely ionic or purely covalent. Such crossing can be avoided by mixing the covalent and the ionic states.

The third reaction type characterized by pushing apart of nonadiabatic surfaces is associated with intersection of the frontier orbitals in the thermal reaction forbidden by the Woodward–Hoffmann rules. A typical example is given by the $(2s + 2s)$ cycloaddition reactions. Figure 1.23b shows that the surface of the double-excited electron state S^2S^2 intersects the PES of the ground state, which in its symmetry coincides with S^2S^2, when these states are described by single electron configurations (one-determinant approximation). Mixing of the configurations and avoiding of the crossing (Fig. 1.23c) can be achieved through inclusion in the Hamiltonian, in addition to the averaged interelectron repulsion, of an operator of the interelectronic interaction which makes allowance for the electron correlation.

It should be noted that the avoided crossing depicted in Fig. 1.23c does not do away with the crossing of the frontier orbitals (Fig. 1.23a). In general, all the foregoing discussion of nonadiabatic transitions concerns the electron states (terms) rather than the orbitals. Fruitfulness of an analysis of the correspondence between the electron states of the reactants and the products was vividly demonstrated by Longuet–Higgins and Abrahamson [129] who developed the most general rules of selection of symmetry-allowed pericyclic reactions. Furthermore, this correspondence served as the basis for a new quite rational classification of the photochemical reactions [127].

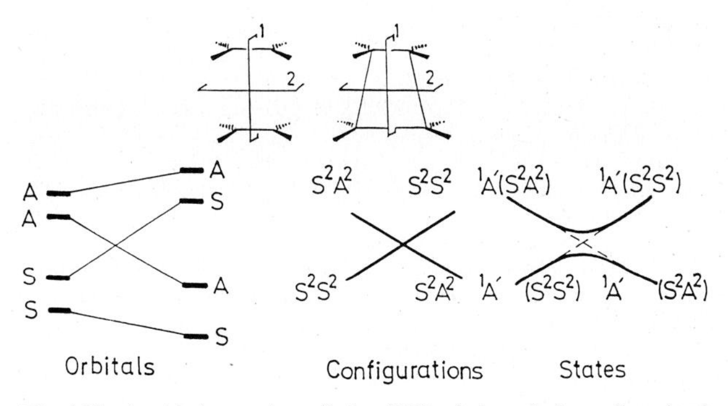

Fig. 1.23. Avoided crossing of the PES of the ethylene dimerization reaction. Classification is performed relative to two symmetry planes retained in the course of reaction

References

1. Ingold CK (1969) Structure and mechanism in organic chemistry. Cornell University Press, Ithaca London
2. Lowry TH, Richardson KS (1981) Mechanism and theory in organic chemistry, Harper & Row, New York
3. Levine RD, Bernstein J (1974) Molecular reaction dynamics. Oxford University Press, New York
4. Karlström G, Engström S, Jönsson B (1978) Chem Phys Lett 57:390
5. Mezey PG (1980) Theor Chim Acta 54:95
6. Stanton RW, McIver JV (1975) J ACS 97:3632
7. Wilson EB, Decius JC, Cross PC (1955) Molecular vibrations. The theory of infrared and raman vibrational spectra. McGraw-Hill, New York
8. Pulay P (1977) in: Schaefer III HF (ed) Applications of electronic structure theory. Modern theoretical chemistry v 4 Plenum Press, New York
9. Hehre WJ, Radom L, Schleyer PvR, Pople JA (1986) Ab initio molecular orbital theory. Wiley and Sons Inc, New York
10. Rauscher G, Clark T, Poppinger D,Schleyer PvR (1978) Angew Chem Int Ed Engl 17:276
11. Ritchie JP (1983) J ACS 105:2083
12. Dewar MJS, Ford GP (1977) ACS 105:1685
13. Dewar MJS, Ford GP, McKee ML, Rzepa HS, Thiel W, Yamaguchi K (1978) J Mol Struct 43:135
14. Eyring H, Lin SH, Lin SM (1980) Basic chemical kinetics. Wiley and Sons, Inc, New York; Klots CE (1988) Acc Chem Res 21:16
15. Flanigan MC, Komornicki A, McIver JW (1977) in: Segal GA (ed) Semiempirical methods of electronic structure calculation. Part B: Applications. Plenum Press, New York
16. Dewar MJS, Ford GP (1977) J ACS 99:7822
17. Wooley RG (1978) J ACS 100:1073
18. Wooley RG (1986) Chem Phys Lett 125:200
19. Bader RFW (1985) Acc Chem Res 18:9
20. Minkin VI, Minyaev RM, Zhdanov YuA (1987) Nonclassical structures of organic compounds. Mir Publ, Moskow
21. Bader RFW, Slee TS, Cremer D, Kraka E (1983) J ACS 105:5061
22. Wiberg KB, Bader RFW, Lau CDH (1987) J ACS 109:985
23. Bader RFW (1986) Can J Chem 64:1036
24. Tang TH, Bader RFW, MacDougall DJ (1985) Inorg Chem 24:2042
25. Tal Y, Bader RFW, Nguyen-Dang TT, Ojha M, Anderson SG (1981) J Chem Phys 74:5162
26. Bader RFW, Tang TH, Tal Y, Biegler-Konig RF (1982) J ACS 104:940, 946
27. Ritchie JP, Bachrach SM (1987) J ACS 109:5909
28. Bersuker IB (1980) Nouv J Chim 4:139, Bersuker IB (1984) The Jahn-Teller effect and vibronic interactions in modern chemistry. Plenum Press, New York
29. McIver JW, Komornicki A (1972) J ACS 94:2625
30. Murrell JN, Laidler KJ (1968) Trans Farad Soc 64:371
31. Dewar MJS (1975) Chem Brit 11:95
32. Rothman MJ, Lohr LL (1980) Chem Phys Lett 70:405
33. Bauschlicher CW, Schaefer III HF, Bender CF (1976) J ACS 98:1653; Ortega M, Lluch JM, Oliva A, Bertran J (1987) Can J Chem 65:1995
34. Ishida K, Morokuma K, Komornicki A (1977) J Chem Phys 66:2153
35. Müller K (1980) Angew Chem, Int Ed Engl 19:1; Dunning TH, Harding LB, Bair RA, Eades RA, Shepard RL (1986) J Phys Chem 90:344
36. Head JD, Weiner B, Zerner MC (1988) Int J Quant Chem 33:177
37. Basilevsky MV, Ryaboy VM (1987) in: Veselov MG (ed) Current problems of quantum chemistry. The quantum chemical methods. The theory of intermolecular interactions and solid state. Khimia, Moskow (in Russian)
38. Himmelblau DM (1972) Applied nonlinear programming. McGraw Hill, New York
39. McIver JW, Komornicki A (1972) J ACS 94:2625
40. Pancíř J (1975) Collect Czech Chem Commun 40:1117
41. Basilevsky MV, Shamov AG (1981) Chem Phys 60:347

42. Cerjan CJ, Miller WH (1981) J Chem Phys 75:2800
43. Simons J, Jørgensen P, Taylor H, Ozment J (1983) J Phys Chem 87:2745
44. Baker J (1986) J Comput Chem 7:385
45. Mezey PG, Peterson MR, Csizmadia IG (1977) Can J Chem 55:2941
46. Halgren TA, Lipscomb WN (1977) Chem Phys Lett 49:225
47. Müller K, Brown LD (1979) Theor Chim Acta 53:75
48. McIver JW (1974) Acc Chem Res 7:72
49. Dewar MJS, Ford GP, Rzepa HS (1977) J Chem Soc Chem Commun 728
50. Minkin VI, Minyaev RM (1977) Zh Org Khim 13:1129
51. Hoffmann R (1982) in: Laidler KJ (ed) IUPAC frontiers of chemistry. Pergamon Press, Oxford
52. Borden WT, Davidson ER (1981) Acc Chem Res 14:69
53. Borden WT, Davidson ER (1979) J ACS 101:3771
54. Glukhovtsev MN, Simkin BYa, Minkin VI (1985) Uspekhi Khim (Russ Chem Fev) 54:86
55. Voter AF, Goddard III WA (1986) J ACS 108:2830
56. Murrell JN (1977) Structure and Bonding 32:93
57. Silver DM, Stevens RW (1973) J Chem Phys 59:3378
58. Melander L, Saunders Jr WH (1980) Reaction rates of isotopic molecules. Wiley and Sons Inc, New York
59. Brown SB, Dewar MJS, Ford GP, Nelson JN, Rzepa HS (1978) J ACS 100 : 7832
60. Jensen A (1983) Theor Chim Acta 63:269
61. Dewar MJS, Kirschner S (1971) J ACS 93:4292
62. Fukui K (1981) Acc Chem Res 14:363
63. Kato S, Morokuma K (1980) J Chem Phys 73:3900
64. Ishida K, Majama S (1983) Theor Chim Acta 62:245
65. Yamashita K, Yamabe T (1983) Int J Quant Chem: Quant Chem Symp 17:177
66. Yamabe T, Zhao CD, Koizumi M, Tachibana A, Fukui K (1985) Can J Chem 63:1532
67. Basilevsky MV (1977) Chem Phys 24:81
68. Mezey PG (1977) in: Csizmadia IG (ed) Progress in theoretical organic chemistry, Elsevier, Amsterdam 2:127
69. Quapp W, Heidrich D (1984) Theor Chim Acta 66:245
70. Eisenhart LP (1963) Riemannian geometry. Prinston Univ Press, Princeton
71. Tachibana A, Fukui K (1978) Theor Chim Acta 49:321
72. Pechukas P (1976) J Chem Phys 64:1516
73. Pearson RG (1976) Symmetry rules for chemical reactions. Orbital topology and elementary processes. Wiley and Sons, Inc, New York
74. Rodger A, Schipper PE (1986) Chem Phys 107:329
75. Valtazanos P, Ruedenberg K (1986) Theor Chim Acta 69:281
76. Valtazanos P, Elbert ST, Ruedenberg K (1986) J ACS 108:3147
77. Minkin VI, Minyaev RM (1975) Zh Org Khim 11:1993
78. Martin JC, Perozzi EF (1976) Science 191:154
79. Burwell RL, Pearson RG (1966) J Phys Chem 70:300
80. Wolfe S, Schlegel HB, Csizmadia IG, Bernardi F (1975) J ACS 97:2020
81. Salem L (1971) Acc Chem Res 4:322
82. Minkin VI, Minyaev RM, Zacharov II (1977) J Chem Soc Chem Commun 213
83. Minkin VI, Minyaev RM (1983) in: Kolotyrkin YaM (ed) Physical Chemistry. Current problems. Khimia, Moskow (in Russian)
84. Ehrenson S (1974) J ACS 96:3778
85. Swanson BI (1976) J ACS 98:3067
86. Altmann JA, Yates K, Csizmadia IG (1976) J ACS 98:1450
87. Swanson BI, Arnord TH, Dewar MJS, Rafalko JJ, Rzepa HS, Yamaguchi Y (1978) J ACS 100:771
88. Rice FO, Teller E (1938) J Chem Phys 6:489
89. Hine J (1977) Adv Phys Org Chem 15:1
90. Hlebnikov AF (1976) in: Current problems of organic chemistry. Leningrad State Univ Press, Leningrad (in Russian)
91. Tee OS (1969) J ACS 91:7144
92. Dunitz JD (1979) X-ray analysis and the structure of organic molecules. Cornell Univ Press Ithaca

93. Bürgi HB (1975) Angew Chem Int Ed Engl 14:460
94. Bürgi HB, Dunitz J (1983) Acc Chem Res 16:153
95. Bürgi HB, Dunitz JD, Lehn JM, Wipff G (1974) Tetrahedron 30:1563
96. Holmes RR (1979) Acc Chem Res 12:257
97. Muetterties EL (1974) Tetrahedron 30:1595
98. Britton D, Dunitz JD (1981) J ACS 103:2971
99. Peterson MR, Csizmadia IG (1982) in: Csizmadia IG (ed) Progress in theoretical organic chemistry. Elsevier, Amsterdam 3:190
100. Chapuisat X, Jean Y (1976) Topics Curr Chem p 3; (1975) J ACS 97:6325
101. Wang ISY, Karplus M (1973) J ACS 95:8160
102. Horsley JA, Jean Y, Moser C, Salem L, Stevens RM, Wright JS (1972) J ACS 94:279
103. Arrighini GP, Guidotti (1983) THEOCHEM 10:49
104. Miller WH, Handy NC, Adams JE (1980) J Chem Phys 72:99
105. Waite BA, Gray SK, Miller WH (1983) J Chem Phys 78:259
106. Gray SK, Miller WH, Yamaguchi Y, Schaefer III HF (1980) J Chem Phys 73:2733
107. Havlas Z, Zahradnik R (1984) Int J Quant Chem 26:607
108. Frisch MJ, Binkley JS, Schaefer III HF (1984) J Chem Phys 81:1882
109. Basilevsky MV (1972) Zh Vsesoyuzn Khim Obszh im DI Mendeleeva 17:322
110. Goldanskii VI (1975) Uspekhi Khim (Russ Chem Rev) 44:2121
111. Goldanskii VI, Trakhtenberg LI, Fleurov VN (1986) Tunneling process in chemical physics. Nauka Press, Moskow (in Russian)
112. Zamaraev KI, Khairutdinov RF (1978) Uspekhi Khim (Russ Chem Rev) 47:992
113. Miller WH (1979) J ACS 101:6810
114. Miller WH (1983) J ACS 105:216
115. Gray SK, Miller WH, Yamaguchi Y, Schaefer III HF (1981) J ACS 103:1900
116. Osamura J, Schaefer III HF, Stepehn KG, Miller WH (1981) J ACS 103:1904
117. Schaefer III HF (1979) Acc Chem Res 12:288
118. Carpenter BK (1983) J ACS 105:1700
119. Dewar MJS, Merz KM, Steward JJP (1984) J ACS 106:4040
120. Dewar MJS, Merz KM (1985) J Phys Chem 89:4739
121. Dewar MJS, Merz KM (1985) THEOCHEM 122:59
122. Bersuker IB, Polinger BZ (1983) Vibronic interactions in molecules and crystals. Nauka Press, Moskow (in Russian)
123. Kondrat'ev VN, Nikitin EE (1975) Kinetic and mechanism of gas-phase reactions. Nauka Press, Moskow (in Russian)
124. Davidson ER (1977) J ACS 99:397
125. Herzberg G (1966) Molecular spectra and molecular structure III. Electronic spectra and electronic structure of polyatomic molecules. Van Nostrand, Toronto
126. Salem L (1974) J ACS 96:3486
127. Salem L, Leforestier C, Segal G, Wetmore R (1975) J ACS 97:479
128. Duben WG, Salem L, Turro NJ (1975) Acc Chem Res 8:41
129. Longuet-Higgins HC, Abrahamson EW (1965) J ACS 87:2045

CHAPTER 2

Quantum Chemical Methods for Calculating Potential Energy Surfaces

This chapter considers briefly the general theoretical principles of modern quantum mechanical methods for calculating the PES's and the pathways of chemical reactions and presents a concise characteristic of some particular methods pointing out the most suitable areas of their application.

2.1 General Requirements Upon the Methods for Calculating Potential Energy Surfaces

Evidently, the shape of the PES of a reacting system, i.e., its topography, the position and nature of stationary points may depend upon the method used for calculating the energy of a given system. Accordingly, the theoretical conclusions as to the reaction mechanisms may differ depending on the type of approximations used in the calculations.

The use of an extended basis set in nonempirical calculations, taking account of the electron correlation as well as of the corrections for relativistic effects and nonadiabaticity would undoubtedly have provided for more exact theoretical predictions. But such a program is not realizable at present for systems containing more than 3–4 atoms of the second and higher periods. In practice, when making theoretical assessment of the mechanisms of organic reactions, one is confronted with the choice between either a rigorous calculational method applicable to rather limited areas of the PES or some simplified approaches, which, however, would permit investigation of a much more extensive area of the PES. Gradually, the most adequate methods are found for the solution to particular problems and, at the same time, typical limitations of these methods become apparent.

A method for calculating a PES has to satisfy certain basic requirements before its results may be used as a basis for theoretical interpretation of a reaction mechanism. Such a method must reliably reproduce: 1) relative energies of the reactants and the reaction products; 2) relative energies of the reactants and the transition state; 3) the PES curvature in those zones of the stationary points which correspond to the structure of the reactants, the products and the transition state of reaction; 4) geometrical characteristics of the reactants, the products and the transition state.

Meeting the conditions 1 and 2 is necessary for correct evaluation of the heat and the activation energy of a given reaction, the condition 3 has to be satisfied in order to reproduce the vibration spectrum and, consequently, the entropy and the isotopic effects while the conditions 3 and 4 are essential for the calculation of the true reaction path.

Many modern quantum-mechanical methods, especially the semiempirical ones, have been devised with the specific aim of calculating some particular properties of molecular systems with the result that they adequately satisfy only one or two of the requirements mentioned. For this reason, the conclusions on reaction mechanisms obtained by such methods should be treated with a degree of caution.

2.2 Nonempirical (ab initio) Methods. The Hartree–Fock Method

The theory and the analysis of calculational schemes of quantum chemistry have been dealt with in detail in a number of books [1–6] and review articles [7–12]. Here we give only a brief account of the main principles of the general theory of molecular orbitals (MO) that provides the basis for constructing the most important nonempirical methods of quantum chemistry.

2.2.1 Closed Electron Schells

The MO approximation is based on the assumption that an individual spin orbital $\varphi_i\alpha$ or $\varphi_i\beta$ (where φ_i is the function of space coordinates and α ($m_s = 1/2$) or $\beta(m_s = -1/2)$ are the functions of spin coordinates) corresponds to each electron. The full wave function of a many-electron system ψ in the Hartree–Fock approximation is written as a Slater determinant whose form provides for the property of antisymmetry of ψ, required by the Pauli principle, with respect to the pairwise permutation of any electron

$$\psi(1,2,\ldots,2n) = [(2n)!]^{-1/2} \begin{vmatrix} \varphi_1(1)\alpha(1) & \varphi_1(1)\beta(1) & \cdots & \varphi_n(1)\beta(1) \\ \varphi_1(2)\alpha(2) & \varphi_1(2)\beta(2) & \cdots & \varphi_n(2)\beta(2) \\ \cdots & \cdots & \cdots & \cdots \\ \cdots & \cdots & \cdots & \cdots \\ \varphi_1(2n)\beta(2n) & \varphi_1(2n)\beta(2n) & \cdots & \varphi_n(2n)\beta(2n) \end{vmatrix}$$

$$\equiv [(2n)!]^{-1/2} |\varphi_1(1)\alpha(1)\varphi_1(2)\beta(2)\cdots\varphi_n(2n)\beta(2n)| \tag{2.1}$$

The wave function of Eq. (2.1) corresponds to the so-called restricted Hartree–Fock method (RHF) for the closed electron shells when $2n$ electrons are located on n MO's pairwise with opposite spins, with the symmetry of each MO corresponding to the point group of symmetry of the nuclear configuration of a molecule. The orbitals φ_i are chosen in such a manner as to minimize the total energy of the system

$$E = \int \psi(1,2,\ldots,2n)\hat{H}\psi(1,2,\ldots,2n)d\tau_1, d\tau_2 \cdots d\tau_{2n} \tag{2.2}$$

where $\hat{H}$ is the electronic Hamiltonian of a molecule representing the sum of the operators of the following energies: the kinetic energy of electrons ($\hat{T}_e$), the potential energy of interaction of the electrons with one another ($\hat{V}_{ee}$) and with the nuclei ($\hat{V}_{en}$) and the potential energy of interaction of the nuclei with one another ($\hat{V}_{nn}$):

$$\hat{H} = \hat{T}_e + \hat{V}_{ee} + \hat{V}_{en} + \hat{V}_{nn}$$

In the existing practical methods of calculation, the linear combination of atomic orbitals (LCAO) approximation is employed where the atomic orbitals χ_μ, centered at each atom, are used as an expansion basis set:

$$\varphi_i = \sum_{\mu=1}^{N} C_{\mu i}\chi_\mu \qquad i = 1,2,\ldots,n \tag{2.3}$$

where χ_μ are the atomic orbitals (AO) and N is their total number.

The coefficients $C_{\mu i}$ in the expansion of Eq. (2.3) can be derived by solving the Roothaan equations:

$$\sum_{\mu=1}^{N} C_{\mu i}(F_{\mu\nu} - \varepsilon_i S_{\mu\nu}) = 0 \tag{2.4}$$

where $F_{\mu\nu}$ are the matrix elements of the Fock matrix:

$$F_{\mu\nu} = H_{\mu\nu} + \sum\sum P_{\lambda\sigma}[(\mu\nu|\lambda\sigma) - \tfrac{1}{2}(\mu\lambda|\nu\sigma)] \tag{2.5}$$

$S_{\mu\nu}$ is the overlap integral between the AO's χ_μ and χ_ν:

$$S_{\mu\nu} = \int \chi_\mu(1)\chi_\nu(1)\,d\tau_1 \tag{2.6}$$

$$H_{\mu\nu} = \int \chi_\mu(1)[-\tfrac{1}{2}\nabla^2]\chi_\nu(1)d\tau_1 + \int \chi_\mu(1)\left(\sum_{A}^{\text{nuclei}} Z_A r_{1A}^{-1}\right)\chi_\nu(1)d\tau_1 \tag{2.7}$$

$$(\mu\nu|\lambda\sigma) = \iint \chi_\mu(1)\chi_\nu(1)r_{12}^{-1}\chi_\lambda(2)\chi_\sigma(2)d\tau_1 d\tau_2 \tag{2.8}$$

$P_{\lambda\sigma}$ is the bond order matrix:

$$P_{\lambda\sigma} = 2 \sum_{i=1}^{\text{occup.}} C_{\lambda i} C_{\sigma i} \tag{2.9}$$

The orbital energies ε_i can be obtained through solving the equation:

$$|F_{\mu\nu} - \varepsilon_i S_{\mu\nu}| = 0 \tag{2.10}$$

In the Roothaan method, the expression for the total energy [Eq. (2.2)] is reduced to:

$$E = 2 \sum_{i}^{\text{occup.}} \varepsilon_i - \sum_{i>j}^{\text{occup. occup.}}\sum (2J_{ij} - K_{ij}) + \sum_{A>B}\sum Z_A Z_B R_{AB}^{-1} \tag{2.11}$$

where J_{ij} and K_{ij} are the Coulomb and the exchange integrals, resp.:

$$J_{ij} = \iint \varphi_i(1)\varphi_j(2) r_{12}^{-1} \varphi_i(1)\varphi_j(2)\, d\tau_1\, d\tau_2 \tag{2.12}$$

$$K_{ij} = \iint \varphi_i(1)\varphi_j(2) r_{12}^{-1} \varphi_i(2)\varphi_j(1)\, d\tau_1\, d\tau_2 \tag{2.13}$$

and the last term represent the energy of repulsion of atomic nuclei.

Equations (2.12) and (2.13) may readily be transformed into a form in which the expansion in the AO [Eq. (2.3)] will be taken into account.

2.2.2 Open Electron Shells

In case a molecular system is characterized by an open (non-closed) electron shell, its wave function will generally be represented more correctly not in one-determinant approximation [Eq. (2.1)] but rather as a linear combination of the Slater determinants:

$$\psi = b_1\psi_1 + b_2\psi_2 + \cdots \tag{2.14}$$

with all possible distributions of the unpaired electron over the partially filled zone of orbitals. However, for the radicals and the lower energy states with the multiplicity $(p - q + 1)$, where p and q are the numbers of the electrons with α and β spin functions $(p > q)$, the wave function may be given by one determinant:

$$^{(p-q+1)}\psi_{\text{RHF}} = |\varphi_1(1)\alpha(1)\varphi_1(2)\beta(2)\cdots\varphi_q(2q)\beta(2q)\varphi_{q+1}(2q+1) \\ \alpha(2q+1)\cdots\varphi(p+q)\alpha(p+q)| \tag{2.15}$$

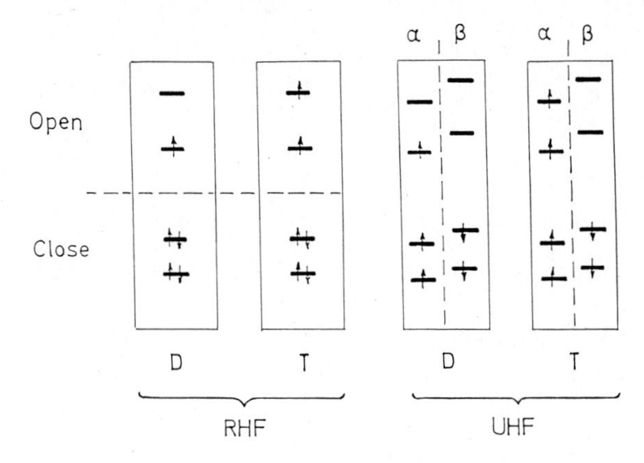

Fig. 2.1. Schematic representation of the electron distribution over the MO's for open shells in the restricted (RHF) and unrestricted Hartree–Fock (UHF) methods (D stands for doublet and T for triplet electronic configuration)

Such a representation corresponds to the restricted Hartree–Fock approximation for open shells: the functions of the $\varphi_1, \varphi_2, \ldots, \varphi_q$ orbitals with the α and β spins are equal (Fig. 2.1). The total energy can be calculated from the equation:

$$
\begin{aligned}
E = {} & \underbrace{2\sum_{i}^{q} \varepsilon_i - \sum_{i>j}^{q}\sum^{q} (2J_{ij} - K_{ij})}_{\text{closed}} \\
& \underbrace{+ 2\sum_{i=q+1}^{p-q+1} \varepsilon_i - \sum_{i=q+1}^{p-q+1}\sum_{j=q+1}^{p-q+1} (2J_{ij} - K_{ij})}_{\text{open}} + \underbrace{\sum_{i=1}^{q}\sum_{j=q+1}^{p-q+1} (2J_{ij} - K_{ij})}_{\text{closed-open}} \\
& + \sum_{A>B}\sum Z_A Z_B R_{AB}^{-1}
\end{aligned}
\tag{2.16}
$$

Since for the open shells the total number of electrons with the β spin is not equal to the number of electrons with the α spin, their electron surroundings must be different and the assumption of equivalence of the space functions describing the spin orbitals of the α- and β-electrons will not be rigorous. In the more general approximation represented by the unrestricted Hartree–Fock (UHF) method, the electrons with the α- and β-spins correspond to different orbitals $\varphi_1^\alpha, \varphi_2^\alpha, \ldots, \varphi_q^\alpha$ and $\varphi_1^\beta, \varphi_2^\beta, \ldots, \varphi_q^\beta$ (Fig. 2.1), and the one-determinant wave function is written as:

$$
\begin{aligned}
{}^{p-q+1}\psi_{\mathrm{UHF}} = {} & |\varphi_1^\alpha(1)\alpha(1)\varphi_1^\beta(2)\beta(2)\varphi_2^\alpha(3)\alpha(3)\cdots \\
& \varphi_q^\beta(2q)\beta(2q)\varphi_{q+1}^\alpha(2q+1)\alpha(2q+1) \\
& \varphi_{q+2}^\alpha(2q+2)\alpha(2q+2)\cdots\varphi_{p+q}^\alpha(p+q)\alpha(p+q)|
\end{aligned}
\tag{2.17}
$$

When the MO's in Eq. (2.17) are represented in the LCAO form, the Roothaan equations of Eq. (2.4) break up into two series of interrelated equations:

$$\sum_{\mu} c_{\mu i}^{\alpha}(F_{\mu\nu}^{\alpha} - \varepsilon_i^{\alpha} S_{\mu\nu}) = 0$$
$$\sum_{\mu} c_{\mu i}^{\beta}(F_{\mu\nu}^{\beta} - \varepsilon_i^{\beta} S_{\mu\nu}) = 0 \tag{2.18}$$

Calculations in the UHF approximation yield as a rule lower values of the total energy and more accurate spin density distributions. However, the wave functions obtained in the UHF approximation are, unlike those in the RHF approximation, not always the eigenfunctions of the operator of the total spin momentum $\hat{S}^2$ and, consequently, do not correspond exactly to the pure singlet, doublet, triplet etc. electronic states. To eliminate admixture of higher multiplicity states, various methods have been devised for designing a required spin component.

2.2.3 Basis Sets of Atomic Orbitals

The simplest level of the nonempirical (ab initio) and semiempirical all-valence calculations is the use of a minimal basis set of AO's where each AO χ_μ in the expansion of Eq. (2.3) is represented by one function, for example, by a Slater-type orbital (STO):

$$\chi_{nlm}(r, \theta, \varphi) = N r^{n-1} e^{-\xi r} Y_{lm}(\theta, \varphi) \tag{2.19}$$

where ξ is the effective charge of the nucleus (the Slater exponent); n, l, m are the quantum numbers and N is the normalization factor.

A considerable improvement in the accuracy of calculations can be achieved by making use of the DZ basis set where each valence orbital corresponds to two functions of the same type [Eq. (2.19)] but with different values of the Slater exponents:

$$H - 1s, 1s'$$
$$Li \cdots F - 1s, 1s', 2s, 2s', 2p_x, 2p_x'$$

Any basis set of the Slater functions that exceeds the DZ-type basis is regarded [2,6] as an extended basis set. Important is the case of a basis in which the functions of the DZ set are supplemented with the AO functions possessing a higher quantum number l, for example, with the d oribitals for the second period atoms and sometimes with the p orbitals for the hydrogen atoms. These additions make it possible to take into account the polarization of the orbitals of the atomic ground state, so when the d orbitals and the p-AO's are mixed we have

The AO's having higher orbital quantum numbers l are termed the polarization (p) functions and the DZ basis set with the polarization functions is designated as DZ + P.

When the basis of the STO functions are employed, by far the greater part of the computer time is spent on calculating the integrals of Eqs. (2.7) and (2.8). As one goes to the Gaussian-type (GTO) basis sets, this time consumption is drastically reduced. In this case, Eqs. (2.19) are approximated by a linear combination of several Cartesian Gaussian functions [6–8, 13].

$$\chi_{pqs}(x, y, z) = NX^pY^qZ^se^{-\alpha r^2} \tag{2.20}$$

where p, q, s are the integers and N is the normalization factor.

If, for instance, $p = s = 0$ and $q = 1$, the GTO corresponds to the function p_y. The minimal orbital basis set in which N Gaussian functions are used to approximate one Slater function is designated as the STO–NG basis set. Usually one sets $N = 3$, seeing that with the further increase the accuracy improves very slowly.

The STO basis set of the DZ type can be approximated by split polynomials of the Gaussian-type functions M–NPG. Each inner AO is replaced by M GTO orbitals, the valence 2s orbital—by N, while the p orbital—by P GTO functions. For example, the 4–31 G basis set describes every inner (1s) orbital by four GTO's, every valence 2s AO by three GTO's and every valence p AO by one GTO. It is important to point out that whereas in the case of the minimal basis set of the NG type the accuracy level of the minimal STO basis set cannot be attained even at great values of N, the use of the split-valence GTO M–NPG basis sets allows the Slater basis set level to be exceeded.

Widely used are the GTO basis sets of the 6–31 G* and 6–31 G** types. They correspond to extended STO basis sets which include polarization function. One asterisk denotes addition of the polarization d–GTO to each p function, while two asterisks mean that, besides that orbital, a p–GTO is added to the 1s orbitals of the hydrogen atoms[1]. There are cases when the 6–31 G**-type basis sets do not satisfy accuracy requirements in the calculation of physical characteristics of molecules. Pople and his co-workers [14] have suggested in this connection the basis sets of the types 6–311 G** (single zeta core, triple zeta valence and polarization functions on all atoms) and 6–311 + + G (3df, 3pd) which differ from the previous ones in further additions of the polarization

[1] The use of the GTO basis sets of the types mentioned in the present-day calculations of organic structures and reactions has been boosted by their inclusion in fast and effective programs GAUSSIAN (GAUSSIAN-70, -76, -80, -82, and GAUSSIAN-85) developed by the Pople group and supplied to the Quantum Chemistry Program Exchange Fund [6]

Table 2.1. Total energies of ethane conformeres (in a.u.) [16]

Structure	E, a.u.				
	STO–3G	4–31G	6–31G*	6–31G**	MP2/6–31G*
D_{3d}	– 78,30618	– 79,11593	– 79,22876	– 79,23823	– 79,49451
D_{3h}	– 78,30160	– 78,11151	– 79,22240	– 79,23321	– 79,48937
D_{2h}	– 77,96889	– 78,85466	– 78,97031	– 78,98786	– 79,25715
D_{4h}	– 77,30344	– 78,32567	– 78,45215	– 79,49100	– 79,79739

and diffusion functions on each atom. The sign + + means that diffusion AO's are added for all atoms, while 3*df* and 3*pd* denote, respectively, that three *d* AO's and one *p* AO are included in the basis set for all the elements of the second period, and three *p* AO's and one *d* AO for the hydrogen atom. A detailed list of the basis sets can be found in the book listed as Ref. [15] as well as in a review [13].

The data of Table 2.1 on the calculations of total energies for ethane in the stable D_{3d} and the eclipsed D_{3h} conformations as well as for its hypothetical bridged structural isomers with D_{2h} and D_{4h} symmetry may give an idea of how the results of nonempirical calculations are refined with the gradual extension of the basis set.

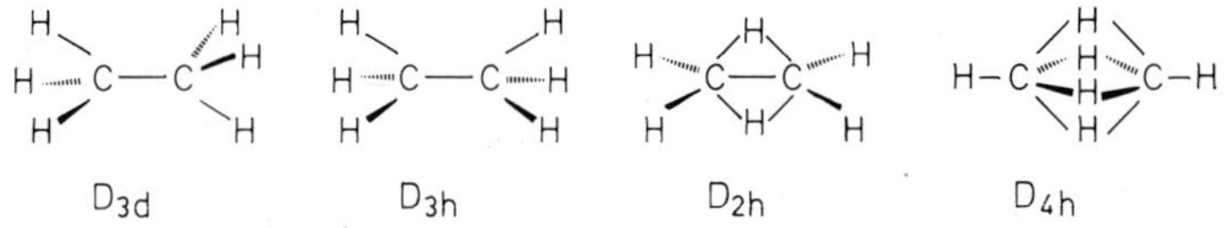

D3d D3h D2h D4h

As one goes from the minimal to an extended basis set, the total energy is lowered even for such simple structures by approximately one a.u., i.e., 625 kcal/mol. As may be seen from the Table, this lowering occurs not uniformly for different geometrical configurations. The relative energies of the two conformers D_{3d} and D_{3h}, which determine the height of the rotation barrier about the C—C bond, lie within the range of 2.7–4.1 kcal/mol, which is in good agreement with the experimental value of 2.93 ± 0.25 kcal/mol [17].

The minimal STO–3G basis set is fairly satisfactory for reproducing the molecular geometry (subject to an error of ~ 0.03 Å in bond lengths and ~ 4° in angles) of most structures with closed shells and the energies of isodesmic reactions. However, with the reactions in which the products do not retain all the bond types of the starting reagents one should pass to at least a 4–31 G-type basis set [7]. A similar basis is also required for a sufficiently good reproduction of the vibration spectrum of molecules, while the barrier of rotation about simple bonds, which include a heteroatom, and the barriers of pyramidal inversion cannot be described without the polarization functions being taken into account, i.e., in this case the use of an extended basis set would be in order.

Table 2.2 lists the basis sets evolved in practical calculations, which have proved the most suitable for calculating specific molecular properties. These

Table 2.2. Types of the basis sets of the AO's for correct description of molecular properties

Properties of molecular system	The least basis set	Notes and exceptions
Molecular geometry	Minimal	Except for calculation of dihedral angles and of geometry of pyramidal structures where polarization functions must be used.
Force constants	DZ	The 3–21 G-type basis set represents satisfactory level. No appreciable improvement is achieved by changing to the extended 6–31G basis.
Rotation barriers	Minimal	Except the molecules with a rotation axis that includes two heteroatoms, such as H_2O_2 and others. For them a DZ + P-type basis set is recommended.
Inversion barriers	DZ + P	—
Energy of reactions	DZ DZ + P	Suitable for reactions in which electronic bond pairs present in reactants are retained in the products (transition states). Otherwise, electron correlation should be taken into account.
Interaction between ions and dipoles; system with H bond.	Minimal	In the calculation of anions and their interactions, inclusion of polarization and diffusion functions is necessary.
Weak intermolecular-interactions	Extended	Polarization and diffusion functions must be included and correction for unbalanced character of the basis set allowed for.

sets represent the lower limit of requirements so that occasionally a higher energy level might be needed for a correct description of certain properties. Such a case may be exemplified by the structure of hexaazabenzene which currently attracts considerable attention in connection with reports that it was observed in experiments with matrix isolation [19].

Even though generally the DZ-type basis sets lead to fairly reliable evaluations of the molecular geometry of systems with a closed shell and permit their

$$N_6 \longrightarrow N_6 \longrightarrow 3\,N_2$$

assignment to a definite type of the critical point on the PES, in this particular case the calculations with, for example, a DZ + P-type set yield results essentially different from the chemical point of view [20]. Whereas with the DZ basis set a structure with D_{3h} symmetry represents the local minimum on the PES and the more symmetrical D_{6h} form corresponds to a transition state, in the case of the extended DZ + P set the theoretical predictions are exactly opposite.

2.2.4 Electron Correlation

The accuracy of the Hartree–Fock method (lowering of the total energy) has a limit when any further extension of the basis set fails to improve the results.

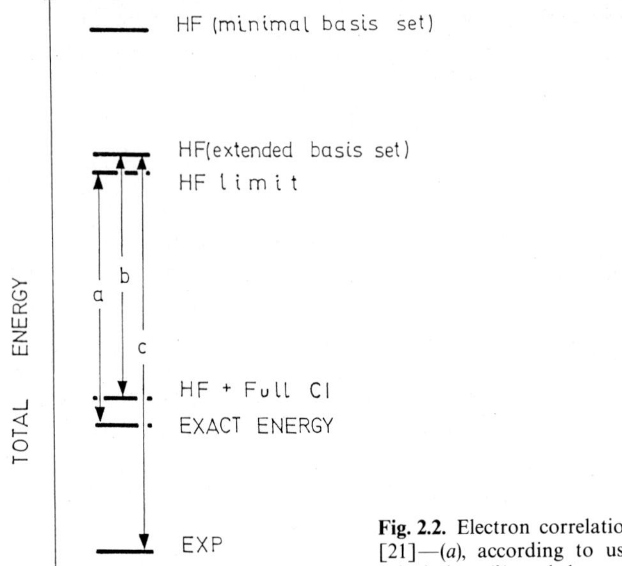

Fig. 2.2. Electron correlation energy as defined in Ref. [21]—(*a*), according to usual estimation in practical calculations (*b*), and the experimental estimation (*c*)

The difference between the energy of the system obtained through exact solution to the nonrelativistic Schrödinger equation and the minimal value derived in the HF approximation is customarily regarded as the energy of electron correlation [21]. The point is that the Hamiltonian in the HF method, in particular Eq. (25), includes an averaged interelectronic potential which does not account for correlated motion of electrons in a molecular system.

The definition proposed in Ref. [21] is not quite satisfactory seeing that an exact solution to the Schrödinger equation cannot be obtained even for a three-atom system. Therefore, a somewhat different definition is adhered to in practical calculations, it is explained by the scheme in Fig. 2.2.

The energy of electron correlation constitutes an insignificant fraction of the total energy of a molecular system (0.5% for the H_2O molecule), on the other hand, the energy of binding in the molecules is, too, a small fraction of the total energy (also 0.5% for H_2O). Moreover, in the general case, the energy correlation is a nonadditive quantity and depends on the geometry of the system [2, 6]. In order to account for the energy of electron correlation, the following procedures are currently widely used: the method of configuration interaction (CI) and the method of the perturbation theory of Møller–Plesset (MP).

In the CI method, the total wave function is written as a linear combination of the Slater determinants, which correspond to different electron configurations:

$$\psi = A_0\psi_0 + \sum_{k=1}^{M} A_k\psi_k \tag{2.21}$$

where ψ_0 is the Slater determinant of the ground state, ψ_k of the excited state and A_k are the expansion coefficients.

Substituting Eq. (2.21) into Eq. (2.2) yields a system of equations analogous in its form to Eq. (2.4):

$$\begin{aligned} \sum_k A_{kl}(H_{kl} - E_l S_{kl}) = 0 \\ |H_{kl} - E_l S_{kl}| = 0 \end{aligned} \tag{2.22}$$

where $H_{kl} = \langle\psi_k|\hat{H}|\psi_l\rangle$, $S_{kl} = \langle\psi_k|\psi_l\rangle$ and $\hat{H}$ is the HF Hamiltonian.

When solving Eqs. (2.22), either the coefficients $C_{\mu i}$ found from Eqs. (2.4) and (2.10) are not varied or A_{kl} and $C_{\mu i}$ are varied simultaneously. The former approach is the configuration interaction method, the latter is called the method of multiconfigurational interaction (MCI). The MCI method is a particularly effective tool for taking an exact account of the correlation energy when a complete enough set of the configurations of Eq. (2.21) is used. In the case of two-atom molecules, it yields potential curves as Eq. (1.2) which practically coincide with the experimental ones [2]. However, for greater size systems this method is at present not applicable in view of exorbitant demands upon the computer.

The principal shortcomings of CI technique are the weak convergence of the configurational sets of Eq. (2.21) and great difficulties in the selection, not amenable to a simple algorithmization, of important configurations. The size of the sets of Eq. (2.21) may be judged from, e.g., the compartively small molecule of sulfurane SH_4 (see Sect. 1.3.4.3). In the calculations [22] where an extended basis set of the TZ type is used (three STO functions which approximate each AO and their subsequent decomposition into the GTO's), the procedure of taking complete account of the configuration interactions comprising all once- and twice-excited configurations involves 7381 configurations. In the case of more complex systems, several tens of thousands of configurations have to be considered so that the diagonalization of the matrix of Eq. (2.22) requires a lot of computer time. Various schemes have been developed intended to expedite convergence of the set [6, 8] of which the so-called coupled electron-pair approximation (CEPA) is the most effective but, at the same time, the most laborious.

This laboriousness of the CI method has led recently to a wider use of the methods of perturbation theory for evaluating the correlation energy. These methods can more readily be algorithmized, their requirements for computer time lie within reasonable bounds and, moreover, they allow one to take into account the influence of various orders of the perturbation theory upon properties to be calculated. A method based on the Møller–Plesset perturbation theory has now found a particularly wide acceptance, its computational scheme included in some GAUSSIAN-series programs [6] is suffuciently effective for practical calculations. The essence of this technique is as follows. The solution to the rigorous Schrödinger equation with the Hamiltonian $\hat{H}$ is sought on the basis of the known solutions to this equation with a model Hamiltonian $\hat{H}^{(0)}$ which is different from $\hat{H}$ by a small perturbation $\hat{W}$. Usually, the Hartree–Fock–Roothaan Hamiltonian, whose matrix elements have the form Eq. (2.5), is taken to be $\hat{H}^{(0)}$, while the operator describing the electron correlation:

$$\hat{W} = V_{ee} - \sum_{i>j}^{\text{occ.}}\sum^{\text{occ.}} (2J_{ij} - K_{ij}) \tag{2.23}$$

represents the perturbation $\hat{W}$. With such a choice of $\hat{H}^{(0)}$ and $\hat{W}$, the total energy of the system may be represented in terms of the perturbation theory as an expansion:

$$E = E^{(0)} + E^{(2)} + E^{(3)} + E^{(4)} + \cdots \tag{2.24}$$

where the first correction to the electron energy of the molecule vanishes and the first nonzero contribution corresponds to the second order of the perturbation theory ($E^{(2)}$).

The calculated data on total energies of the ethane isomers given in Table 2.1 show the importance of inclusion of the correlation effects even in regard to the relatively small second order of the Møller–Plesset perturbation (MP2). The correlation energy calculated for different geometrical configurations is different with the highest value belonging to those configurations in which the bonds have a multicenter character as, e.g., in the structures III and IV. Thus, when account is taken of the correlation energy, essential corrections can be introduced into the calculation of the thermodynamic stability of nonclassical structures whose stability is primarily determined by multicenter bonds [23]. So a HF/6–31G* calculation [24] of the relative stability of classical (open) V and nonclassical (bridged) VI forms of the acetyl ethyl cation:

V, C_s VI, C_{2v}

shows that V is more stable than VI by 2 kcal/mol. When the correlation energy is taken into account in terms of MP2 using the 6–31G* basis set, the bridged form VI becomes more stable by 6 kcal/mol. Further refinement of the calculation does not change this trend.

An even more impressive example of the effect of correlation energy on the form of the PES is given by the calculation of the isomerization reaction of the vinyl cation VII ⇄ VIII [25]:

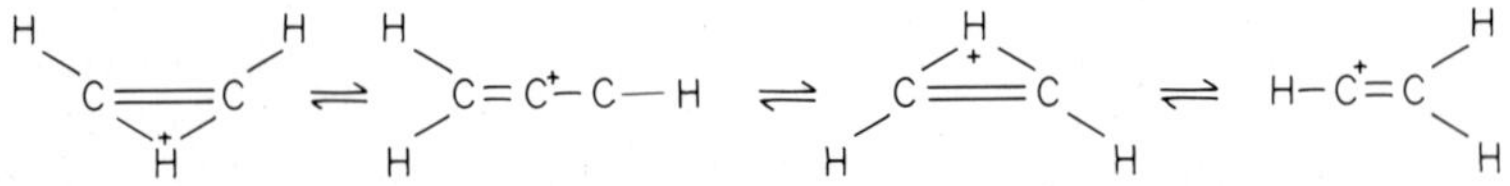

VIIa ⇌ VIIIa ⇌ VIIb ⇌ VIIIb

Calculations on the HF level show that both forms VII and VIII correspond to a minimum on the PES with the classical form VIII possessing a greater (7 kcal/mol) thermodynamic stabilty. The inclusion of the correlation energy in terms of MP2 indicates a nearly complete energy equivalence of both forms with a small activation barrier between them. Further refinement of the calculation up to MP4 has brought an extremely interesting result. The saddle point on the PES, corresponding to the transition state in the isomerization reaction VII $\rightleftarrows$ VIII, has vanished, while the minimum corresponding to the classical form VIII has deformed into the saddle point of the first rank ($\lambda = 1$). In other words, on the MP4 level, the structure VIII turns wholly unexpectedly into the transition state of a truly strange reaction of rotation of hydrogen atoms about the C=C fragment VIIa $\rightleftarrows$ VIIIa $\rightleftarrows$ VIIb with the barrier equalling 4.9 kcal/mol. Further refinement in the calculation and inclusion of the zero vibration energy does not alter this last result.

The inclusion of the electron correlation energy in the theoretical analysis of chemical reactions becomes a "must" when electron bond pairs of starting structures are destroyed during reaction. This concerns in the first place the reactions of bond breaking since a calculation in terms of the HF approximation leads to incorrect dissociated states, such as $F_2 \rightarrow F^+ + F^-$ rather than 2F. Also the reactions proceeding via biradical type structures (see Sect. 1.6 and 8.3) and those forbidden by the orbital symmetry conservation rules fall within the category of such transformations. A typical example is provided by the electrocyclic-type reaction:

IX → X

The data on nonempirical calculations [26] presented in Fig. 2.3 show that, when the electron correlation is allowed for through the use of the CI technique, not only the barriers and the transformation energetics (the reaction is exothermal) are described more correctly, but also the shortcoming inherent in the HF calculation in regard to disrotatory mechanism is removed. The HF

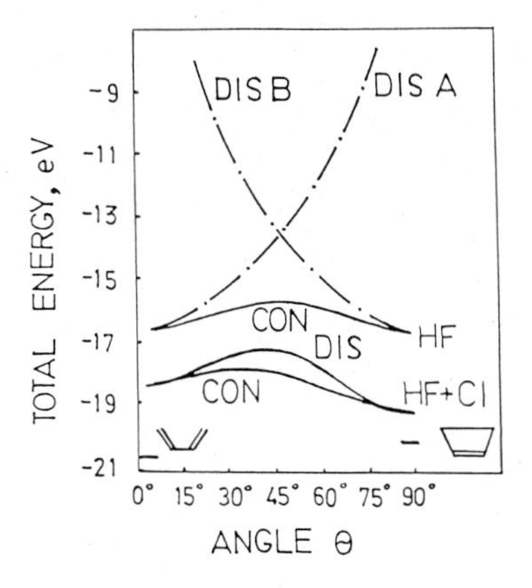

Fig. 2.3. Minimal energy reaction paths of the electrocyclic butadiene-cyclobutene reaction in the HF and HF + CI approximations [26]. The reaction coordinate is represented by the angle θ of synchronous rotation of the methylene groups. The value of 154.0 hartrees is taken to be the zero energy; DIS denotes the disrotatory and CON—the conrotatory mode of the cycloreversion. (Reproduced with permission from the American Chemical Society)

approximation produces a result which contradicts the microscopic reversibility principle since the thermally prohibited disrotatory reaction gives rise to two different electronic states (different filling of orbitals) depending on whether the reaction proceeds from butadiene or cyclobutene (for a more detailed description see Chap. 10).

Recently Spellmeyer and Houk [27] conducted a systematic study of the influence exerted by the size of basis sets and by correlation effects upon the activation energy of the reaction cyclobutene-butadiene. Parallel with the nonempirical calculations, this reaction was studied by means of the most frequent semiempirical methods MINDO/3, MINDO, and AM1 (see below Sect. 2.3.2). The authors think, quite correctly, that the experimental value of the activation energy 32.9 ± 0.5 kcal/mol should be compared with the results of only those empirical methods whose parameterization includes corrections for thermal vibrations and the energy of zero vibrations. At the same time, nonempirical calculations have to be compared with the value 34.5 kcal/mol, which is the result of the subtraction of the difference between the energies of zero vibrations and the energies of thermal vibrations of the ground and the transition states from the experimental value. Figure 2.4 shows some results from the work [27].

The most important result is, apparently, the nonmonotonical improvement of the results with the extension of the basis set and with a fuller inclusion of the correlation energy. It may easily be seen that the semiempirical methods

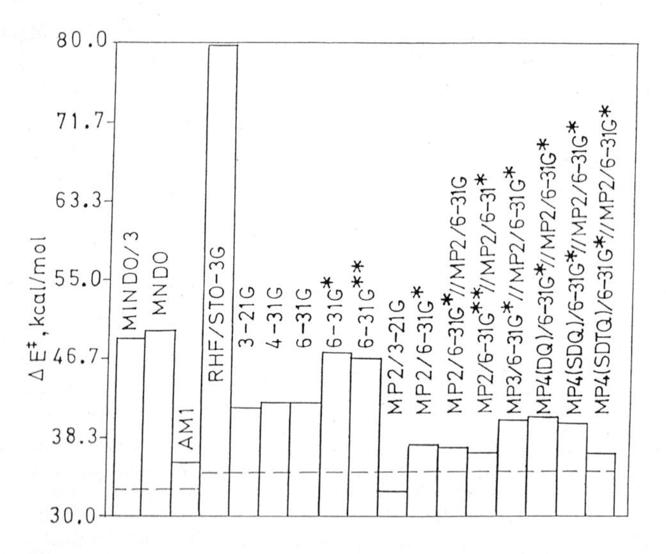

Fig. 2.4. Comparison of calculated activation energies at several levels of approximation. The *dashed line* indicates the value of experimental activation energy for comparison purposes. The semiempirical values are compared to $\Delta H^{\neq} = 32.9$ kcal/mol and ab initio values to $\Delta E^{\neq} = 34.5$ kcal/mol. (Adapted from Ref. [7] with permission from the American Chemical Society)

successfully compete with the ab initio schemes. Thus, the AM1 results are, in effect, as accurate as the nonempirical calculation using the scheme MP4 (SDTQ)/6–31G*/MP2/6–31G*, which requires a several orders of magnitude greater computer time expenditure.

Another important case where the electron correlation must be taken into account is the calculation on structures with degenerate or pseudodegenerate electron configurations. A typical example is given by the cyclobutadiene molecule. When the correlation energy is not allowed for, the calculations, even those with an extended basis set, lead to an erroneous conclusion in favor of the preferability of a triplet electron configuration and a square rather than rectangular geometry of the molecule [27–30].

The inclusion of the correlation effects is indispensable in the calculation of intermolecular interactions, since the important effect of dispersional attraction is of purely correlational nature [31].

2.2.5 The Problems of Stability of Hartree–Fock Solutions

The HF equations, in particular, the equation of Roothaan are nonlinear with respect to the one-electron functions φ_i. This creates the problem of additional solutions to the HF equations and of an analysis for the type of the extremum of the results obtained for φ_i (points in the functional space of the test functions), similarly to how the character of a stationary point in the configurational space of the PES nuclear variables is determined. To this end, one needs to investigate the environment of this point in the functional space, i.e., to derive the second energy variation in the test function space:

$$\delta^2 E = \delta^2 \frac{\langle \psi | \hat{H} | \psi \rangle}{\langle \psi | \psi \rangle} \tag{2.25}$$

where ψ is defined according to (2.1).

The problem of determining the sign before $\delta^2 E$ is reduced to one of finding the signs of the eigenvalues of a certain Hermitian matrix $\| A \|$ [32] whose elements are determined from the known formulas for matrix elements of the HF Hamiltonian between the singly excited states as well as between the ground and the doubly excited states [33, 34].

If the matrix $\| A \|$ is positive definite, the HF solution obtained (a set of φ_i) is stable (the local minimum). The negative eigenvalue indicates instability of the solution (the local maximum or the saddle point in the functional space). The form of the eigenvector, which corresponds to a negative eigenvalue, predetermines the procedure of constructing the initial wave function that would allow one to arrive in the SCF MO scheme at a stable solution. There exists a ramified hierarchy of instability types of the HF solutions [35]. In the RHF method a singlet and a triplet instability are distinguished whose appearance may roughly be estimated from Eqs. (2.26) and (2.27) [36]:

$$(\varepsilon_j - \varepsilon_i) - J_{ij} + 3K_{ij} < 0 \tag{2.26}$$

$$(\varepsilon_j - \varepsilon_i) - J_{ij} - K_{ij} < 0 \tag{2.27}$$

where the subscript i refers to the occupied and j to the vacant orbital. Thus:

$${}^1\Delta E_{ij} < -K_{ij} \tag{2.28}$$

$${}^3\Delta E_{ij} < K_{ij} \tag{2.29}$$

The quantities ${}^{1,3}E_{ij}$ are the energies of, respectively, the singlet and the triplet transition of the electron from the i- to the j-orbital, and K_{ij} is the exchange integral of Eq. (2.13).

Equations (2.26) and (2.27) mean that in terms of the method under discussion there occur, respectively, a singlet or a triplet excited states lying close to the ground state.

The singlet instability involves the existence of a solution with the electron distribution of lower symmetry, which gives rise to the trend towards distortion of a fully symmetrical geometry configuration (lattice instability). A characteristic example is given by the instability of HF solutions for the fully symmetrical D_{2h} structure of propalene [29]. The singlet instability should be viewed as an indication that the initially assumed form of the nuclear configuration of a system needs correction.

The triplet instability of the RHF solutions is a necessary, but insufficient, condition for the conclusion as to the biradical character or the triplet ground state of a given system, which would be important for an analysis of the internal mechanism of a number of reactions (see Sect. 5.1). Usually, a reliable result may be achieved in such cases by passing to the UHF approximation.

Generally, instability of the HF solutions signifies drawing together of the PES's of different electron states and shows the need for the inclusion of correlation effects in order that a more exact description of the molecular system may be achieved. Since the effects of instability of the HF solutions are easy to diagnose, their detection should be viewed as procedure valuable for the prediction of structural features and reaction mechanisms.

2.3 Semiempirical Methods

In nonempirical methods of calculation, usually about 70% of the computer time is spent on computing the integrals of the interelectron interaction $(\mu\nu|\lambda\sigma)$ in Eq. (2.5). As the size of a molecular system is increased, the number of such integrals grows roughly proportionally to N^4 (N is the size of the AO basis set of

Table 2.3. Dependence of the number of integrals and of the time for their calculation on the number of atoms in a molecule and the basis set

Molecule	Gaussian basis set	Number of integrals	Time, min (CDC 3300)
CH_2	3G	189	1.0
CH_2	6–31G*	6358	12.5
CH_3	3G	369	134.0
CH_3	6–31G*	13160	21.1
$CH_3—CH_3$	6–31G*	194205	45.8
$CH_3—OH$	6–31G*	215572	147.4
$CH_3—NH_2$	6–31G*	230766	181.6

Eq. (2.3)). Accordingly, the time and costs of the calculation grow, which is well illustrated by the data in Table 2.3.

There are various ways of reducing this inconvenience. One of these consists in neglecting a certain part of the integrals contained in Eq. (2.5), though retaining the overall scheme of the nonempirical calculation. Such an approach has been implented in the PRDDO method, where the one-, two-, and three-center Coulomb integrals as well as the one- and two-center exchange integrals are retained, but no more, so that out of N^4 integrals only N^3 remain [37]. Another promising approach that is currently gaining acceptance is associated with the use of the effective core potential (ECP) or pseudopotential as it is also called. This approach is based on the observation that the atomic core orbitals react weakly to the formation of a chemical bond so that the influence of the core electrons on the valence electrons can be expressed through introduction of the corresponding potential functions, in other words the core electrons can be frozen into the nuclear core and then the so-called "frozen core approximation" may be used. In this case, the Hartree–Fock–Roothaan scheme is wholly retained for the valence electrons, while the core electrons are replaced by the pseudopotential. As a result, calculation time is reduced by almost a factor of 10 and no longer grows catastrophically with the increase in the atomic number of the elemens contained in the molecule [38]. By now, the ECP's have been developed for nearly all the elements of the Periodic Table [39, 40].

Finally, there is one more approach which involves replacement of most integrals by the experimental parameters (such as atomic ionization potentials in orbital valence states, and others) and the use of various approximate expressions, which include these parameters, for the evaluation of the integrals. The methods based on such an approach are called semiempirical.

Nearly all semiempirical methods used for calculating the PES's are the valence approximation methods; in contrast to the nonempirical procedures they take account of only the valence electrons and the atomic orbitals of valence shells. The influence of the non-valence (core) electrons is included in empirical parameters.

The overall schemes and the details of parametrization of various semiempirical methods have been thoroughly described and analyzed [1, 4, 5, 41–43], so we confine ourselves to brief characterization of the frequently used methods pointing out their areas of application as well as their limitations.

2.3.1 The Extended Hückel Method

The extended Hückel method (EHM) developed by Hoffmann [44] is the simplest noniterative semiempirical method. Formally, the EHM equations coincide with the equations of Roothaan [Eqs. (2.4) and (2.10)], however, only the overlap integrals $S_{\mu\nu}$ are calculated in them exactly, while the matrix elements $F_{\mu\nu}$ are replaced by empirical parameters:

$$F_{\mu\mu} = -I_{\mu}, \quad F_{\mu\nu} = k(F_{\mu\mu} + F_{\nu\nu})S_{\mu\nu} \tag{2.30}$$

where I_{μ} is the ionization potential of an electron from the μ-th orbital and k is a certain empirical constant. In consequence of this approximation, the matrix elements $F_{\mu\nu}$ are not the functions of the coefficients $c_{\mu i}$ being sought and the method is not selfconsistent.

Nowadays the EHM is regarded as mainly a qualitative method not claiming to satisfy the demands listed in Sect. 1 of this chapter. Its chief advantage is a quite fair reproduction of the relative energies and the form of MO's, particularly, in the case of not very strongly polarized molecules. In view of its extreme simplicity, it may be useful in calculations on the systems with a practically unlimited size of the basis set. When DZ-type basis sets are used, this method is sufficiently effective for qualitative analysis of structures and reactions of organometallic compounds [45–47].

2.3.2 Semiempirical Selfconsistent Field Methods

In the SCF methods, the matrix elements $F_{\mu\nu}$ depend on the coefficients $C_{\mu i}$ in Eqs. (2.4), therefore the solutions are sought iteratively. Various semiempirical SCF methods differ from one another in parametrization procedures as well as in the character and number of the integrals in Eq. (2.5) whose calculation is neglected. The set of parameters for each method is dependent upon what property (or properties) of a given molecular system has been chosen for a parameter calibration. For this reason, the semiempirical methods yield the most reliable results in, as a rule, rather narrow areas of application.

2.3.2.1 The CNDO/2 Method

The complete neglect of differential overlap (CNDO) method [1] rests on the zero differential overlap (ZDO) approximation, which means that all the products of

$\chi_\mu \chi_\nu$ are set to zero and:

$$S_{\mu\nu} = \delta_{\mu\nu} \tag{2.31}$$

The CNDO method admits of several variants of parametrization of which the CNDO/2 scheme is the one which is used more frequently than others. The ZDO approximation drastically reduces the number of two electron integrals since all the three-center, four-center, and exchange integrals are set to zero:

$$(\mu\nu|\lambda\sigma) = \delta_{\mu\nu}\delta_{\lambda\sigma}(\mu\mu|\nu\nu) \tag{2.32}$$

When Eqs. (2.31) and (2.32) are taken into account, the matrix elements of the Fock operator take in the CNDO/2 method the form:

$$F_{\mu\mu} = -\tfrac{1}{2}(I_\mu + A_\mu) + (P_{AA} - Z_A) \tag{2.33}$$

where μ belongs to the atom A:

$$F_{\mu\nu} = \tfrac{1}{2}(\beta_A^0 + \beta_B^0)S_{\mu\nu} - \tfrac{1}{2}\rho_{\mu\nu}\gamma_{AB} \tag{2.33}$$

where μ and ν belong to AO's and A, B to the atoms; I_μ and A_μ are the ionization potential and the electron affinity of the orbital, respectively; γ_{AA} and γ_{AB} are the interaction repulsion integrals which are calculated exactly over the Slater S orbitals for any AO.

The chief calibration parameter is the resonance integral β_A^0 which depends only on the type of the atom A. It is chosen in such a way as to ensure that the relative order of the energy levels of the occupied MO's and the expansion coefficients of the MO's in the LCAO of Eq. (2.3) optimally coincide with the ab initio calculations using the minimal basis set of the AO.

The CNDO/2 method gives the most reliable results in the calculations of the electron distributions and of those properties which are determined on the basis of these electron distributions, such as the dipole moments. It underestimates the valence bond length by on average 10%, particularly sizeable is this underestimation in regard to the hypervalence bonds, such as the axial bonds in trigonal-bipyramidal structures of the transition states of the S_N2 reactions (see Table 5.1 in Chap. 5). This inaccuracy follows from the drawback of the method, which is strongly reflected in the character of the PES's obtained, namely the overestimation (by about 20%) of the covalence bonding. This and a number of other shortcomings of the method (underestimation of the conjugation energy and of the stability of bridged structures) stem from the ZDO approximation and are hard to remove by reparametrization.

For example, unlike the nonempirical methods of calculation, which predict, in complete agreement with experiment, stability of protonated benzene in the form of a σ-complex XI [48], the CNDO/2 calculation [49] gives considerable preference to a bridged isomer XII where the proton is bound to both C atoms of the π-bond.

STO-3G	0 kcal/mol	28 kcal/mol
4–31G	0 kcal/mol	21 kcal/mol
CNDO	46 kcal/mol	0 kcal/mol

At the same time, the CNDO/2 method describes fairly accurately those reactions in which electrostatic interactions rather than covalent bonding is the moving force. The explanation is simple: the interactions of this type are governed by the character of the electron distribution. Thus, this method reproduces quite satisfactorily the systems with hydrogen bonding, and the protonation reactions of heteroatomic compounds where the reaction course is determined by the form of the electrostatic potential. The example of the model peptide given in Fig. 2.5 shows how well the CNDO/2 method can reproduce the picture of the electrostatic potential.

2.3.2.2 *The MINDO/3 Method*

The methods of the MINDO family are based on the INDO approximation suggested by Pople [1]. Unlike the CNDO, the INDO approximation is characterized by inclusion in Eqs. (2.30) of the one-center exchange integrals $(\mu\nu|\lambda\sigma)$ when the AO's μ and ν belong to the same atom.

The parametrization of the MINDO methods, of which the MINDO/3 scheme [51] is the most important, is carried out differently than in the case

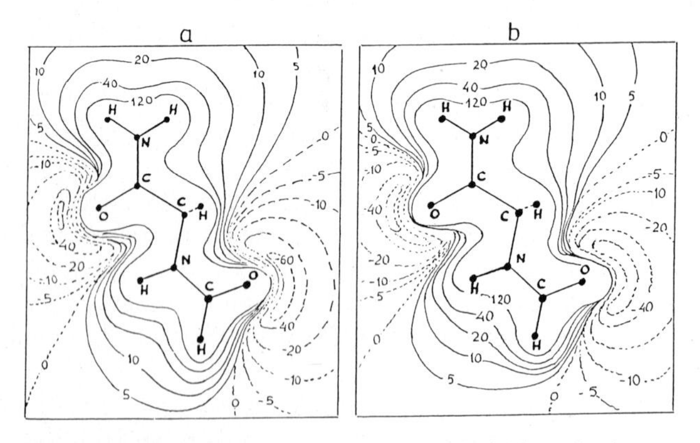

Fig. 2.5a, b. Maps of the electrostatic potential of 2-formylamino acetamide obtained in **a** ab initio and **b** CNDO/2 calculations [50]. (Reproduced with permission from Kluwer Academic Publishers)

of the CNDO/2. The empirical parameters introduced into the terms of the Fock operator with a two-center interaction are chosen in such a manner as to reproduce possibly best the experimental characteristics (geometry, heat of formation, ionization potentials etc.) of a large number of standard compounds. The overall number of the parameters in the MINDO/3 method for the hydrogen atom and the second-period elements alone amounts to 102. The procedure of parameter optimization takes a lot of computer time, but this investment pays off, as a rule, by providing fairly good results when evaluating the heat of formation, activation barriers and vibration spectra for compounds containing the H, C, N, and O atoms. On the whole, these results satisfy the principal requirements placed upon the methods for calculating the PES's [41, 52–54], however, one should keep in mind also a number of shortcomings inherent in the MINDO/3 method which have become apparent in practical calculations [54–56].

1. The MINDO/3 method overestimates the valence angles by on average 6–8°. This feature is particularly pronounced with the double bonds, for example, in the case of 1,3-butadiene the angle CCC comes out at 131° [51] while the experimental value is 123.6°. The accumulation of such errors in the calculation of the compounds of type XIII leads to a lengthening of the X···X distance by 1–1.5 Å. As a consequence, not only the migration barrier for XIII ⇄ XIIIa is sharply increased but also the reaction mechanism may change. This difficulty may be circumvented by taking in the calculation the

XIII ⇄ XIIIa

magnitudes of the angles α and β to be constant and equal to their experimental values when optimizing the remaining geometrical parameters [57].
2. The method considerably overestimates the stability of small cycles, particularly those containing two neighboring heteroatoms.
3. Similarly to CNDO/2, the MINDO/3 method overestimates stability of the three-center two-electron bonds and reduces a good deal their length (see Table 5.1 in Chap. 5).
4. The method is not suitable for describing the systems with hydrogen bonding since it cannot reproduce the stabilization energy of, e.g., such dimers as $(H_2O)_2$, $CH_3OH, \ldots, NH_3$ and others.
5. The method does not yield accurate results in the description of the compounds that contain neighboring heteroatoms with electron lone pairs, the organic boron compounds etc. Since organic compounds with fixed valence of the atoms that form them are used for the parametrization of this method, it underestimates, as a rule, the stability of nonclassical skeleton and polyhedral structures.

6. In virtue of its parametrization, the method underestimates interaction of the atoms which lie at a noncovalent distance. This may result in distortions of the mechanism of addition reaction [58]. (Addition reactions are described in greater detail in Chap. 6).

A UHF scheme of the MINDO/3 method [59] and a simpler "half-electron" scheme based on the RHF approximation [60] have been developed with the aim to describe radical reactions and structures. A detailed list of various areas where the MINDO/3 method yields good results is given in the review article [54].

2.3.2.3 *The MNDO Method*

The MNDO method [61] is based on a more rigorous approximation NDDO [1] which takes into account in Eq. (2.5) the interelectron-repulsion integrals that include the one-center overlaps. The integrals $(\mu\mu|\nu\nu)$ between any orbitals on the A and B atoms are not approximated by the corresponding integrals over the S orbitals but rather calculated for the corresponding functions χ_μ and χ_ν. An important advantage of the MNDO over the MINDO/3 method consists in the rejection of parametrization of the resonance integral $\beta_{\mu\nu}$ in accordance with the bonding type, and a transition to the relationship:

$$\beta_{\mu\nu} = S_{\mu\nu}(I_\mu + I_\nu) f(R_{AB}) \tag{2.34}$$

As a result, even though the number of interelectronic interaction integrals of various types increases appreciably, the overall number of the parameters for the second-period elements is reduced from 102 to 41. Using this method, the computer calculation takes 1.5 times longer than with the MINDO/3 method. As may be seen from Table 2.4 compiled from the data presented by Dewar [53, 62] for over 200 compounds, the MNDO method achieves a better agreement with experiment than the MINDO/3 method.

Notwithstanding this better agreement concerning physical characteristics of the molecules in the singlet ground state, particularly of the molecules containing heteroatoms (this case is listed above as representing one of the shortcomings

Table 2.4. Average deviations of calculated lengths obtaned by the MINDO/3,MNDO and AM1 methods from experimental values for compounds containing the C, H, O, and N atoms

Method	Heat of formation kcal/mol	Bond length Å	Valence angle	Vibration frequency
MINDO/3	11	0.022	5.6°	5%
MNDO	5	0.014	2.8°	< 5%
AM1	5.8	0.012	2.5°	< 5%

Table 2.5. Heats of activation (kcal/mol) [63]

Reaction	Obs	AM1	MNDO	MINDO/3
$CH_3^{\cdot} + HC{\equiv}CH \rightarrow CH_3CH{=}CH^{\cdot}$	7.7	6.83	16.7	7.3
$CH_3^{\cdot} + CH_2{=}CHCH_3 \rightarrow CH_3CH_2\dot{C}HCH_3$	7.4	1.31	13.5	7.8
$CH_3^{\cdot} + CH_3{-}CH_3 \rightarrow CH_4 + C_2H_5^{\cdot}$	11	11.96	27.2	6.1
$:CHCH_2 \rightarrow CH_2{=}CH_2$	1–3	14.92	21.8	0.7
XIIIa → XIII (R = H)	4–5	22.17	39	28.2
cyclobutene → butadiene	32.9	36.0	51.7[a]	49[b]

[a]See Ref. [64]; [b]see Ref. [65]

of the MINDO/3 method), many of the drawbacks inherent in the MINDO/3 method are present in the MNDO method also. So the MNDO method is unsuitable for describing the H-bond, moreover, in contrast to the MINDO/3 method, it strongly overestimates activation barriers of the reactions, as may be seen from the data of Table 2.5. The recently developed parametrization of the MNDO method, which covers the I–III period elements, almost all halogens as well as a number of heavy metals, such as Zn, Hg, Sn, Pb, and Ge, makes it possible to apply successfully this method in studying the mechanisms of organic, organometallic and biochemical reactions. Specific studies aimed at comparing the structures of transition states and energetics of the corresponding reactions obtained by the MNDO method, on the one hand, and by nonempirical calculations, on the other, Refs. [66–68] have shown that the MNDO method often gives reuslts not inferior to those of the nonempirical calculations using the 4–31G basis set [66–69] with the obvious advantage of reducing the computer time by a factor of nearly one thousand! [63].

However, even though thanks to parametrization the MNDO method partially takes account of the electron correlation energy, the common drawback of a one-determinant approximation, namely, its inability to correctly evaluate the bond-breaking reactions, is not thereby removed as was pointed out, for example, in Ref. [70] using as an illustration a number of dissociation reactions. In such cases, a more complete inclusion of the correlation energy can remedy the situation. In this connection, Thiel suggested a MNDOC scheme [71] in which the correlation corrections are calculated explicitly in the second order of perturbation from the Epstein–Nesbet formula [72, 73]:

$$\Delta E^{(2)} = -\sum_{K} \frac{V_{ko}^2}{E_k - (E_o + \Delta E^{(2)})} \tag{2.35}$$

where V_{ko} is the matrix element of the perturbation between ground ψ_o and

Table 2.6. Comparison of semiempirical MNDO, MNDOC, and ab initio results for triplet $^3B_1$) and singlet ($^1A_1$) methylene

State		MNDO[a]	MNDOC[a]	SCF[b] DZ	SCF + CI[b]	Experimental
$^3B_1$	C—H, Å	1.052	1.066			1.078[c]
	HCH, deg	149.9	134.6			136[c]
$^1A_1$	C—H, Å	1.091	1.106			1.11[d]
	HCH, deg	111.1	101.5			102.4[d]
$^3B_1 \rightarrow {}^1A_1$	ΔE_{ST} kcal/mol	3.01	7.3	32	11	9[e]

[a]Ref. [77]; [b]Ref. [78]; [c]Ref. [79]; [d]Ref. [80]; [e]Ref. [81]

doubly excited configurations ψ_k and E_o and E_k are the Hartree–Fock energies of the configuations ψ_o and ψ_k. Explicit formulas for V_{ko} and E_k in terms of MO integrals and orbital energies are given in Ref. [74]. Equation (2.35) is solved iteratively.

The correlation energy taken into account by means of Eq. (2.35) has necessitated a reparametrization of the MNDO method. In his MNDOC scheme, Thiel introduced parameters for the atoms H, C, N, and O only. A comparison between the results of the MNDO and the MNDOC calculations on the molecules with closed electron shell in the electron ground state has revealed their nearly complete coincidence. At the same time, the MNDOC method has been shown to be more adequate for investigating intermediates and transition states [75, 76] as well as excited states and photochemical reactions [77]. An illustrative example is provided by a good agreement between the results of calculations of geometry in the $^3B_1$ and $^1A_1$ states of methylene and of the singlet-triplet splitting (E_{st}) using the MNDOC method [77] and the experimental data [79–81] (see Table 2.6). Both the geometry and the energy difference between the singlet $^1A_1$ and the triplet $^3B_1$ states have for a long time been a subject of sharp debates between theoreticians and experimentalists [82]. Initial discussions as to whether the CH_2 molecule existed in angular or linear geometry ended to the advantage of the theoreticians who had predicted angular geometry. Calculations and experiments conducted prior to 1970 indicated that the triplet $^3B_1$ state lay lower than the singlet state $^1A_1$. However, the experiment showed the difference between them to be 2.5 kcal/mol whereas the theoretical calculation gave a much higher value of 20 kcal/mol. Then Hase and his co-workers [83] obtained for E_{ST} the experimental value of 10 ± 3 kcal/mol while the calculations carried out by groups at Caltech [84] and at University of California, Berkeley [85] gave 11.5 and 13 kcal/mol, resp., which was a remarkable agreement. This harmony between the theoreticians and the experimentalists lasted for nearly five years. Next, in the late seventies, extremely rigorous calculations produced for ΔE_{ST} the value of 10.4 kcal/mol and new experiments gave 8 ± 1 kcal/mol. Though the difference between theory and experiment was not at all significant, the theoreticians would insist on their result being right alleging that photoelectron spectra of the CH_2 ion had been interpreted incorrectly giving thus rise to underestimation of the experimental value of ΔE_{ST}. Experimentalists felt wounded in their pride and wagered several bottles of best French champagne that it was their result of 8 ± 1 kcal/mol, which was correct. In the course of three years, the Colorado experimentalists were meticulously preparing a new experiment expecting to put the theoreticians to shame. March 29, 1984 arrived and the first results portended experimentalists's defeat—they got ΔE_{ST} equal to 9.0 kcal/mol. Thus, in the end, the theoreticians proved right and exquisite French wine went to Caltech. (For greater detail see Ref. [82]).

2.3.2.4 The AM1 Method

The AM1 method (Austin Model 1) [63] is a novel semiempirical scheme. It has been developed under Dewar's guidance and, like the MNDO method, is based on the NDDO approximation. Apart from original MNDO parametrization, the AM1 method differs from the MNDO method in that the function:

$$E_{AB} = Z_A Z_B \gamma_{AB}^{SS}(1 + f(R_{AB})) \tag{2.36}$$

is in its case replaced by a new multiparameter function of the energy of core repulsion:

$$E_{AB} = Z_A Z_B \gamma_{AB}^{SS}(1 + f_A(R_{AB}) + f_B(R_{AB})) \tag{2.37}$$

Introduction of Eq. (2.37) permits the H-bond to be satisfactorily described as may be seen from Table 2.7. The accuracy of the AM1 calculations is increased by 20% over the MNDO method [87] (see Table 2.4). An obvious advantage of the

Table 2.7. Intermolecular hydrogen bonding (energies in kcal/mol) [86]

	ΔE, kcal/mol		$R_{X\cdots H\cdots Y}$, Å	
System	AM1	Exp	AM1	Exp
Acrylic acid (dimer)	6.17	14.10	3.06	2.76
Acetic acid (dimer)	6.29	14.20	3.06	—
HF(dimer)	3.17	5-7	2.93	—
H_2O(dimer)	2.74	5.0	3.10	2.98
NH_3(dimer)	0.94	4.5	3.68	—
$NH_3 \cdot CH_3CN$	1.70	2.43	3.68	—
$HF \cdot H_2O$	4.07	6.22	2.94	2.69
CH_3OH·pyridine	1.24	3.2–4.4	3.67	—
$CH_3OH \cdot N(CH_3)_2$	4.32	5.8–7.5	3.62	—
NH_3·pyridine	2.16	3.75	3.77	—

AM1 method, as contrasted with the MINDO/3 and MNDO schemes, is that it can much better describe the magnitudes of activation barriers of organic reactions (see Table 2.5).

This account of the types and characteristics of quantum chemical calculation methods is not complete and reflects certain predilections of different authors. Nevertheless, it describes more or less objectively the current state and development trends of these methods.

References

1. Pople JA, Beveridge DL (1970) Approximate molecular orbital theory. McGraw-Hill, New York
2. Schaefer HF (1972) The electronic structure of atoms and molecules. A survey of rigorous quantum mechanical results. Addison-Wesley, Massachusetts
3. Csizmadia IG (1976) Theory and practice of MO calculation on organic molecules. Elsevier, Amsterdam
4. Minkin VI, Simkin BYa, Minyaev RM (1979) Theory of the molecular structure. Vysshaya Shkola, Moscow (in Russian)
5. Zhidomirov GM, Bagaturyants AA, Abronin IA (1979) Applied quantum chemistry. Khimia, Moscow
6. Hehre WJ, Radom L, Schleyer PvR, Pople JA (1986) Ab initio molecular orbital theory. Wiley, Inc, New York
7. Hehre WJ (1976) Acc Chem Res 9:399
8. Carsky P, Urban M (1980) Ab initio calculations. Lecture notes in chemistry. Springer-Verlag, Berlin
9. Clementi E (1985) J Phys Chem 89:4426
10. Goddard III WA (1985) Science 227:917
11. Schaefer HF (1983) Chem Future Proc 29th JUPAC Cong Cologne, 5–10 June
12. Davidson ER (1984) Faraday Symp Chem Soc 19:7
13. Davidson ER, Feller D (1986) Chem Rev 86:681
14. Frisch MJ, Pople JA, Binkley JS (1984) J Chem Phys 80:3265
15. Husinaga S (ed) (1984) Gaussian basis sets for molecular calculations. Elsevier, Amsterdam
16. Kos AJ, Jemmis ED, Schleyer PvR, Gleiter R, Fischbach U, Pople JA (1981) J ACS 103:4996
17. Orville-Thomas WJ (ed) (1974) Internal rotation in molecules. Wiley and Sons, New York
18. Pople JA (1977) in: Schaefer III HF (ed) Applications of electronic structure theory v 4 Modern theoretical chemistry. Plenum Press, New York
19. Vogler A, Wright RE, Kunkley H (1980) Angew Chem Engl Intern Ed 19:717
20. Saxe P, Schaefer III HF (1983) J ACS 105:1760
21. Löwdin PO (1959) Adv Chem Phys 2:207
22. Yoshioka Y, Goddard IIIWA, Schaefer III HF (1981) J Chem Phys 74:1855
23. Minkin VI, Minyaev RM, Zhdanov YA (1987) Nonclassical structures of organic compounds. Mir Publ, Moscow
24. Wong MW, Baker J, Nobes RH, Radom L (1987) J ACS 109:2245
25. Pople JA (1987) J Chem Phys 137:10
26. Hsu K, Buenker RJ, Peyerimhoff SD (1971) J ACS 93:2117
27. Spellmeyer DC, Houk KN (1988) J ACS 110:3412
28. Bally T, Masamune S (1980) Tetrahedron 36:343
29. Dewar MJS, Komornicki A (1977) J ACS 99:61
30. Voter AF, Goddard III WA (1986) J ACS 108:2830
31. Kaplan IG (1982) Introduction to the theory of intermolecular interactions. Nauka, Moscow
32. Cizek J, Paldus J (1967) J Chem Phys 47:3676
33. Prat RF (1975) in: Chalvet O, Daudel N (eds) Localization and delocalization in quantum chemistry v 1 Atoms and molecules in the ground state. Reidel Publ Co, Dordrect, Boston
34. Bouman H (1978) J ACS 100:7196

35. Seeger R, Pople JA (1977) J Chem Phys 66:3045
36. Hirao K, Nakatsui H (1978) J Chem Phys 69:4535
37. Halgren TA, Keier DA, Hall JH, Brown LD, Lipscomb WN (1978) J ACS 100:6595
38. Leasure SC, Martin TP, Balint-Kurti GG (1984) J Chem Phys 80:1186
39. Bachelet GB, Hamann DR, Schuler M (1982) Phys Rev B 26:4199
40. Rappe AK, Smedley TA, Goddard III WA (1981) J Chem Phys 85:1662
41. Segal GA (1977) Semiempirical methods of electronic structure calculation. Part A: techniques and Part B: Applications. Plenum Press, New York
42. Schembelov GA, Ustynyuk YA, Mamaev VM (1980) Quantum chemical methods of molecular calculations. Khimia, Moscow
43. Gubanov VA, Zhukov VP, Litinskii AO (1976) Semiempirical methods of molecular orbitals in quantum chemistry. Nauka, Moscow
44. Hoffman R (1963) J Chem Phys 39:1397
45. Hoffman R, Albright TA, Thorn DL (1978) Pure Appl Chem 50:1
46. Albright TA (1982) Tetrahedron 38:1339
47. Hoffmann R (1987) Angew Chem 99:871
48. Hehre WJ (1977) in: Schaefer III HF (ed) Applications of electronic structure theory. Plenum Press, New York
49. Heidrich D, Grimmer M, Köhler (1976) Tetrahedron 32:1193
50. Tomasi J (1979) in: Daudel R, Pullman A, Salem L, Veillard (eds) Quantum theory of chemical reactions. 1:191. Reidel Pub Co, Amsterdam
51. Bingham RC, Dewar MJS, Lo DH (1975) 97:1285
52. Dewar MJS (1975) Chem Brit 11:95
53. Dewar MJS (1985) J Phys Chem 89:2145
54. Lewis DFV (1986) Chem Rev 86:1111
55. Jug K (1980) Theor Chim Acta 53:148
56. Bantle S, Ahlrichs R (1978) Chem Phys Lett 53:148
57. Simkin BYa, Golyanskii BV, Minkin VI (1981) Zh Org Khim (USSR) 17:3
58. Dewar MJS, Olivella S, Stewart JP (1987) 108:5771
59. Bischof P (1977) J ACS 99:8145
60. Dewar MJS, Hashmall JA, Venier CG (1968) J ACS 90:1963
61. Dewar MJS, Thiel W (1977) J ACS 99:4899
62. Dewar MJS (1978) in: Further perspectives in organic chemistry. Elsevier, Amsterdam
63. Dewar MJS, Zoebisch EG, Healy EF, Stewart JP (1985) J ACS 107:3902
64. Jensen A (1983) Theor Chim Acta 63:269
65. Dewar MJS, Kirschner S (1974) J ACS 96:6809
66. Schröder S, Thiel W (1985) J ACS 107:4422
67. Schröder S, Thiel W (1986) J ACS 108:7985
68. Schröder S, Thiel W (1986) Theochem 138:141
69. Voityuk AA (1988) Zh Strukt Khim 29:138
70. Engelke R (1981) Chem Phys Lett 83:151
71. Thiel W (1981) J ACS 103:1413
72. Epstein PS (1926) Phys Rev 28:695
73. Nesbet RK (1955) Proc R Soc London, Ser A 230:312, 322
74. Ostlund NS, Bowen MF (1976) Theor Chim Acta 40:175
75. Thiel W (1981) J ACS 103:1420
76. Maier WF, Lau GC, McEwen AB (1985) J ACS 107:4724
77. Schweig A, Thiel W (1981) J ACS 103:1425
78. Davidson ER (1982) in: Borden WT (ed) Diradicals, Wiley, New York
79. Herzberg G, Johns JW (1971) J Chem Phys 54:2276
80. Herzberg G (1961) Proc R Soc London, Ser A 262:291
81. Leopold DG, Murray HU, Lineberger WC (1984) J Chem Phys 81:1048
82. Goddard III WA (1985) Science 227:917
83. Hase WL, Phillips RJ, Simons JW (1971) Chem Phys Lett 12:161
84. Hay PJ, Hunt WJ, Goddard III WA (1972) Chem Phys Lett 13:30
85. Bender CF, Schaefer HF, Franceschetti DR, Allen LC (1972) J ACS 94:6888
86. Buemi J, Zuccarello F, Raudino A (1988) Theochem 164:379
87. Dewar MJS, O'Connor BM (1987) Chem Phys Lett 138:141

CHAPTER 3

Effects of the Medium

Most organic reactions occur in solution and, naturally, any comprehensive analysis of the mechanism, kinetics, and thermodynamics of a given reaction must take account of effects of the medium, in other words, of the solvent effects. The number of publications is very great devoted to experimental research into the multifarious effects a solvent has on the rate and mechanism of organic reactions which were first detected about one hundred years ago by Menshutkin. How essential the effect of a solvent can be shows, for example, a comparison between the reaction of alkali hydrolysis of alkylhalogenides and esters in water and in the gas phase. The rate of the gas-phase reaction is by 20 orders of magnitude faster than in solution (see Sect. 5.1.1.4).

$$OH^-(H_2O)_n + CH_3Br \rightarrow Br^-(H_2O)_n + CH_3OH$$

In the gas-phase hydrolysis of esters, a competitive channel appears associated with an attack not on the carbonyl but rather on the alkyl carbon atom [1]:

$$R{-}C\genfrac{}{}{0pt}{}{{=}O}{{\smallsetminus}OCH_3} + {}^{18}OH^- \longrightarrow RCOO^- + CH_3{}^{18}OH$$

This channel cannot practically arise in solution. On the other hand, there are processes whose development does not in fact depend on properties of the medium. These interesting facts will explain why the study of the dependence of rates and mechanisms of chemical reactions on the medium has grown into one of the most rapidly developing branches of physical chemistry. The current state of this problem in the field of physical organic chemistry is represented in several reviews [2–4] and in a number of earlier publications [5–7].

A strict quantum mechanical analysis of the solvation effects on the reactivity, in which a direct calculation of the solvent-solute molecules system would be performed for all stages of its transformation during a reaction, cannot, in view of obvious mathematical difficulties, be realized as yet. In actual calculations, various simplified techniques are employed for assessing the solvent effects. The most important of these will be briefly considered in this chapter; for more detail see our reviews [8–10].

3.1 General Scheme for Calculating the Solvation Effects

In order to calculate the solvation effects, it is necessary, on the one hand, to find the wave function and the energy of the solute molecule in the solvent field, i.e., to solve the corresponding Schrödinger equation, and, on the other, to apply statistical methods, namely, to perform averaging of the energy and structural parameters of the solution weighted over all its configurations.

The most correct approach to the solution of the first problem involves the so-called supermolecular approximation in which the molecules of the reactants and the greatest amount of the solvent molecules admissible by the calculational scheme and computer capacity are merged into the unified system of a "supermolecule" (Sect. 3.4). The reliability of this approximation is determined by the accuracy of the quantum chemical scheme of calculation and also by the number of the solvent molecules taken into account. Of course, by far not all the potentialities of this method can currently be implemented because of calculational difficulties, in particular, excessive demands on computer time.

For calculating the energy of interaction between a solute molecule and the molecules of the solvent using the supermolecular approximation, it is necessary to study carefully the PES of the corresponding cluster. This surface has as a rule several or even many local minima the getting into which is primarily determined by the "starting" structure of the solvation shell. Therefore, the Boltzmann mean is calculated over the different energies corresponding to the local minima. This is, however, a rather crude substitute for the statistical methods (see Sect. 3.5).

The following effects can be described in terms of the supermolecular approximation: the degree of charge transfer between a solute molecule and the molecules of the solvent, changes in the electron and geometry structure of the molecule incorporated into solution, the influence a solvent has on the mechanism of chemical reaction.

Only statistical methods can be of use when studying the thermodynamic and structural characteristics of solution, such as heat of dissolution, energy of solvation, energy of solvent reorganization and radial distribution functions (probabilities of solvent molecules being located at a given distance from the solute molecule). In the statistical theory of solution, the energy of interaction with the medium is written as [11]:

$$E = C \int V(R) g(R) R^2 dr \tag{3.1}$$

where C is the constant that depends partly on the density of solution, $V(R)$ is the function describing the dependence of the energy of pairwise interactions on the distance R between the molecules, and $g(R)$ is the radial distribution function.

These two approaches, the supermolecular and the statistical, have up to now been used separately to study distinct manifestations of solvation effects. The model methods for their description may be divided into the discrete (microscopic) and the continual (macroscopic) ones.

3.2 Macroscopic Approximation

3.2.1 General Theory

The medium is assumed to exert weak polarizing effect upon the electron structure of a solute molecule or a reaction complex. This restriction is best fulfilled in the case of the molecules without conjugated bonds and of the solutions in which no specific solvation takes place (no formation occurs of hydrogen bonds, complexes with charge transfer or of other stabilized adducts of the solute molecule and the solvent).

In this case, the total energy of a molecule in solvent is:

$$E = E_o + E_s \tag{3.2}$$

where E_o is the energy of the isolated molecule calculated by the quantum chemical method and E_s is the solvation energy of the solute molecule.

Using the fundamental continuum theories (of Born, Onsager, Kirkwood), a direct calculation is in fact made not of the solvation energy but rather of the free solvation energy. Since, however, in most publications on this theme the calculated free solvation energy is stubbornly called the solvation energy, we shall retain this customary term.

The solvation energy can be represented as a sum of individual contributions. This decomposition into contributions, like any other, is rather aribtrary admitting alternative variants. The most common of these is that of Sinanoglu [12] in which E_s is determined as the sum of the following contributions:

$$E_s = E_{es} + E_{disp} + E_{cav} + E_{rep} \tag{3.3}$$

where E_{es} is the electrostatic energy of interaction of the constant and induced dipole moments of the solute molecule with those of the solvent molecules, E_{disp} is the dispersion energy, E_{cav} is the energy of formation of a cavity in the dielectric where the solute molecule is inserted and E_{rep} is the energy of repulsion of the valence-non-bound atoms.

Clearly, various components of E_s can be evaluated in terms of both the classic and the quantum mechanics. Commonly the former is used invoking also the theory of dielectrics. The value of E_{rep} is large only for the intermolecular distances less than the sum of van der Waals atomic radii. For this reason, the repulsion energy may, as a rule, be ignored in view of its smallness for actual intermolecular distances in solution and for lack of information on the distribution of the solvent molecules around the solute molecule. Usually, E_{es} and E_{disp} are negative while E_{cav} is positive.

When ions are solvated by polar solvents, the electrostatic component is predominant. In a somewhat softer case, when, for example, polar molecules are solvated in polar solvents, the components E_{es}, E_{disp} and E_{cav} become comparable. At the same time, some calculation results [13] indicate that it is precisely the changes in E_{es} which determine the observed variations in E_s. In other words, for qualitative estimates of E_s for a series of similar structures, particularly conformers, it would sometimes suffice to assess E_{es}.

In terms of the general theory of Kirkwood, the electrostatic energy of interaction of a solute molecule with the solvent, assumed to be an isotropic dielectric with the dielectric permittivity ε will be:

$$E_{es} = \frac{1}{2} \sum_{n=0}^{\infty} \frac{(n+1)(1-\varepsilon)}{n+(n+1)\varepsilon} \sum_{j=1}^{N} \sum_{k=1}^{N} Q_j Q_k \frac{(\mathbf{r}_j \mathbf{r}_k)^n}{a^{2n+1}} P_n(\cos\theta_{jk}) \tag{3.4}$$

where a is the radius of the spherical cavity, Q_j are the point changes of each of N atoms of the solute molecule, P_n are the Legendre polynomials that describe, respectively, the monopole, dipole, octupole interactions and the effects of higher orders.

A choice of the size of the spherical cavity is not, generally speaking, unambiguous. In practical terms, a good approximation is represented by the volume of the molecule calculated by means of molecular refraction.

In most cases, molecules do not possess spherical symmetry, but the choice of the cavity of spherical form does not ordinarily introduce any substantial error into the estimate of the electrostatic energy. Moreover, even this error may be reduced by selecting the shape of the cavity according to molecular configuration. In a number of publications, Ref. [14] being one of the latest, the calculation of the electrostatic energy was carried out under the assumption of a cavity having the shape of an ellipsoid of revolution.

The terms of the series in Eq. (3.4) rapidly decrease in value so that in practical calculations it is sufficient to take into account no more than four terms [14]. The values of the electrostatic energy calculated from Eq. (3.4) cannot be used as absolute seeing that the variation of the sizes of the cavity may change these values several times. Equation (3.4) may appropriately be employed to estimate relative values of the electrostatic energy for a series of structurally similar molecules (with roughly equal volume, such as conformers, tautomers etc.). A review concerned with the application of the Kirkwood theory to calculations of the electrostatic energy was written by Tapia [15].

A particular case of Eq. (3.4) for $n = 0$ is the Born equation for the ion solvation (in the case of the neutral molecules this contribution is equal to zero):

$$E_0 = -\frac{1}{2}\frac{Q^2}{a}\left(1 - \frac{1}{\varepsilon}\right) \tag{3.5}$$

Because of the simplicity, Eq. (3.5) has gained fairly wide acceptance for the interpretation of experimental facts. Furthermore, it served as a basis for the construction of a number of model Hamiltonians (see below).

At $n = 1$ Eq. (3.4) becomes an analog of the Onsager dipole energy (a model of the reactive field)

$$E_1 = -\frac{1}{2}\left[\frac{2(\varepsilon - 1)}{2\varepsilon + 1}\frac{\mu^2}{a^3}\right] \tag{3.6}$$

A detailed review of the uses and limitations of the theory of reaction field as applied to studies of intermolecular effects was made by Linder [16] nearly 20 years ago. This area of the theory was already then fairly well developed and no essential additions have been made since.

The calculation of the dispersion component and the energy of formation of the cavity in Eq. (3.3) is made, according to Ref. [12], by means of simplified relationships analyzed in Ref. [10].

3.2.2 Model Hamiltonians in the Macroscopic Approximation

A considerable shortcoming of Eq. (3.2) consists in the assumption of invariability of the electron structure of the solute molecule when it passes from "free space" into solution, i.e., in the assumption of constancy of E_0. However, E_0 can change significantly, particularly for the systems of high polarizability. This limitation of the calculational scheme may be overcome by constructing a model Hamiltonian of the solute molecule. The general form of this Hamiltonian $\hat{H}$ is:

$$\hat{H} = \hat{H}_0 + \hat{U} \tag{3.7}$$

where $\hat{H}_0$ is the Hamiltonian of the isolated molecule and $\hat{U}$ is the operator describing interaction of the solute molecule with the medium.

Strictly speaking, when constructing the operator $\hat{U}$ it is necessary to take into account all types of the intermolecular forces. However, such a scheme that would actually represent the supermolecular approach (see below) requires unjustifiably excessive expenditure of computer time. For this reason, when selecting the form of the model Hamiltonian, only the electrostatic interactions are commonly taken into consideration.

3.2.2.1 *Model Hamiltonian in the Kirkwood Approximation*

The most rigorous method for constructing the operator $\hat{U}$ in the continuum model is based on the Kirkwood theory. In this case, the distribution of point charges is replaced in the classic expression of Eq. (3.4) with the corresponding quantum mechanical distribution. In other words, in Eq. (3.4) r_j and r_k and the angle θ_{jk} between these vectors will have to be regarded not as constants but rather as variable quantities. Then, using the atomic units, $\hat{U}$ can be written as a sum of the nuclear-nuclear, nuclear-electron, and electron-electron terms:

$$\hat{U} = \sum_{n=0}^{\infty}\left[\frac{(n+1)(1-\varepsilon)}{(n+1)\varepsilon+n}\left(\frac{1}{a}\right)\left\{\frac{1}{2}\sum_{\alpha,\beta}^{P} Z_A Z_B\left(\frac{\mathbf{R}_\alpha \mathbf{R}_\beta}{a^2}\right)^n P_n(\cos\theta_{\alpha\beta})\right.\right.$$
$$\left.\left. - \sum_{\alpha}^{P}\sum_{i}^{s} Z_\alpha\left(\frac{\mathbf{R}_\alpha \mathbf{r}_i}{a^2}\right)^n P_n(\cos\theta_{\alpha i}) + \frac{1}{2}\sum\left(\frac{r_i r_j}{a^2}\right)^n P_n(\cos\theta_{ij})\right\}\right] \tag{3.8}$$

It is assumed that in this system there are P nuclei and s electrons, and the positions of the nuclei $\mathbf{R}_\alpha$, $\mathbf{R}_\beta$ are fixed. After solving the Schrödinger equation with the Hamiltonian in Eq. (3.7), we obtain the wave function of the solute molecule depending on the dielectric permittivity of the solvent.

The complete continuum approach was employed in the Kirkwood model on an ab initio level with the basis set of the floating Gauss functions in 1976 [17]. Around that time, a similar formalism for taking the solvent into account was included in the CNDO/2 method [18]. However, such calculational schemes did not gain wide acceptance by reason of excessive expenditure of computer time, difficulties in evaluating some integrals and overall drawbacks inherent in the macroscopic approximation. Eventually some simplified techniques were developed, each of which takes usually one of the components in Eq. (3.8) into account. Next the simplest of these will be considered.

3.2.2.2 *A Model Hamiltonian Based on the Born Formula. Scheme of Solvatons*

Strictly speaking, the Born equation of Eq. (3.5) can be used only for calculating the solvaton energy of an ion. However with some additional assumptions, it is possible to construct from Eq. (3.5) a model Hamiltonian for neutral molecules. Such a scheme was first suggested by Klopman [19], while its nonselfconsistent version had already been employed in Ref. [20].

Let each charge Q_A of the solute molecule induce in the medium with the polarizability α a counter-charge:

$$Q_A^{\mathrm{pol}} = -\alpha Q_A \tag{3.9}$$

The charges induced in the solvent determine the electric field that has an effect on the electron density of the solute molecule and, consequently, causes its reorganization. One may find the charge distribution of a solute molecule in the polarizable medium by solving modified Hartree–Fock equations in which the Fockian has the form:

$$F_{\mu\mu}(\alpha) = F^0_{\mu\mu} - \sum_B Q_B^{\text{pol}} \gamma_{AB} \tag{3.10}$$

The polarizability α is a parameter of the calculational scheme. Clearly, it must be related with the dielectric permittivity of the medium. Actual form of the function $\alpha(\varepsilon)$ can be quite diverse [19, 21, 22]. The first suggested dependence $\alpha(\varepsilon)$ see Ref. [19]:

$$\alpha(\varepsilon) = \frac{1}{2}\left(1 - \frac{1}{\varepsilon}\right) \tag{3.11}$$

was based on the Born equation and did not satisfy the obvious limiting conditions:

$$\lim_{\varepsilon \to 1} \alpha(\varepsilon) = 0 \quad \text{and} \quad \lim_{\varepsilon \to \infty} \alpha(\varepsilon) = 1 \tag{3.12}$$

Equation (3.11) did not in fact take into account the polarization energy of the medium. The later models [21, 22] corrected this shortcoming. The total energy of the solute molecule is, in accordance with Eqs. (3.10), (3.11), given by:

$$E(\varepsilon) = E_0 - \frac{1}{2}\left(1 - \frac{1}{\varepsilon}\right)\sum_{A > B}\sum Q_A Q_B \gamma_{AB} \tag{3.13}$$

Note that our interpretation of the method represented by Eqs. (3.10), (3.11), (3.13) differs from the original [19]. According to Klopman, each atom of the solute molecule generates a particle of the medium, the "solvaton", whose charge equals the atomic charge but is opposite in sign. No interaction is assumed between the solvatons, while the interaction between the atomic charge and the solvatons is described by the equation:

$$E = \frac{Q_A Q_s}{2 r_{As}}\left(1 - \frac{1}{\varepsilon}\right) \tag{3.14}$$

A detailed analysis of the solvaton scheme may be found in [8–10, 23–25]. There are instances of its succesful application in the solution of thermodynamic and kinetical problems. An illustrative example is provided by the theoretical treatment of the reaction of nucleophilic addition of the hydroxyl to CO_2:

$$OH^- + CO_2 \rightleftarrows HCO_3^- \tag{3.15}$$

The activation energy of this reaction in water was experimentally found to be 13 kcal/mol [26]. The ab initio calculations (27) with solvaton not taken into account has shown this process to develop without a barrier with the heat effect of 56 kcal/mol. The experimental value of the heat effect is much lower being a mere 11 kcal/mol. Ab initio (DZ basis set) and MINDO/2 calculations, with the solvent included in accordance with the solvaton scheme, made it possible to adequately reproduce the form of the PES (presence of the activation barrier) and the value of the heat effect [28].

Of course, the solvation method should not be applied before its possibilities in regard to a given problem have been explored. Thus, an attempt to reproduce, in terms of this scheme, the energy surface of the reaction:

$$NH_3 + HF \rightarrow NH_4^+ + F^- \tag{3.16}$$

in water failed to produce an even qualitatively correct energy profile [29].

3.2.2.3 *The Scheme of Virtual Charges*

The idea of the method of virtual (imaginary) charges developed in [21, 30, 31] is close to that of the Klopman scheme. Clearly, Eq. (3.14) is unsuitable for the description of the interaction between the charge A and the solvaton generated by it. Therefore, Constanciel and Tapia worked out a scheme of virtual charges under which the polarized solvent forms on the surface of the cavity that surrounds a given atom an external charge [Eq. (3.9)]. Whereas in the solvation scheme $\alpha = 1$ and Eq. (3.14) should be regarded as being a rule based on the Born formula, in the virtual charges method the choice of α rests on meaningful physical assumptions according to which the energy of interaction between the charge Q_0 and the medium:

$$E_{\text{int}} = -\frac{Q_0^2}{2a}(\alpha^2 - 2\alpha) \tag{3.17}$$

So as to describe correctly the asymptotic behavior of $\alpha(\varepsilon)$, Constanciel and Tapia [21] have taken the following empiric relationship:

$$\alpha = 1 - \frac{1}{\sqrt{\varepsilon}} \tag{3.18}$$

Then Eq. (3.17) will coincide with the rigorous Born expression:

$$E_{\text{int}} = -\left(1 - \frac{1}{\sqrt{\varepsilon}}\right)\frac{Q_0^2}{2a} \tag{3.19}$$

The virtual charges scheme is better substantiated than the Klopman model since in it both the charge and the energy depend on ε in a noncontradictory manner.

3.2.2.4 *The Theory of Selfconsistent Reactive Field*

The schemes of solvatons and of virtual charges represent particular cases of the general theory of selfconsistent reactive field (SCRF) [21, 22, 32–35]. The model Hamiltonian has in the general theory of SCRF the form:

$$\hat{H} = \hat{H}_0 - \hat{\boldsymbol{\mu}} g \langle \psi | \hat{\boldsymbol{\mu}} | \psi \rangle \tag{3.20}$$

where g is the tensor determining the susceptibility of the reactive field; ψ is the wave function of the solute molecule; $\hat{\boldsymbol{\mu}}$ is the operator of the dipole moment vector.

The second term in Eq. (3.20) is a quantum mechanical analog of the classic electrostatic energy of interaction:

$$E_{\text{es}} = -\boldsymbol{\mu}\mathbf{R} \tag{3.21}$$

where $\boldsymbol{\mu}$ is the sum of the permanent dipole moment and the moment induced by the medium and $\mathbf{R}$ is the Onsager reactive field.

In the general case, the tensor g depends on the macroscopic parameters of the medium and on the structure of the first solvation shells. Commonly, information on the structure of the surrounding of the solute molecule is ignored in view of its scarcity and the tensor g is taken to be (on analogy of the Onsager theory):

$$g = \frac{2}{a^3}\frac{\varepsilon - 1}{2\varepsilon + 1} \tag{3.22}$$

Thus, the problem of finding the electronic and geometrical structure of a solute molecule comes down to solving the nonlinear Schrödinger equation:

$$\hat{H}(\psi)\psi = \{\hat{H}_0 - \hat{\boldsymbol{\mu}} g \langle \psi | \hat{\boldsymbol{\mu}} | \psi \rangle\}\psi = E(\mathbf{R})\psi \tag{3.23}$$

The SCRF model has been included in the semiempirical (CNDO/2, MINDO/3) [32, 35] and ab initio [36] calculational methods. Its simplicity allows a good deal of static and dynamic problems to be solved.

One more scheme, also based on a reactive field model developed recently [37–40], is noteworthy. This scheme makes it possible to calculate both the free energy and the solvation enthalpy. In Refs. [39, 40] it has been applied in an analysis of the effect a solvent has on the energy profile and mechanism of chemical reaction.

3.3 Discrete Representation of Solvent Molecules. Model Hamiltonians in the Microscopic Approximation

Let the operator $\hat{U}$ of Eq. (3.7) of the electrostatic interaction with the medium be written for the case when the solute molecule is surrounded by the solvent molecules regarded as a set of electrons and nuclei (discrete description of the medium):

$$\hat{U} = \sum_{n=1}^{S} \left\{ \sum_{i=1}^{N} \sum_{\alpha=1}^{N'} \frac{1}{r_{i\alpha n}} - \sum_{i=1}^{N} \sum_{a=1}^{M'} \frac{Z_{an}}{r_{ian}} + \sum_{A=1}^{M} \sum_{a=1}^{M'} \frac{Z_{an} Z_A}{r_{Aan}} - \sum_{A=1}^{M} \sum_{\alpha=1}^{N'} \frac{Z_A}{r_{A\alpha n}} \right\} \tag{3.24}$$

The following notation is used here: i are the electrons of the solute molecule; N is their number; ε are the electrons of the solvent molecule; N' is their number; A are the nuclei of the solute molecule; M is their number; a are the nuclei of the solvent molecule; M' is their number; Z_a and Z_A are the nuclear charges of the solvent and the solute molecule, respectively; S is the total number of the solvent molecules under consideration.

The solution of the problem with the perturbation Hamiltonian in Eq. (3.24) is quite a laborious task for which reason it is usually simplified to:

$$\hat{U} = \sum_{k=1}^{L} \sum_{i=1}^{N} \frac{q_k}{r_{ik}} - \sum_{k=1}^{L} \sum_{A=1}^{M} \frac{q_k Z_A}{r_{kA}} \tag{3.25}$$

Here q_k are taken to be point charges whose positions are determined by the solvation shell structure and the electron density distribution in the solvent molecules; L is the number of the point charges. The values of q_k are commonly assumed fixed and localized in certain points of space. The best variant of the point charges method is obtained when the values of q_k and their localization places are chosen in accordance with the electron structure of solvent molecules and the true structure of the solvation shells. Information on their respective structures is, however, quite limited so that frequently the nearest surrounding of the solute molecule is modelled. Using Eq. (3.25), one may study the effect the position of charges and their magnitude have on properties of the solute molecule or the solvation complex.

The use of the microscopic model Hamiltonian of Eq. (3.25) involves, as compared to the macroscopic version, both advantages and drawbacks. The advantages are these: 1) inclusion of the real structure and the electron density distribution of solvent molecules; 2) rejection of macroscopic characteristics of the medium that grow quite vague near the solute molecule; 3) absence of such uncertain parameters as, e.g., the cavity radius. Among the shortcomings one may note; 1) no allowance is made for the polarizing effect of the solute molecule on the electron distribution of solvent molecules which is assumed fixed in the solution of the Schrödinger equation; 2) repulsion between the solute molecule and the solvent molecules is not taken into account, which, in terms of the scheme discussed, leaves no room for the question as to the construction of solvation shells, nor does it provide for optimization of the geometry parameters of the solute molecule since opposite charges may draw together to a distance less than the equilibrium distance. In this case, the electrostatic description of intermolecular interactions is liable to inaccuracy seeing that the magnitudes of other contributions to the total energy may substantially grow.

The point charge method has been primarily developed in Refs. [41–45]. Some problems of its application were examined and solved in Ref. [45]. Use was made of the MINDO/3 method. The point charges model of Eq. (3.25) was supplemented by inclusion of two additional contributions into the core energy, namely, the dispersion energy and the energy of exchange repulsion.

The dependence of calculation results on the number and the localization places of the point charges was analyzed in Refs. [43–45]. According to ab initio calculations, covalent bonding is predominant in the gas-phase in the ammonia-hydrogen fluoride (NH_3—HF) complex and the PES has one minimum. Then two hydrate shells were constructed. In the first of these, six water molecules were located in positions of minimal energies of interaction between one water molecule and the complex (for monosolvation scheme see below). The second hydrate shell is constructed by adding water molecules linked to the first six through a hydrogen bond.

Perturbation by the point charges localized at the first shell atoms did not lead to any substantial changes in the form of the PES of the complex. On the other hand, inclusion of the second solvation shell resulted in a qualitatively correct pattern of the PES with two minima, one of which corresponded to the structure of $NH_4^+ \cdots F^-$. This finding shows the need for taking into account of the possibly greatest number of point charges and seems to indicate that their localization sites influence the perturbation effects.

The operator of perturbation of a solute molecule by the point dipoles that surround it may be written similarly to Eq. (3.25):

$$\hat{U} = \sum_{k=1}^{L} \sum_{i=1}^{N} \frac{\boldsymbol{\mu}_k \mathbf{r}_{ki}}{r_{ki}^3} - \sum_{k=1}^{L} \sum_{A=1}^{N} \frac{\boldsymbol{\mu}_k \mathbf{r}_{kA} Z_A}{r_{kA}^3} \tag{3.26}$$

where $\boldsymbol{\mu}_k$ are the vectors of dipole moments of the solvent molecules.

The methods of perturbation by point dipoles was chiefly developed in Refs. [42, 46–48]. There are different ways in which the values of the point dipoles may be selected. The medium may be described by permanent dipoles whose influence upon the solute molecule is then analyzed. In this case, the point dipoles scheme has no advantage over the scheme of the point charges. Considering, however, the fact that, owing to polarization, induced dipoles arise in the solvent, one may find Eq. (2.26) quite attractive. Indeed, the quantities $\boldsymbol{\mu}_k$ may be regarded as consisting of two parts: a permanent and an induced dipole. Thus, one may "switch on" the influence of the solute molecule upon the electron characteristics of the solvent molecules. Note that the interaction of induced dipoles with the electrostatic field of a solute molecule invariably diminishes the energy of the system since the induced dipoles are orientated along the field. There exists a possibility of combing Eqs. (3.25) and (3.26). In this case, the point charges approximate the electron distribution of an unperturbed solvent, while the point dipoles describe only the polarization of the medium [42, 49, 50].

Using Eq. (3.26), the effect of solvent on various classes of chemical reactions has been studied [42, 45–50, 51, 52].

As noted earlier, successful application of the point charge and the point dipole methods is essentially predetermined by the manner in which the solvation shells are constructed. The most rigorous approach consists in combining the statistical methods for studying solutions (see Sect. 3.5), permitting calculation of atom–atom distribution functions and, consequently, providing information on the average structure of solvation shells, with the model Hamiltonians in microscopic approximation. As it often happens, the more rigorous the approach, the greater must be calculational efforts. Thus, whereas perturbation by the point charges or dipoles in terms of the Hartree–Fock method does not practically increase computer time consumption as compared to calculations on an isolated molecule, the construction of a liquid phase with the aid of computer modelling requires computer resources greater by up to two orders of magnitude. Keeping this in mind, one is commonly forced to confine oneself to modelling the solvation shells or even their parts only.

The model solvation shells are preferably constructed by means of simple calculational schemes, occasionally without any calculations at all, based on intuitive assumptions of specific interactions in solution. The following calculation procedure may be regarded as the simplest. Using the method of molecular electrostatic potential (MEP) the regions are determined of maximal attraction between the solute molecule and the molecules of the solvent. The MEP method rests on the scheme for calculating the interaction $V(r_i)$ of a single positive charge with the molecule [53, 54]:

$$V(r_i) = -\int \frac{\rho}{r_i} dr + \sum_A \frac{Z_A}{r_{Ai}} \tag{3.27}$$

where the first and the second terms describe, respectively, the electron attraction and the nuclear repulsion; ρ is the function of the electron density distribution and Z_A are the nuclear charges.

Having calculated the values of the MEP for the whole region surrounding the solute molecule, one may determine the minima of $V(r_i)$ and place there the molecules of the medium. The structure of the nearest surrounding may be built by means of the monosolvation scheme [55, 56]. By this scheme, using supermolecular calculations (see Sect. 3.4), all possible minima may be found of the interaction between the solute molecule and one molecule of the solvent. Solvent molecules are introduced into each minimum and refining optimization of the total energy is carried out. Sometimes use is made [57, 58] of the so-called cluster approximation, also referred to as the polysolvation scheme, which consists in a direct search for the minimum on the hypersurface of the system solute molecule $+ n$ molecules of the solvent (for a more detailed description see Ref. [10]).

Noteworthy is also a relatively simple procedure of determining the localization sites of solvent molecules based on a certain scheme of their disposition which is independent of the properties of the solute molecule. For example, solvent molecules may be placed in cubic lattice sites, with part of the molecules removed upon incorporation of the solute molecule. Such a scheme was, for instance, implemented in terms of the point dipoles model in Refs. [42, 46–48].

3.4 Specific Features of the Supermolecular Approach in Studies of Solvation Effects

Only the supermolecular approach to the analysis of solvation effects makes it possible to take into account a) the intermolecular charge transfer; b) interactions of all types among molecules; c) that the solvent molecules may play an active role in determining chemical reactivity. The chief advantage of the supermolecular method is that in it the solvent is not assigned a passive role of merely following the motion of the reacting system, as in the above described schemes for taking the medium into account, in which an obviously inexact assumption is present as to an equilibrium solvation of a nonequilibrium system (for example, in a transition state). The restructuring of the solvation shell proceeds not as a relaxation process but rather as an integral part of this process associated with motion of the reactants. An ever greater number of computation data have been accumulating that support precisely such an interpretation of certain chemical reactions [59–63]. Sometimes, even only one solvent molecule directly taken into account helps ascertain the reason for often basic differences between the reactions in the gas phase and in solution.

Thus, the electrophilic addition reaction:

$$CH_2{=}CH_2 + Cl_2 \rightarrow Cl{-}CH_2{-}CH_2{-}Cl$$

has thoroughly been studied theoretically both without inclusion of a solvent and with one water molecule [64] by the ab initio method with an extended basis set (the dispersion part of the energy was also taken into account). The energy of the ionic structure $C_2H_4Cl^+Cl$ was so high that it could not lie on the reaction coordinate. The addition in the supermolecular calculation of only one water molecule opens up a reaction channel of the ionic mechanism. Moreover, a calculation in the absence of a solvent did not produce a transition state between the cations of the π- and the σ-types (see Sect. 6.1) which was revealed upon addition of a solvent molecule.

At the same time, a potentially high accuracy of calculations inherent in this method may be in a stark contrast with considerable errors caused by incorrect determination of the structure of a solvation shell due to the limited number of the solvent molecules taken into account and laboriousness of the optimization of the geometry of the complex. As a rule, the supermolecular approach makes use of model conceptions in regard to the structure of the solvation shell. For example, when studying the hydration effect, the water molecules are intially arranged in such a fashion that they form linear hydrogen bonds with the active centers of the solute molecule.

A more complicated but still realizable procedure consists in the use of the monosolvation and polysolvation scheme, however, with only few solvent molecules. A certain progress was made in the application of the supermolecular approach by the Pullmans (see the reviews listed as Refs. [55, 65]). They distinguished three types of solvent molecules depending on the degree of their interaction with the solute molecule. Such differentiation reduces calculational difficulties, since in this case not the same precision is required in the localization of the solvent molecules depending on their disposition with respect to the solute molecule or the adduct.

There are practically no studies on the dependence of the results of supermolecular calculations on the number of the medium molecules taken into account and of their localization sites. The solution of this problem is, in terms of an ab initio approach, a difficult task. It is for that reason that the MINDO/3 and CNDO/2 methods were applied in Ref. [66]. The shortcomings of these schemes (see Chap. 2) restricted the analysis to the dependence of the intra- and intermolecular charge transfer in solution on the choice of structure of the solvation shell. The intramolecular redistribution of charges does not practically depend on the structure of the hydrate shell. Changes in the charges in solution, even for relatively strongly polarizable molecules, such as H-pyridone, are small as compared to the atomic charges. This confirms the approximation of rigidity of the electron structure of molecules, standard for numerical modelling of solutions (see next section).

3.5 Statistical Methods for Studying Solutions

A correct calculation of solvation thermodynamics and solution structure is conceivable only in terms of the methods of statistical physics, in particular, the computer experiment schemes, including, in the first place, the molecular dynamics (MD) and the Monte-Carlo (MC) methods [10]. By means of the MD method Newton's classic equations of motion are solved numerically with the aid of a computer assuming that the potential energy of molecular interaction is known. In this manner, the motion of molecules of the liquid may be "observed", the phase trajectories found and then the values of the necessary functions are averaged over time and determined. This method permits both the equilibrium and the kinetical properties of the system to be calculated.

A basically different approach is used in the MC method. The states of a system of particles are taken to be stochastic and one has to select the most probable configurations over which various properties are then averaged. Therefore, this method is suitable only for calculating equilibrium quantities since it cannot, in principle, give an answer to the question how the system achieves its equilibrium because, instead of a genuine dynamic evolution of the system, an artificial stochastic process is modelled. A possibility arises of taking into account by statistical methods the effect of the temperature on the properties of a solution.

The methods of the computer experiment require simplified representation of the potential energy $V(R)$, so in most calculations only the two-particle contributions are taken into account:

$$V(R) = \sum_{i,j} V_{ij} \tag{3.28}$$

Moreover, the two-particle contributions V_{ij} are, as a rule, calculated as a sum of the atom-atom interactions:

$$V_{ij} = \sum_{m,n} U_{ij}(r_{mn}) \tag{3.29}$$

where r_{mn} is the distance between the atom m of the i-th molecule and the atom n of the j-th molecule.

The simplified representation of the potential energy makes it possible to consider in the MD and MC methods a great number of the solvent molecules (up to 1000). It should, however, be kept in mind that whereas the medium molecules may, as a rule, be regarded as electronically and geometrically rigid, in the case of reactants this assumption is clearly incorrect. In the course of reaction, a strong distortion occurs of the electronic and geometric structure of these molecules which cannot be described by means of the standard potentials. In order to do that, the dependence must be found between the intermolecular potentials and the structure of reacting molecules. The first

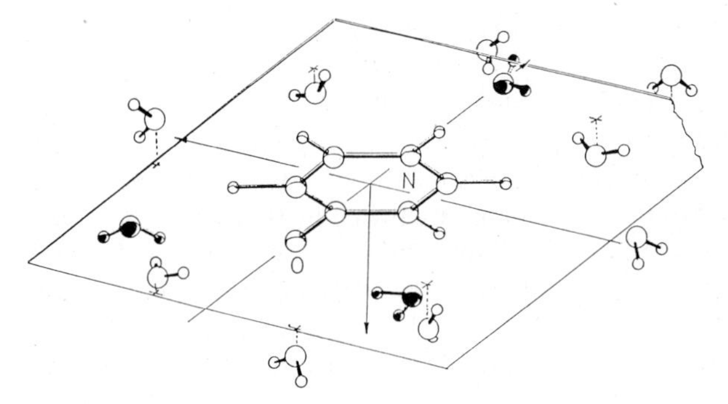

Fig. 3.1. Comparison of the structure of 4-pyridone solution in water calculated by the MC method [8] to that of the solvation shell (*black circles* representing water molecules) obtained by the monohydration method

attempts have already been made at calculations in which the internal degree of freedom of a reactant is equivalent with respect to the degrees of freedom of the motion of medium molecules [67–70].

Unlike the monohydration schemes, the computer experiment methods account in solutions both for interactions of the solute molecule with molecules of the solvent and for those of solvent molecules among each other. In some cases, particularly for aqueous solutions, the latter component is comparable to and may even exceed the water-solute molecule interaction energy. This results in considerable differences in the structure of solvation shells. Below (Fig. 3.1), solvation shells are given of 4-pyridone obtained by calculations using the MC method [8] and the monohydration scheme. The data presented show that for correct determination of localization sites of the point charges and dipoles of solvent molecules, the calculations should be employed based on statistical methods.

One more methodological problem should be touched upon whose solution is necessary for correct description of mechanisms of reactions in solutions. This problem has to do with the relaxation of the solvent during the dynamic process. In all methods, except the MD scheme, it is assumed that, regardless of the velocity of the process, the medium is equilibrated in each point of the PES of reaction. This assumption is actually one of the necessary conditions for applying the theory of transition state. Clearly, in the case of fast reactions the time of reorganization of the solvent molecules is comparable to the time of realization of these reactions. This justifies the conclusion that the equilibrium of the medium is not always fulfilled. The model calculations by van der Zwan and Hynes [71, 72], later extended to more realistic cases of the S_N2 reactions in polar media [73, 74], bear witness to the dependence of the reaction rate constants upon the degree of nonequilibrium of a given solvent.

References

1. Takashima K, Riveroc JM (1978) J ACS 100:6128
2. Menger FM (1983) Tetrahedron 39:1013
3. Kurtz AL (1964) Zh Vsesoyuzn Khim Obszh im DI Mendeleeva 29:530
4. Thea S, Williams A (1985) Chem Soc Rev 15:125
5. Reichardt C (1979) Solvent effects in organic chemistry. VCH, New York
6. Entelis SG, Tiger RP (1973) The kinetic of reactions in liquid phase. Khimia, Moskow (in Russian)
7. Moelwyn-Hughes EA (1971) The chemical statics and kinetics of solutions. Academic Press, New York
8. Simkin BYa, Sheykhet II (1983) J Mol Liquids 27:79
9. Simkin BYa, Scheykhet II (1983) in: Kolotyrkin YaM (ed) Physical chemistry. Current problems. Khimia, Moskow (in Russian), p 148
10. Simkin BYa, Sheykhet II (1989) The quantum chemical and statistical theories of solutions. The calculational methods and their applications. Khimia, Moskow (in Russian)
11. Pierotti RA (1976) Chem Rev 76:717
12. Sinanoglu O (1974) Theor Chim Acta 33:279
13. Krebs C, Hoffman HJ, Weiss C (1981) Z Chem 227
14. Gesten JT, Spsae AM (1985) J ACS 107:3786
15. Tapia O (1982) in: Ratajczak H, Orville-Thomas WJ (eds) Molecular interactions. Wiley, New York, p 47
16. Linder B (1967) Advan Chem Phys 12:283
17. McGreery JH, Chrystoffersen RE, Hall GG (1976) J ACS 98:7191
18. Rivail J-L, Rinaldi D (1976) Chem Phys 18:233
19. Klopman G (1967) Chem Phys Lett 1:200
20. Hoijtink GJ (1956) Rec Trav Chim Pays Bas 75:487
21. Tapia O, Goscinski O (1975) Mol Phys 29:1653
22. Constanciel R, Tapia O (1978) Theor Chim Acta 48:75
23. Simkin BYa (1978) Khim Geterocycl Soed N1 (without volume):94
24. Simkin BYa, Minkin VI, Kosobutski VA, Zhdanov YuA J. Mol. structure (1975) 24:237
25. Gorb LG, Arbonin IA, Kharchevnikova NV (1984) Zh Phys Khim 58:9
26. Pinsent BRW, Pearson L, Roughton FIW (1956) Trans Farad Soc 52:1512
27. Jonsson B, Karlstrom G. Wennerstrom H (1978) J ACS 100:1658
28. Miertus S, Kysel O, Krajcik K (1981) Chem Zvesti 35:3
29. Miertus S, Bartos J (1980) Coll Czech Chem Comm 45:2308
30. Constanciel R (1980) Theor Chim Acta 54:123
31. Constanciel R (1986) Theor Chim Acta 69:505
32. Tapia O, Silvi B (1980) J Phys Chem 84:2646
33. Tapia O, Johannin G (1981) J Chem Phys 75:3624
34. Sanhueza JE, Tapia O (1982) THEOCHEM 89:131
35. Tapia O, Sussman F, Poulain E (1978) J Theor Biol 71:49
36. Karelson MM (1980) React Sposopbn Org Soed 17:363
37. Miertuš S, Scrocco E, Tomasi J (1981) Chem Phys 55:177
38. Bonaccorsi R, Cimiraglia R, Tomasi J (1983) J Comput Chem 4:567
39. Sheykhet II, Edelstein LL, Levchuk VN, Simkin BYa (1989) Khim Phys (in press)
40. Bonaccorsi R, Cimiraglia R, Miertuš S, Tomasi J (1983) THEOCHEM 94:11
41. Abronin IA, Burstein KYa, Zhidomirov GM (1980) Zh Struct Khim 21:145
42. Warshel A, Levitt M (1976) J Mol Biol 103:227
43. Noell JO, Morokuma K (1976) J Phys Chem 80:2675
44. Noell JO, Morokuma K (1977) J Phys Chem 81:2295
45. Simkin BYa, Sheykhet II, Levchuk VN (1984) Zh Struct Khim 25:55
46. Warshel A (1979) J Phys Chem 83:1640
47. Levchuk VN, Sheykhet II, Simkin BYa (1988) J Mol Liquids 38:35
48. Burshtein KYa (1987) THEOCHEM 153:195
49. Warshel A (1981) Acc Chem Res 14:284
50. Warshel A, Russell ST (1984) Quart Rev Biophys 17:283

51. Burshtein KYa (1987) THEOCHEM 153:203
52. Burshtein KYa (1987) THEOCHEM 153:209
53. Politzer P, Truhlar DC (eds) (1981) Chemical applications of atomic and molecular electrostatic potentials, Plenum Press, New York
54. Tomasi J (1979) in: Daudel R (ed) Quantum theory of chemical reactions. Reidel, Amsterdam, p 191 (v 1)
55. Pullman A (1980) in: Daudel R (ed) Quantum theory of chemical reactions. Reidel, Amsterdam, p 1 (v 2)
56. Demontis P, Ercoli R, ES (1981) Thoer Chim Acta 58:97
57. Pullman A Perahia D (1978) Theor Chim Acta 48:29
58. Pullman A, Miertuš S, Perahia D (1979) Theor Chim Acta 50:317
59. Williams IH (1987) J ACS 109:6299
60. Bertran J, Lledos A (1985) THEOCHEM 123:211
61. Zielinski ThJ, Poirier RA, Peterson MR, Csizmadia IG (1983) J Comput Chem 4:419
62. Ventura ON, Lledos A, Bonaccorsi R, Bertran J, Tomsai J (1987) Theor Chim Acta 72:175
63. Field MJ, Hiller IH (1987) J Chem Soc Perkin Trans 11:617
64. Kochanski E (1981) in: Pullman B (ed) Intermolecular forces. The Jerusalem symposium on quantum chemistry and biochemistry. Reidel, Amsterdam, p 15 (v 14)
65. Pullman A (1976) in: Pullman B, Parr R (eds) The new world of quantum chemistry. Proceedings of the second international congress of quantum chemistry. Reidel, Amsterdam, p 149
66. Sheikhet II, Simkin BYa (1988) Zh Struct Khim 29:84
67. Fresier BC, Jolly DL, Nordholm S (1983) Chem Phys 82:369
68. Chandrasekhar J, Smith SF, Jorgensen WL (1985) J ACS 107:154
69. Evanseck JD, Blake JF, Jorgensen WL (1987) J ACS 109:2349
70. Levchuk VN, Sheykhet II, Simkin BYa, Dorogan IV (1989) Zh Struct Khim (in press)
71. van der Zwan G, Hunes JT (1983) J Chem Phys 78:4174
72. van der Zwan G, Hynes JT (1984) Chem Phys 90:21
73. Bergsma JP, Gertner BJ, Wilson KR, Hynes JT (1987) J Chem Phys 86:1356
74. Gertner BJ, Bergsma JP, Wilson KR, Lee S, Hynes JT (1987) J Chem Phys 86:1377

CHAPTER 4

Orbital Interactions and the Pathway of a Chemical Reaction

4.1 The Role of Frontier Orbitals

The above-considered methods for determining the pathway of a chemical reaction are based on the calculation of critical parts of the PES of a reacting system taken as a supermolecule. This approach is the most rigorous, however, its implementation requires a large amount of calculations. Moreover, the results obtained need explanation based on current models of theoretical chemistry. Of primary importance is in this connection the concept of orbital interactions in reacting systems which determine the stereochemistry and the height of energy barriers of concerted reactions. Based on qualitative and semiquantitative arguments of the theories of conservation of the orbital symmetry [1], and of orientation and stereoselection [2, 3], this concept establishes a clear-cut interconnection among the symmetry properties, the amplitudes of interacting orbitals and the optimal pathways for the approaching reactants.

A special role belongs in this case to the interaction between frontier orbitals of the reactants which grows stronger with the increase in orbital overlapping. By analyzing the character of changes in the orbital interactions when a reacting system is moving along the reaction path described by the intrinsic reaction coordinate (IRC), Fukui [3, 4] has shown that the distortion of the initial system occuring on this pathway is largely predetermined by its adaptation to geometrical requirements for maximal overlapping of the highest occupied and the lowest vacant MO's (HOMO and LUMO) of the interacting reactants. This situation may be illustrated by an example, already considered earlier, of nucleophilic addition to a carbonyl group (see Sect. 1.3.5.3). Figure 4.1 features the tendency, which characterizes the change in the shape of the π^*-type LUMO of formaldehyde when it is approached by the attacking nucleophile, the hydride ion. At the final stage of reaction, the carbon atom must possess a localized vacant orbital of the sp^3-type so as to be able to form the σ-bond C—H. The pyramidalization of this atom, which rapidly develops as the reactants draw nearer to each other, is just that type of deformation which transforms and localizes the initial π^*-orbital to insure its maximal overlapping with the s orbital of the hydride ion being added.

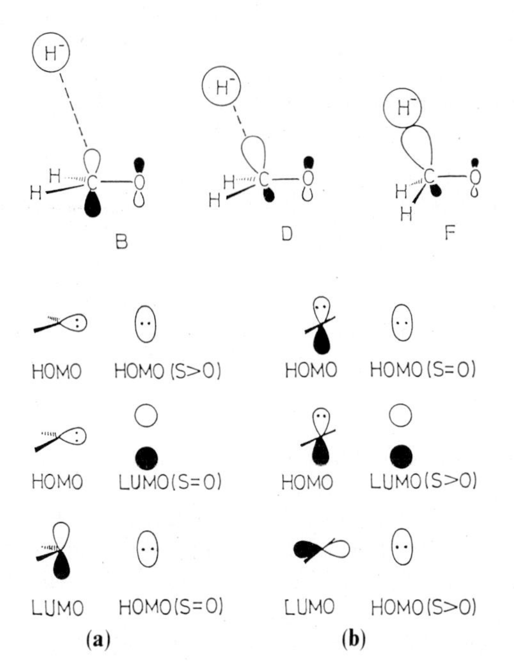

Fig. 4.1. Increase in the $S\pi^*$ overlap and rehybridization of the π^* MO of formaldehyde in the course of the hydride-ion addition reaction according to the calculations in Ref. [5]. B, D, and F correspond to geometry configurations in Fig. 1.18

Fig. 4.2a, b. Frontier orbital interactions in the course of the least-motion addition (insertion) of bent (**a**) and linear (**b**) singlet methylene to the hydrogen molecule, as adapted from Ref. [7]. Note: the degeneracy of two frontier orbitals of the isolated linear methylene molecule is broken down as the hydrogen molecule approaches it

Important is also the fact that, as a rule, the localization of frontier orbitals at the interacting centers leads to the narrowing of the gap between energy levels of the orbitals and thereby to the enhancement of orbital interactions [3]. If the symmetry (nodal) properties of the frontier MO's of reactants are such that an effective overlapping of the orbitals is impossible, then the mechnism described cannot be realized for the given geometry of approach. This conclusion constitutes the basis for the principle of stereoelectronic selection of appropriate reaction pathways.

The optimization of orbital interactions between reacting species is responsible for some drastic changes in reactant geometries occurring in the transition state and sometimes at the earlier stages of reaction. This may be illustrated by the results of ab initio calculations of the least-motion addition of methylene to the hydrogen molecule [6, 7]. It has been found that the HCH bond angle of the methylene in the vicinity of the saddle point is much larger than that for either the reactants or the products, taking the value close to 180°. The explanation for this is straightforward in terms of the frontier orbital interactions. As Fig. 4.2 shows, the symmetry of the HOMO orbital of bent methylene provides for its strong repulsive interaction with the $1\sigma_g$ HOMO orbital of the hydrogen molecule, whereas that of the HOMO orbital of linear methylene excludes such an interaction. On the other hand, the stabilizing pairwise HOMO–LUMO interactions (see the next section of this chapter) are cancelled in the case of bent methylene but are strongly favored with linear methylene. These stereoelectronic effects lie at the origin of the opening of the methylene angle which is produced during the least-motion approach of methylene to the hydrogen molecule.

Let us now examine the nature of orbital interactions at a more quantitative level.

4.2 Theory of Orbital Interactions

The interaction between the frontier MO's of the reactants, i.e., the HOMO of the donor component and the LUMO of the acceptor component, is usually the most important but by no means the unique orbital interaction arising during a reaction. The total interaction energy can be represented by the sum of all the pairwise interactions each of which is either of attractive or of repulsive type.

The mathematical formalism employed in the theory of orbital interactions has been elaborated both in general terms [8–12] and at different levels of the MO theory approximation, including the Hückel [13] and extended Hückel molecular orbital [14] as well as the SCF MO [15] and configuration interaction [16, 17] methods. In what follows, only the principal general relationships and conclusions will be dealt with.

When two reactants A and B are drawing together, their oribitals φ_a and φ_b with the energy levels of ε_a and ε_b, respectively, interact to produce two new orbitals φ'_a and φ'_b of the forming molecular system A–B (Fig. 4.3). Through this interaction the energy level of the orbital φ'_a is lowered relative to that of the orbital φ_a by the value of:

$$\Delta\varepsilon_a = \varepsilon'_a - \varepsilon_a = \frac{(H_{ab} - \varepsilon_a S_{ab})^2}{\varepsilon_a - \varepsilon_b} \tag{4.1}$$

while the higher energy orbital φ'_b is destabilized with respect to φ_b by:

$$\Delta\varepsilon_b = \varepsilon'_b - \varepsilon_b = \frac{(H_{ab} - \varepsilon_b S_{ab})^2}{\varepsilon_b - \varepsilon_a} \tag{4.2}$$

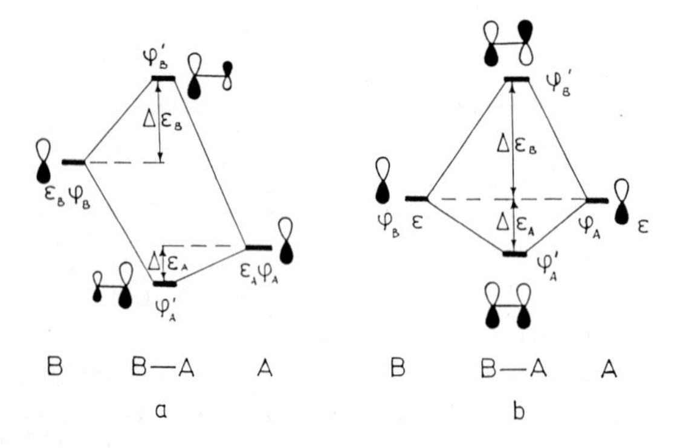

Fig. 4.3a, b. Formation of MO's of the A–B system (φ'_a, φ'_b) from the orbitals of the fragments A and B (φ_a and φ_b): **a** interaction between non-degenerate orbitals; **b** interaction between degenerate orbitals

where H_{ab} is the energy of interaction associated with two overlapping orbitals φ_a and φ_b and S_{ab} is the overlap integral for these. It is usually assumed that:

$$H_{ab} = kS_{ab} \tag{4.3}$$

The proportionality coefficient k is varied depending on the initial approximation for calculating the orbitals φ_a and φ_b (-39.7 eV in the SCF MO approximation).

Equations (4.1), (4.2) make it easy to determine what factor affects the extent of orbital interactions. In the first place, it is the orbital overlap integral whose value is defined by the symmetry and the mutual orientation of orbitals. The second factor is the close proximity of energy levels of the orbitals φ_a and φ_b. The strongest interaction corresponds to the case of the degenerate ($\varepsilon_a = \varepsilon_b = \varepsilon$) orbitals φ_a and φ_b (Fig. 4.3b). In this case, the energy of stabilization of the lower orbital φ'_a and the destabilization of the upper orbital φ'_b of the composite system A–B can be calculated from perturbation theory using the relationships:

$$\Delta\varepsilon_a = \varepsilon'_a - \varepsilon_a = \frac{H_{ab} - \varepsilon S_{ab}}{1 + S_{ab}} = \frac{(k - \varepsilon)S_{ab}}{1 + S_{ab}} \tag{4.4}$$

$$\Delta\varepsilon_b = \varepsilon'_b - \varepsilon_b = \frac{-H_{ab} + \varepsilon S_{ab}}{1 - S_{ab}} = -\frac{(k - \varepsilon)S_{ab}}{1 - S_{ab}} \tag{4.5}$$

The orbitals φ'_a and φ'_b of the composite system A–B may be represented as linear combinations of the orbitals of the reactants A and B:

$$\varphi'_a = N(\varphi_a + \lambda\varphi_b) \tag{4.6}$$

$$\varphi'_b = N(\varphi_b + \mu\varphi_a) \tag{4.7}$$

with the coefficients λ and μ being less than 1, i.e., the orbital φ'_a contains the orbital φ_a with a larger weight, the same holds for the orbitals φ'_b and φ_b. This means that the shape of the bonding combination φ'_a is closer to that of the lower reactant orbital, i.e., to $\varphi_a(\lambda > 0$ with S_{ab} being positive), while the antibonding combination φ'_b is nearer in its shape to that of $\varphi_b(\mu < 0)$.

By substituting Eq. (4.3) into Eqs. (4.1) and (4.2) the following expressions, convenient for practical calculations, are obtained:

$$\Delta\varepsilon_a = \frac{(k - \varepsilon_a)^2 S_{ab}^2}{(\varepsilon_a - \varepsilon_b)} \tag{4.8}$$

$$\Delta\varepsilon_b = \frac{(k - \varepsilon_b)^2 S_{ab}^2}{(\varepsilon_b - \varepsilon_a)} \tag{4.9}$$

Since $\varepsilon_a < \varepsilon_b$, it is clear that in its absolute value $\Delta\varepsilon_a < \Delta\varepsilon_b$. Consequently, if both orbitals of the reactants φ_a and φ_b are occupied by electron pairs which populate the MO's of the A–B system, then the overall effect of the orbital interaction will be destabilization. The effect of the four-electron destabilization $\Delta E_{ab}^{(4)}$ is nothing other than an exchange repulsion of filled electron shells. Its description by means of Eqs. (4.1) and (4.2) is made possible thanks to taking implicit account of the overlap integral S_{ab} between the interacting orbitals.

If, however, only one of the interacting orbitals φ_a and φ_b is occupied by a pair of electrons or if each of these orbitals contains only one electron, then both electrons will be placed in the composite molecule A–B in the lower orbital φ'_a. In this case, the interaction results in the stabilization of the system, which is referred to as the two-electron stabilization $\Delta E_{ab}^{(2)}$ or the charge transfer effect.

The total interaction energy can be calculated as a sum of all stabilizing two-electron and destabilizing four-electron contributions:

$$\Delta E = \sum_{ab} \Delta E_{ab}^{(2)} + \sum_{a'b'} \Delta E_{a'b'}^{(4)} \tag{4.10}$$

The use of Eq. (4.10) in practical calculations implies that the total energy of a system may be represented by the sum of the orbital energies $E = \sum_i n_i \varepsilon_i$, where n_i is the number of the electrons in the i-th orbital. This is equivalent to an assumption of a mutual compensation of the energies of the electrostatic nuclear repulsion and the electronic interaction. In actual fact, this compensation is not complete and the sum of the interactions mentioned is quite close to 1/3 E [18]. Thus, the relationship:

$$E = \frac{3}{2} \sum_i n_i \varepsilon_i \tag{4.11}$$

must hold. Its validity for the molecular structures not significantly distorted from their equilibrium configurations has been verified by a number of theoretical investigations. Calculations on some real molecular systems have shown that it is satisfied within the accuracy of up to 11% [19], which is quite sufficient for a qualitative examination. In case interacting reagents are strongly polarized, the electrostatic term $\sum_{\mu<\nu} q_\mu q_\nu R_{\mu\nu}^{-1}$ may be added to Eq. (4.10) [20].

Calculations of MERP by means of the scheme described take into account both the attractive and the repulsive terms of interaction between the reacting species. The latter fact represents a considerable improvement on the simple scheme of interaction between the frontier orbitals. The point is that quite often the selection of intermediate and even of the stable structures arising during a reaction or conformational transformation is to a greater extent governed by a favorable lowering of the exchange repulsion rather than by an increase in attractive interaction of active regions of molecules participating in a reaction. However, since all the above-given relationships rest on the perturbation theory, it is necessary to emphasize that their applicability, even though it extends to the case of a quite close proximity between the reactants ($S = 0.1$–0.5) [15], is none the less restricted to that part of the MERP, which contains rather small deformations of the initial structure.

4.3 Components of the Interaction Energy of a Reacting System in a Transition State

The above restriction obviously deters investigation of the most important part of the reaction pathway in the vicinity of the transition state whose structure significantly differs from that of the reactants. In order to evaluate rigorously enough the contributions made by the deformation of the molecular structure and those by other components to the total energy of a composite system for any point of a reaction path, Morokuma and Kitaura [21, 22] have developed a method, based on the Hartree–Fock one-determinant approximation, allowing one to decompose the interaction energy into various physically meaningful energy components. Components of the interaction energy can be determined in this approximation unambiguously by successively singling out the corresponding matrix elements of the Hartree–Fock equations.

The interaction energy ΔE during reaction between A and B can be decomposed into two components viz., the energy of deformation E_{def} of the structure of starting reactants in a given point of the reaction path, which is a destabilizing term, and the energy of interaction E_{int} between the deformed structures of the reactants, which is a stabilizing term:

$$\Delta E = (E_{A \cdots B} - E_A - E_B) = E_{\text{def}} + E_{\text{int}} \tag{4.12}$$

The latter term represents the sum of the following individual energy components: the electrostatic interaction E_{es}; the polarization effect E_{pol}, due to polarization of the electronic distribution of A by the system of the electric charges of B, and vice versa; the exchange repulsion term E_{ex} and the charge transfer term E_{ct}:

$$E_{\text{int}} = E_{\text{es}} + E_{\text{pol}} + E_{\text{ex}} + E_{\text{ct}} + E_{\text{mix}} \tag{4.13}$$

The term E_{mix} in this equation represents the high order interactions between the energy components given above, it is usually relatively small.

The relationship between the energy components of this expansion and the scheme of orbital interactions described above can be easily established. This relationship is featured by Fig. 4.4. The components E_{pol} and E_{ct} are always of attractive character, whereas E_{ex} represents in all cases the repulsive term. The electrostatic term E_{s} can change the sign depending on the position of the system along the reaction path.

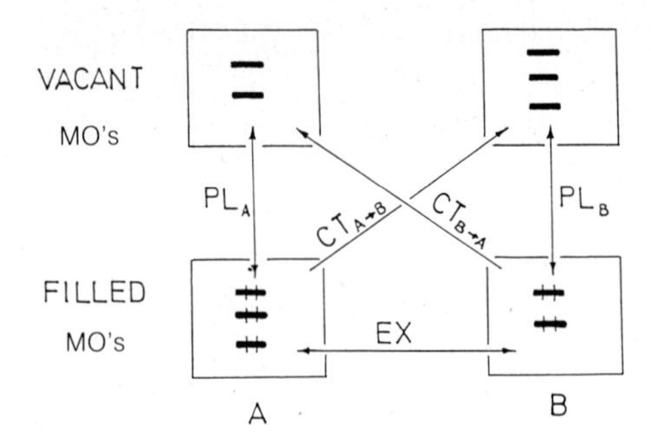

Fig. 4.4. Relationship between the orbital interactions and the energy components of Eq. (4.13). Polarization is due to admixture of vacant MO's

The calculations carried out for various radical and heterolytic addition and substitution reactions have led to the conclusion that generally one connot single out one component of the interaction energy which would control the course of a reaction and determine the height of the activation barrier. In the early stages of a reaction (at the beginning of the reaction path), the interaction energy is mainly related to the energy E_{int}, see Eq. (4.12), and the reactants choose their initial orientation by optimizing their orbital interactions. As the reacting system approaches the region of the transition state, there occurs a sharp, often jumpwise rise in the deformation energy E_{def}, which becomes the principal component of the energy barrier.

Table 4.1 lists the data obtained in the calculations [22, 23] using the energy decomposition analysis according to the above-described scheme for certain transition state structures of typical organic reactions. For comparison, the data are also given regarding the formation of the H-complex of the water dimer in its stable configuration (the value of $\Delta E < 0$ indicates the complex to be stable).

Table 4.1. Energy components (kcal mol^{-1}) according to Eqs. (4.12), (4.13) for the transition state structures of some reactions as calculated using ab initio (4–31G) method [22, 23]

E component	$CH_4 + H(D_{3h})$	$CH_4 + H^-(D_{3h})$	$CH_2{=}CH_2 + HCl$	$^3CH_2 + H_2 \rightarrow H + CH_3$	$(H_2O)_2$ [a]
E_{es}	−34.8	−44.7	−86.5	−25.3	−8.8
E_{ex}	87.2	74.1	181.6	72.2	4.3
E_{pol}	−6.0	−16.3	−25.6	−10.3	−0.5
E_{ct}	−50.9	−32.3	−76.0	−40.6	−2.2
E_{mix}	12.0	−9.3	26.3	12.0	−0.2
E_{int}	7.6	−21.5	19.6	7.2	−7.6
E_{def}	46.4	72.4	28.7	9.6	–
ΔE	62.1	48.8	48.3	17.4	−7.6

[a] For the water dimer at optimized intermolecular distance

4.4 Isolobal Analogy

A theoretical interpretation of the reactions of organometallic compounds of transition metals is quite a complex problem since the atoms of transition elements make use, for the formation of their bonds, not only of the valence *s*- and *p*-orbitals, but also of the *d*-orbitals close to them in energy. Such an enlargement of the orbital basis set in transition metal compounds makes it difficult to perform rigorous calculations because of computer-time-consuming problem and thus enhances the role of qualitative approaches to the analysis of the reactivity of organometallic and coordination compounds. An important advancement in this area is the introduction by Hoffmann and his co-workers [24–26] of the powerful concept of isolobal (from "lobe") groups, i.e., such groups for which the number, the symmetry properties, the form and the energy of frontier orbitals are roughly the same. This definition allows one to establish simple correspondence between the energy and the spatial (nodal) characteristics of the groups formed by the main-group and transition elements as well as to analyze in an approximate manner the structure and the reactivity of inorganic and organometallic compounds of the transition elements by analogy with those of the simple organic compounds [12, 26, 27].

The orbitals of the group ML_n ($n = 2$–5), where M is the transition metal atom and L is the two-electron ligand (CO, NH_3), can be formed from the orbitals of the octahedral ML_6 or the square-planar ML_4 complexes [12, 24]. The transition metal atom M participates with the following nine orbitals in the formation of the M—L bonds: five md, one $(m+1)s$ and three $(m+1)p$. Three *d*-orbitals of the axially symmetrical fragment ML_n do not mix with the orbitals of the ligands, they form a low-lying, weakly split set of energy levels which corresponds to the degenerate t_{2g} level in octahedral complexes. The orbitals of n two-electron ligands give rise, by interacting with remaining orbitals of the central atom, to n bonding (Fig. 4.5) and n antibonding (not shown in the figure) MO's of the group ML_n, which are localized at the M—L bonds. Thus, $(6-n)$ orbitals of the group ML_n filled with $x - 2(6-n)$ electrons remain in the valence zone (x is the number of the valence metal electrons).

These $(6-n)$ valence orbitals are used by the group ML_n for bonding with other ligands. Even though these group orbitals contain considerable contributions from *d*-orbitals of the metal, their form and symmetry properties (see Fig. 4.5) are analogous to those of the lobes of CH, CH_2, CH_3 and other similar groups. Electron counting is a necessary step to determine the population of the frontier orbitals on the metal and thus to establish the electron-equivalence relationship with the organic isolobal analog. The basic idea is the metal's quest to attain 18 valence electrons in octahedral (usually 16 in square planar) complexes. The atomic groups typical of organic structures and isolobally related to electron-equivalent groups of transition metals are summarized in Table 4.2. A more complete catalog of the isolobal groups can be found elsewhere [12].

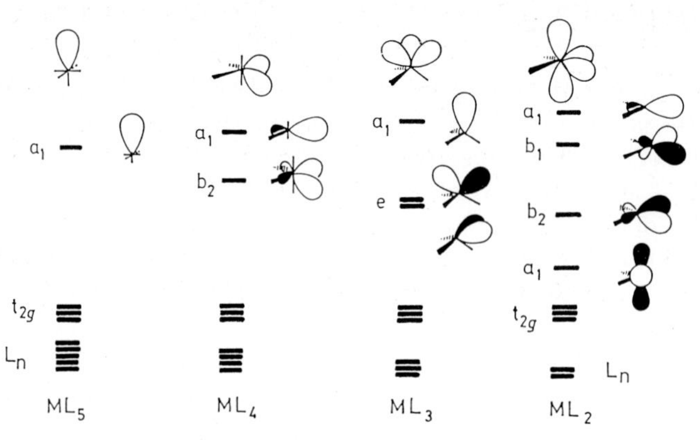

Fig. 4.5. Group orbitals of ML_n fragments, with M for a transition element atom and L for a two-electron ligand (CO, Cl, PX_3, etc.). The a_1 and b_2 levels for ML_n as well as the a_1 and e levels for ML_3 are inverted to the respective levels of the CH and CH_2 groups due to the d-orbitals contributing to the ML_n group MO's ensuing alterations of their nodal properties.

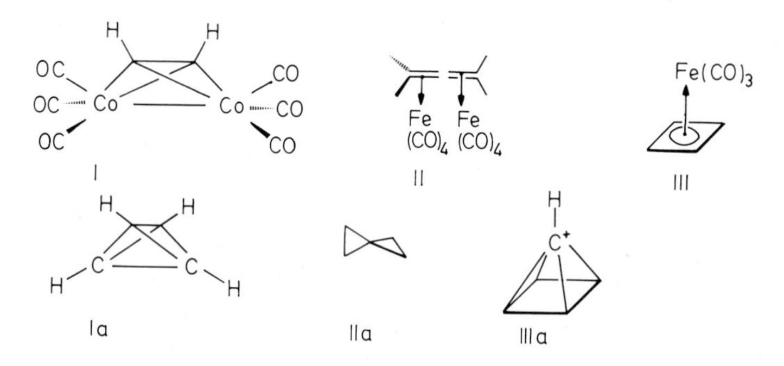

Table 4.2. Isolobal correlations

Organic group	Number of valence electrons	Number of framework electrons	Organometallic group
CH_3	7	5	$Mn(CO)_5$, $Fe(CO)_2(\eta^5\text{-}C_5H_5)$, $Zn(\eta^5\text{-}C_5H_5)$, $Mo(CO)_3(\eta^5\text{-}C_5H_5)$, $Mn(CO)_2(\eta^6\text{-}C_6H_6)$
CH_3^+	6	4	$Cr(CO)_5$, $Mn(CO)_2(\eta^5\text{-}C_5H_5)$
CH_2	6	4	$Fe(CO)_4$, $Cr(CO)_5$, $CoCO(\eta^5\text{-}C_5H_5)$, $Cu(\eta^5\text{-}C_5H_5)$, $Rh(CO)_4^+$, $Ni(PR_3)_2$
CH, N, O^+	5	3	$Co(CO)_3$, $Ni(\eta^5\text{-}C_5H_5)$, $W(CO)_2(\eta^5\text{-}C_5H_5)$, $Re(CO)_4$ $Rh(\eta^6\text{-}C_6H_6)$
CH^+, C, BH	4	2	$Fe(CO)_3$, $Co(\eta^5\text{-}C_5H_5)$, $O_3(CO)_3$
CH^{2+}, C^+, BeH	3	1	$Mn(CO)_3$, $Fe(\eta^5\text{-}C_5H_5)$
CH^{3+}			$Cr(CO)_3$, $Fe(CO)_2$, $Cr(\eta^6\text{-}C_6H_6)$

The valence orbitals of the chain, cyclic, and cluster structures constructed from the isolobal groups also have the similar shapes. For example, the valence orbitals of $Co_2(CO)_6$ are similar to those of acetylene C_2H_2, the orbitals of the cyclic clusters $Fe_2(CO)_{12}$ and $Fe_3(CO)_9$ are analogous to the MO's of the cyclopropane and cyclopropenium cation. Some other examples of isolobally related organic and organometallic compounds are represented by the structures I–III:

Similarity of the orbital structure causes the similarities in the types and the mechanisms of organometallic and organic compounds constructed of isolobal fragments.

Some typical examples will be considered below (see Sects. 6.2, 10.3).

References

1. Woodward RB, Hoffmann R (1970) The conservation of orbital symmetry. VCH Acad Press Weinheim
2. Fukui K (1975) Theory of orientation and stereoselection, Springer-Verlag, Berlin
3. Fukui K (1982) Angew Chem 94:852
4. Fukui K, Kato S, Fujimoto H (1975) J ACS 97:1
5. Kletskii ME, Minyaev RM, Minkin VI (1980) Zh Org Khim 16:686
6. Bauschlicher Jr CW, Schaefer III HF, Bender CF (1976) J ACS 99:3610
7. Ortega M, Lluch JM, Oliva A, Bertran J (1987) Canad J Chem 65:1995
8. Hoffmann R (1971) Acc Chem Res 4:71
9. Whangbo M-H, Wolfe S (1980) Isr J Chem 20:36
10. Epiotis ND (1977) Topics Curr Chem 70:250
11. Fukui K (1971) Acc Chem Res 4:57
12. Albright TA, Burdett JK, Whangbo MH (1985) Orbital interactions in chemistry. Wiley Interscience, New York
13. Salem L (1968) J ACS 90:453, 553
14. Imamura A (1970) Mol Phys 15:225
15. Whangbo MH, Schlegel HB, Wolfe S (1977) J ACS 99:1296
16. Fujimoto H, Koga N, Hatane J (1984) J Phys Chem 88:3539
17. Inagaki S, Fujimoto H, Fukui K (1976) J ACS 98:4054
18. Ruedenberg K (1977) J Chem Phys 66:375
19. Sen KD (1980) Intern J Quant Chem 18:907
20. Bernardi F, Bottoni A (1982) in: Csizmadia (ed) Progress in theoretical organic chemistry. Elsevier, Amsterdam 3:65
21. Kitaura K, Morokuma K (1976) Intern J Quant Chem 10:325
22. Morokuma K, Kitaura K (1981) in: Politzer P, Truhlar DC (eds) Chemical applications of atomic and molecular electrostatic potentials. Plenum Press, New York, p. 215
23. Nagase S, Morokuma K (1978) J ACS 100:1666
24. Hoffmann R, Albright TA, Thorn DL (1978) Pure Appl Chem. 50:1
25. Elian M, Chen MML, Mingos DMP, Hoffmann R (1976) Inorg Chem 15:1148
26. Hoffmann R (1982) Angew Chem Intern Ed Engl 21:711
27. Minkin VI, Minyaev RM (1982) Usp Khim (Russ Chem Rev) 51:586

CHAPTER 5

Substitution Reactions

In this and the following chapters we shall consider how quantum mechanical calculations of the PES's and the pathways of reactions may be used to analyze their mechanism. We shall explore the potentialities of theory and try to show the significance of the calculation data for understanding the essence of a reaction mechanism using as examples some typical reactions of organic chemistry. In our presentation we stick to the conventional classification of reaction mechanisms which takes into account the following three factors: 1) stoichiometric mechanism; 2) characterization of the processes resulting in breaking of the old and making of the new bonds at the limiting elementary stage of the complete reaction and 3) molecularity of this stage [1–4].

As regards the first factor, all reactions are divided into three types: substitution, addition-elimination, and rearrangement reactions. The rearrangement reactions have an independent status only when they proceed intramolecularly.

In relation to the mechanism of bond-breaking and bond-making, the reactions can also be divided into three groups. If during the formation of the X—Y bond electrons come from one of the atoms, e.g., from: X (X is a nucleophile and Y is an electrophile), the reaction is classified under the heterolytic, i.e., ionic type. In case, however, X and Y contribute each one electron, this would be homolytic or free-radical reaction. The process of the bond breaking is described likewise. A special class of reactions are the transformations in which the bond-breaking and bond-making processes are the result of an electron reorganization of the whole molecular system and cannot be reduced to cleavage and formation of the electron pair of one bond. Pericyclic reactions are assigned to just this class of transformations, which are sometimes called "reactions without a mechanism". In contrast to the heterolytic or homolytic reactions, which are usually characterized by the formation of intermediate ions, radicals, ion-radicals or ionic and radical pairs, in pericyclic reactions the breaking and the making of the bonds occur concertedly[1], i.e., with

[1] This does not mean that these processes must proceed synchronously. At a certain stage of a reaction one may be strongly predominant over the other, see Chap. 10 for more detailed explanation of how to distinguish between concerted and synchronous transformations.

no intermediate compounds emerging. The concerted reactions are highly stereospecific, their direction and velocity depend but weakly on solvent. The driving force of these reactions is the need for the conservation of orbital symmetry [5].

In regard to molecularity, the reactions are divided into mono- and bimolecular ones depending on whether their limiting stage is represented by the breaking of one of the bonds in the reactant molecule or by the formation of a transition state structure upon drawing together of two molecules of the reactants. Lately, these reactions have also been termed dissociative and associative, respectively. Hereafter we shall stick to the classification presented above although we are aware that some other more rigorous and detailed schemes have been suggested both for the specification of the mechanisms of organic reactions [6, 7] and for their classification [8, 9], which are specially intended for the quantum chemical analysis but so far have not gained sufficient acceptance among chemists.

5.1 Nucleophilic Substitution at a Tetrahedral Carbon Atom

Reactions of nucleophilic substitution at a tetrahedral carbon atom is a heterocyclic process in which a leaving (nucleofugal) group X in substrate I is displaced by a nucleophile Y through transferring a pair of electrons from Y to the reaction site and from this to X. They belong to the classical domain of organic chemistry [1, (Chap. 7), 10, 11] and were the first known type of reactions with a definite stereochemical outcome predetermined by a given path along which the reactants draw together.

Mechanisms of reactions of this type are broadly varied: from the classical associative mechanism of bimolecular substitution S_N2 (I → II → III) to the classical dissociative mechanism S_N1 which requires the formation of the free carbocation VI, including the intermediate mechanisms of Winstein and Sneen where the contact IV or the solvent-separated V ion pairs serve as the active forms, see Ref. [12] for a detailed review.

$$\underset{\text{II}}{[\text{Y}-\text{R}-\text{X}]^{\ddagger}} \xleftarrow{\text{Y}} \underset{\text{I}}{\text{R}-\text{X}} \rightleftharpoons \underset{\text{IV}}{\text{R}^{+}\text{X}^{-}} \rightleftharpoons \underset{\text{V}}{\text{R}^{+}\|\text{X}^{-}} \rightleftharpoons \underset{\text{VI}}{\text{R}^{+} + \text{X}^{-}}; \quad \text{II},\ \text{IV}\ (\text{Y}),\ \text{V}\ (\text{Y}),\ \text{VI}\ (\text{Y}) \longrightarrow \underset{\text{III}}{\text{R}-\text{Y}} \tag{5.1}$$

The geometry of the nucleophilic attack and the stereochemical outcome of a reaction are dictated by the type of the mechanism which in its turn, depends to a great extent on the solvent effects. The role of these factors has been the subject of a number of theoretical calculations which have led to a deeper understanding of the reactions in question.

5.1.1 The S_N2 Reactions

5.1.1.1 Stereochemistry of the Reactions

The reactions S_N2 are accompanied by the inversion of configuration of the tetrahedral carbon atom—Walden inversion, discovered as early as 1895 [13]. This result constituted the basis for the firmly established view regarding the concerted reaction mechanism whose important features are the rear-side approach of the nucleophile Y to the breaking bond C—X and the emergence of a trigonal-bipyramidal transition state structure II in Eq. (5.2):

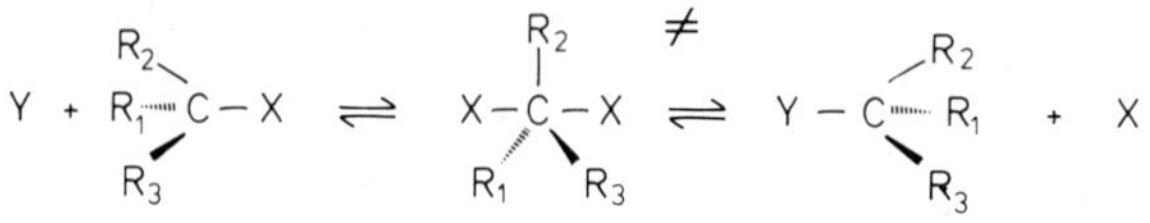

$$\underset{\text{I}}{Y + R_1R_2R_3C{-}X} \rightleftharpoons \underset{\text{II}}{\left[X{-}C(R_1)(R_2)(R_3){-}X\right]^{\neq}} \rightleftharpoons \underset{\text{III}}{Y{-}CR_1R_2R_3} + X \qquad (5.2)$$

Energetical preferability of such a reaction path over any other has been confirmed by all semiempirical—CNDO/2 [14], INDO [15], MINDO/3 [16], MNDO [17, 18]—as well as nonempirical [19–28] calculations.* The principal reason for this is that the rear-side approach VII of the nucleophile provides for optimal possibilities of forming a new bond through overlapping of its lone-pair orbital with the lowest unoccupied MO of the substrate, which is, as follows from the calculations, primarily represented by the σ^*-orbital of the C—X bond .

VII  VIII

By contrast, the front-side attack VIII cannot be supported by an effective $n - \sigma^*$ interaction, which is why it is energetically unfavorable, as also are other variants of approach of a nucleophile to the central atom shown in Scheme IX, which have carefully been studied in ab initio (4–31G) calculations of the reaction $F^- + CH_3F$ [24]. The difference between the energies of the structures, which are formed along the paths of the front-side and the rear-side attacks, amounts in the reaction zone to approximately 50 kcal/mol, which is an evidence in favor of the rear-side attack.

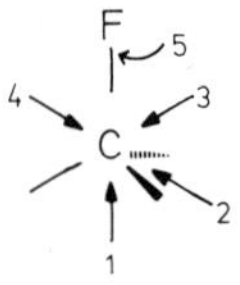

IX

*A fairly comprehensive list of reference (over forty) on reaction calculations is given in Ref. [23]. The results of theoretical studies on the S_N2 reactions were also reviewed in Refs. [29, 30].

This result throws also some light on the reasons for the stereochemical outcome of the S_N2 reaction. Whereas the approaches of an attacking nucleophile along the pathways 3–5 must, after the departure of the leaving group, lead to retention of the initial bond configuration at the central atom, the front-side approach 2 from an edge (rather than a face) of the tetrahedron must result in its inversion. This was pointed out for the first time by Westheimer [31] who noted that the transition trigonal-bipyramidal structure X in which the entering and the leaving groups are not in diaxial, as in Eq. (5.2II), but rather in diequatorial positions is also consistent with the observed stereochemical outcome of Eq. (5.2).

It should be noted that the structure X as a possible transition state of the reaction may be rejected also on the basis of the above presented symmetry rules of selection of transition states of reactions (see Sect. 1.3.3.2).

$$Y + C(R)(R_1)(R_2)X \rightleftharpoons [Y{-}C(R)(R_1)(R_2)X]^{\ddagger} \rightleftharpoons C(R)(R_2)(Y)R_1 + X$$

I   X   III

Indeed, at R=R′=R″ three equivalent pathways exist which can lead to the formation of X (for the approach of type 2 to the edges H—H in IX). Otherwise, the rotation about the C_3-axis in X, which interchanges the positions of the leaving and the entering group in the equatorial plane, would result in inversion of configuration, i.e., in the transformation of the reactants into products, which is prohibited by the given rules.

5.1.1.2 *Reaction Coordinate and the Structure of the Transition State*

Thus, the rear-side approach of the nucleophilic reagent is the only possible path for the S_N2 reaction which explains inversion of configuration at the attacked carbon center. The transition state of this reaction is represented by the trigonal-bipyramidal structure with both the entering nucleophilic and the leaving nucleofugal groups being in axial positions.

A correct description of the reaction coordinate (IRC) has been obtained by applying the method of the steepest descent from the saddle point to the neighboring minimum of the PES [19, 27, 32] (see Sect. 1.3.4.2) for the reaction $H^- + CH_4$ ($X = Y = H$). The calculations confirm that along all the path of Eq. (5.2) the C_{3v} symmetry of the reacting system is retained. Therefore the PES in the reaction zone is defined by only four independent parameters (Fig. 5.1). In earlier calculations of this reaction, for example in Ref. [22], where an extended basis of orbitals was employed and polarization functions were included, the reaction path and the transition state structure characteristics were computed in the reaction coordinate regime. The distance r_2 (Fig. 5.1) was chosen as such a coordinate. The calculation led to the erroneous conclusion

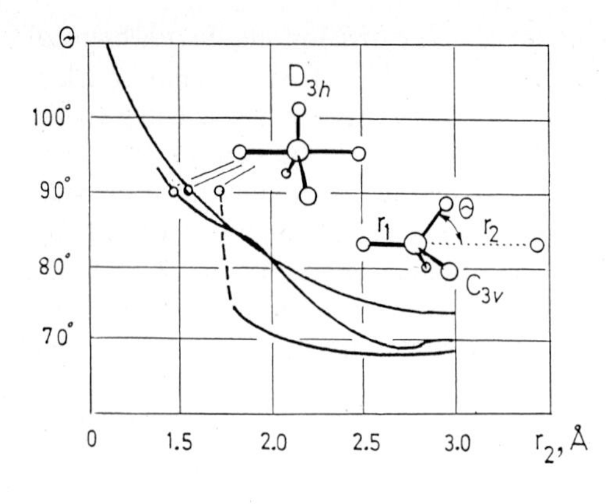

Fig. 5.1. Pathways of the S_N2 reaction $H^- + CH_4$. **1** Calculated by the gradient method [19]; **2** calculated in the reaction coordinate regime by varying r_2 with the rest of parameters optimized; **3** according to the data of an X-ray structural mapping of the five-coordinate complexes of cadmium XI as described in Refs. [27, 32]. The PRDDO and STO-3G methods yield consistent results

that all independent coordinates reacted weakly to changes in r_2 and only at close proximity to the saddle point they, almost jumpwise, adjusted to the geometry II of Eq. (5.2).

This result was explained in terms of the Hammond postulate, however, in actual fact it is an artifact of the reaction coordinate approach to finding the MERP. As may be seen from Fig. 5.1, where the depyramidalization of the methyl group occurring in the course of the reaction is described through correlation of the angle θ with r_2, all the coordinates vary during the reaction quite smoothly. This conclusion is consistent with the data on X-ray structural mapping of the pathway of the model reaction of nucleophilic substitution at the tetrahedral cadmium atom in compounds XI [33]:

RS
Y·········Cd–X
RS
RS
XI

-0.148
H
0.670
H – C – H
-0.613 -0.613
H H
XII

0.089
H
0.639
F – C – F
-0.864 -0.864
H H
XIII

Based on these data, a logarithmic relationship of type of Eq. (1.34) between r_2 and θ has been derived.

Thus, the reaction coordinate of the S_N2 reaction cannot be described by the change of any single geometrical parameter, rather, it is a function of all independent parameters of reacting system.

Of particular interest is the structure of the transition state II in Eq. (5.2). Data of both the semiempirical and the nonempirical calculation methods bear witness to a considerable lengthening of axial over the equatorial bonds. In virtue of inadequacies connected with the differential overlap being neglected (see Sect. 2.3.2) the first group of methods strongly underestimates this effect. Its magnitude can be reproduced to the full extent in nonempirical calculations using an extended basis of orbitals. The results of some important calculations are listed in Table 5.1.

Table 5.1. The axial and equatorial bond distances (Å) calculated for transition state structures II of reactions $H^- + CH_4$ and $F^- + CH_3F$

Method basis set	$H \cdots CH_3 \cdots H$		$F \cdots CH_3 \cdots F$		Ref.
	CH_{ax}	CH_{eq}	CF_{ax}	CF_{eq}	
CNDO/2	1.13	1.21	1.439	1.158	[14]
MINDO/3	1.140	1.225	—	—	[20]
PRDDO	1.54	1.10	—	—	[27]
STO-3G	1.481	1.090			[19]
3–21G	1.702	1.061	1.776	1.062	[26, 34]
4–31G	1.730	1.059	1.830	1.060	[26, 34]
6–31G*			1.820	1.061	[26, 34]
$(8, 4/4) \rightarrow (4, 2/2) +$	1.740	1.06	1.810	1.06	[22]
$+ p(H) + d(F, C)$					
$(11, 7/6) \rightarrow (5, 3/3) +$	1.737	1.062	1.878	1.06	[21]
$+ p(H) + d(F, C)$			—	—	
Extended, CEPA	1.736	1.064			

In the case of nondegenerate reactions as in Eq. (5.2) a sizeable variation in the length of axial bonds is observed. The angle YCX in the transition state X may slightly deviate from 180°. These changes correlate with the energy barriers of the reaction.

Charge diagrams show that in the case of the transition state structures CH_5^- [20] XIII and FCH_3F^- [21] XIII, found in ab initio calcultions, the negative charge concentrates at axial centers. This fact and the looseness of the markedly lengthened axial bonds reflect the sequences of the so-called polarity rule, according to which the most electronegative groups always occupy in trigonal-bipyramidal structures axial positions [35]. This can be explained as follows: the highest occupied MO of these structures is practically fully localized at the axial centers, therefore the energy level of such a HOMO is much lowered with the increase in electronegativity of the groups X and Y which belong in Eq. (5.2II) to the nucleophile and the leaving group, respectively. This, evidently, results in stabilization of that of the trigonal-bipyramidal permutational isomers which bears both most electronegative substituents in axial position.

5.1.1.3 *Energetics and Stoichiometric Mechanisms of the Gas-Phase S_N2 Reactions*

Only the use of an extended basis set in nonempirical calculations allows one to reproduce reliably the character of the critical point related to the pentacoordinate carbon structure having trigonal-bipyramidal bond configuration of II in Eq. (5.2). When a minimal basis set, such as STO-3G, is used, the D_{3h} structure FCH_3F^- corresponds not to a transition state but rather to the minimum on the PES, which lies 39 kcal/mol lower than the initial energy level of the nonreacting reactants [24]. Passing to the 4–31G basis set of DZ type puts the situation right: the D_{3h} structure II in Eq. (5.2) (X = Y = F) represents in this approximation the true transition state of the reaction although the calculation persistently shows that its energy level still lies 14.5 kcal/mol lower than that of the reactants [23, 24, 34].

This unexpected situation, when the energy level of the transition state structure with D_{3h} symmetry turns out to be lower than that of the initial system of the reactants, is predicted for a number of other S_N2 reactions even with the correlation effects taken into account [22]:

$H^- + CH_3Cl$ $\Delta E' = -6.8$ kcal/mol [22]
$F^- + CH_3Cl$ $\Delta E' = -15.0$ kcal/mol [22]
$NH_2^- + CH_3F$ $\Delta E' = -15.2$ kcal/mol, 4-31G [34]
$F^- + CH_3SH$ $\Delta E' = -15.9$ kcal/mol, 4-31G [34]

This result can have only one explanation: the trigonal-bipyramidal structure of the type II in Eq. (5.2) serves as a transition state not for the concerted reaction of the direct transformation of I to II but for a reaction between intermediately forming complexes of the reactant and the product. Indeed, nonempirical calculations [21, 22, 24, 26] using an extended basis as well as semiempirical calculations at the level of the approximations INDO [15, 36], MINDO/3 [37], and MNDO [17] predict the formation of stable ion-molecular complexes when the anion and the substrate draw together to a distance of 2.5–3.5 Å. The structures XIV and XV exhibit geometrical characteristics of two such complexes—intermediates in the reactions $F^- + CH_3F$ [24, 34] and $Cl^- + CH_3CN$ [38]—which have been calculated using the 4–31G basis set:

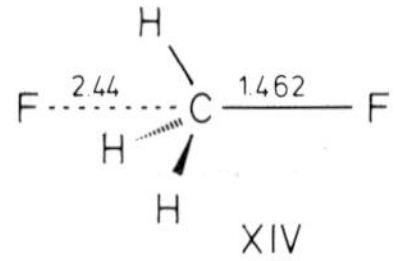

F···2.44···C(H)(H)(H)—1.462—F
XIV

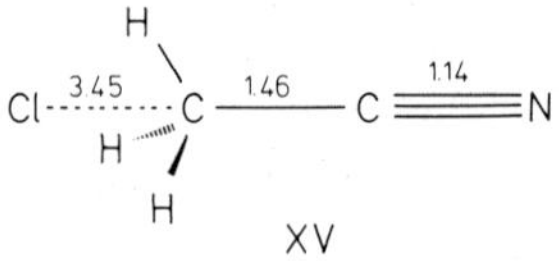

Cl···3.45···C(H)(H)(H)—1.46—C≡1.14≡N
XV

The emergence of ion-molecular adducts of halogenide ions with alkyl halogenides was verified experimentally by high-pressure spectrometric measurements [39–41]. The appreciation of the role of the type XIV and XV adduct formation stage in the overall mechanism of the S_N2 reactions came primarily from the studies of the gas-phase ion-molecular reactions of this type by the method of ion cyclotron resonance (ICR) carried out by Brauman et al. They observed, as a result of the gas-phase S_N2 reactions, inversion of configuration at the carbon atom being attacked [42]. The reaction rate was found to be strongly dependent on what combination of the entering and the leaving groups in Eq. (5.2) had been used [43]. Although the rate of the gas-phase reactions with an anion as a nucleophilic species is higher by 17–20 orders of magnitude than the rate of the same reactions in solution, it is still several times lower than the frequency of collisions of the reacting particles. This was explained [44, 45] by invoking a kinetic model based on the shape of the PES

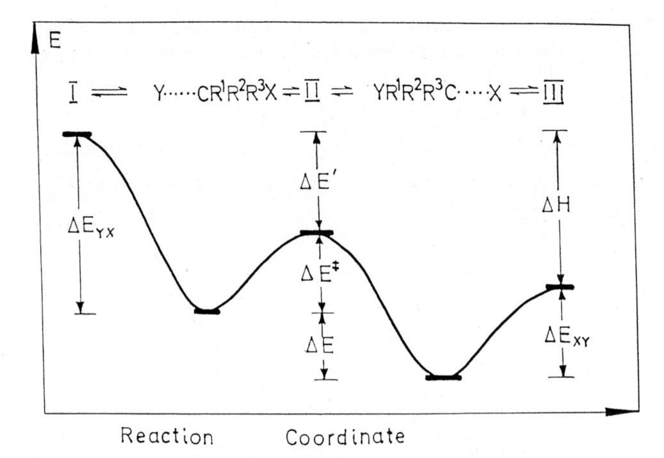

Fig. 5.2. Potential energy profile for the gas-phase S_N2 reactions at the tetrahedral carbon atom. The case of negative $\Delta E'$ is shown, i.e., $\Delta E^{\neq} < \Delta E_{yx}$

shown in Fig. 5.2 which takes into account the lowering of the observed effectiveness of reaction in consequence of the formation of ion-molecular adducts. By using the RRKM calculational scheme, which requires certain assumptions concerning vibrational frequences and moments of inertia, it became possible to assess the internal barriers ($\Delta E^{\neq}$ in Fig. 5.2) of some S_N2 reactions [45]. The data obtained are listed in Table 5.2 along with the calculated principal characteristics of energy parameters of the potential curve in Fig. 5.2. These results were obtained by nonempirical methods as well as using the semiempirical method λ-INDO [36] whose parametrization was adapted so as to ensure reliable computation of the reactions involving stretching and rupture of bonds.

More complete lists of calculated energies of the gas-phase S_N2 reactions of $Y^- + CH_3X$ general type are available in Refs. [30, 37] (the λ-INDO method) and [26, 34] (4–31G). The data obtained make it possible to check how the well-known relationships among the structural, kinetic and thermodynamic characteristics, such as the Bell–Evans–Polanyi principle, the Hammond postulate, etc., are obeyed in the case of the reactions which proceed without a solvent.

Calculations [26] of the heats ΔH of the reactions $Y^- + CH_3F$ yield a sequence of exothermicity which represents very well the experimental trend: $H^- > H_2N^- > HO^- > HCl^- > CH_3O^- > F^- > NC^- > HOO^- > CN^- > FO^-$. The length of the bond C—F in II [Eq. (5.2)] increases in this order from 1.725 Å to 1.986 Å while the angle θ (see Fig. 5.1) is reduced from 96.3 to 84.2°. Thus, the results of the calculations are in agreement with the above-noted relationships, according to which the more exothermic a reaction, the closer to the starting reactants the structure of a transition state will be. Furthermore, they

Table 5.2. Energy characteristics (kcal/mol) of potential curves for S_N2 reactions as shown in Fig. 5.2. All energy values are given relative to the energy of initial reactants which is taken to be zero

X, Y	Method of calculation[a]	ΔE_{YX}	$\Delta E^{\neq}$	$\Delta E'$	ΔE_{XY}	ΔH
H, H	ab initio [46]	− 3.1	60.3	60.2	− 3.1	0
	INDO	− 3.6	51.6	47.8	− 3.6	0
	ab initio [21]	−13.1	21.0	7.9	−13.1	0
F, F	INDO	−16.7	22.0	5.3	−16.7	0
	Experimental	—	26.3 [45]	—	—	—
	ab initio [21]	−12.2	16.0	3.8	—	−68.1
H, F	INDO	−14.8	31.3	15.8	—	−50.0
	Experimental	—	—	3.6 [47]	—	−60.7 ± 7.9 [22]
	ab initio [22]	—	—	− 6.7	—	−88.0
H, Cl	INDO	−16.3	6.0	−10.3	−1.9	—
	Experimental	—	—	0.7 [47]	—	−88.0 ± 9.1 [22]
	ab initio [22]	—	—	−15.1	—	−35.9
F, Cl	INDO	−10.3	17.4	7.2	1.4	−35.9
	Experimental	−10.0	—	3.1	—	−32.0 ± 11.0 [22]

[a] Ab initio calculations were performed using an extended basis set of orbitals with polarization functions included (Table 5.1). INDO calculations were carried out according to the λ-INDO scheme (36) employing special parametrization for adequate reproduction of bond rupture processes

indicate that structural characteristics of a transition state of the S_N2 reaction depend to a greater degree on exothermicity of reaction than on electronic interactions between the entering and the leaving nucleophilic groups. A correlation may also be established between the magnitudes of the energy barriers of S_N2 reactions and the transition state structure geometries [50].

Complex character of the energy profile of the gas-phase reactions S_N2 (see Fig. 5.2) explains the lack of direct correlations between the experimentally assessed activation barriers $\Delta E'$ (when they are positive) and the exothermicity of the reaction ΔH. A better-grounded correlation should be sought between the values of the "intrinsic" barriers $\Delta E^{\neq}$ and the quantity $\Delta E = \Delta H + (\Delta E_{XY} - \Delta E_{YX})$. Wolfe et al. [26] treated, using the 4–31G basis set, the results of their calculations on over twenty S_N2 reactions and found that the data obey quite satisfactorily the well-known Marcus equation [48] which describes reactions of electron and proton transfer as well as those of the alkyl groups' transfer [49] in solution.

In place of the values of free energies (for solutions), potential energies are used, and the energy barrier of the reactions $Y + CH_3Y$ and $X + CH_3X$:

$$\Delta E_0^{\neq} = 1/2(\Delta E_{XX}^{\neq} + \Delta E_{YY}^{\neq}) \tag{5.3}$$

The calculations show that the equality (5.3) holds quite well, as also does the following relationship, which formally coincides with the Marcus equation:

$$\Delta E^{\neq} = \Delta E_0^{\neq}(1 + \Delta E/4\Delta E_0^{\neq})^2 \tag{5.4}$$

Pellerite and Brauman [45] also gave several examples of experimental verification of validity of the Marcus equation for the gas-phase reactions of the S_N2 type.

The significance of the above-described findings consists in that they make it possible to determine to what extent the extrathermodynamic relationships found to be useful for structure-property correlations in solution chemistry are applicable to the description of intrinsic characteristics of reactions which are not connected with the solvent effects.

5.1.1.4 *Effect of the Solvent*

The effect which a solvent has on the rate and character of the S_N2 reactions is enormous [1, 11, 15, 52]. Some important theoretical results were obtained in the supermolecular approximation with various degrees of completeness in regard to reproduction of the solvation shell.

Calculations of the reaction $F^- + CH_3F$ in water, which take into account surrounding of the reactants with eleven water molecules, allow one to obtain, even when the CNDO/2 method is used, a qualitatively correct pattern of the energy profile. The hydrated D_{3h} structure FCH_3F^- was found to represent a transition state rather than to be located in a deep minimum of the PES, which is the case with the reaction in vacuum analyzed in the same approximation [14]. The most detailed calculations of the pathways of the S_N2 reactions:

$$Cl^-(H_2O)_n + CH_3Cl \rightarrow ClCH_3 + Cl^-(H_2O)_n \quad n = 0, 1, 2 \qquad (5.5)$$

which permit an assessment of the hydration effect upon energetics and the mechanism of the reaction have been carried out recently by Morokuma [53] who applied the gradient technique in searching for a reaction path using the ab initio 3–21G technique. These calculations can be directly compared with the experimental data on an analogous reaction:

$$OH^-(H_2O)_n + CH_3Br \rightarrow HOCH_3 + Br^-(H_2O)_n \quad n = 0,1,2,3 \qquad (5.6)$$

Thanks to newly evolved experimental possibilities of generating solvated ions in gas phase, the rates of the gas-phase reactions of Eq. (5.6) could be determined for various degrees of the hydroxyl anion hydration [54]. The calculations [53] have shown a substantial rise of the barrier height ($\Delta E^{\neq} + \Delta E'$ in Fig. 5.2) with the increase in the degree of anion hydration in Eq. (5.5), which explains the observed fall by four orders of magnitude in the rate of the analogous reaction of Eq. (5.6) upon passing from the unhydrated hydroxyl anion ($n = 0$) to the hydrate $n = 3$.

An interesting result of the calculations consists in the selection of two possible ways of the solvation shell reorganization in Eq. (5.5), $n = 1, 2$. The hydrated ion-molecular complex XVI, which is strongly stabilized with respect to separate reactants, rearranges either, owing to synchronized motions of both water molecules of the anion's hydrate shell, to the transition state XVII or to the energetically somewhat more favored transition state XX:

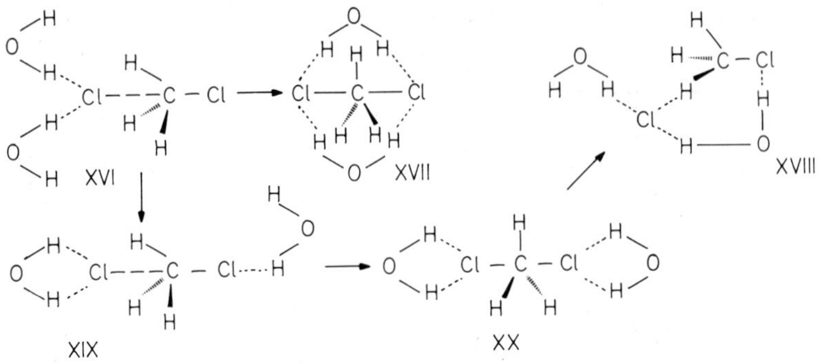

XVI → XVII; XVI → XIX → XX; XVIII

The path to XX is provided through the transfer of one water molecule to the leaving group, which is associated with a temporary deviation of the hydrated nucleophilic reactant from the C_3-axis XVIII (transition state)→XIX (local minimum). It is clear from even this simplified scheme (only two solvent molecules are taken into account) that the geometry parameters of the solvent are an important component of the reaction coordinate in solution. The calculated energy of the solvent reorganization accompanying the formation of the transition state structure is higher than the central energy barrier of Eq. (5.5) in vacuum at $n = 0$. On the other hand, the calculations carried out for the hydrated reactants do not, in principle, alter the form of the potential curve which remains the same as in Fig. 5.2.

The most rigorous calculations of the mechanisms of S_N2 reactions in solution were conducted by Jorgensen and co-workers [55, 56]. They examined as an example the process $Cl^- + CH_3Cl$ using the ab initio (6–31G* basis set) and the Monte Carlo (MC) methods. So as to carry out the MC calculations of the reaction in question in water and dimethyl formamide (DMF), the following problems had to be solved.

1) Approximation of the gas phase profile of the energy surface by means of an analytical function with the reaction coordinate clearly singled out (such an approximation should allow inclusion of the reaction coordinate as a fully equal degree of freedom, along with the degree of freedom of the solvent).
2) Determination of the relationship governing the change in intermolecular potentials of interaction of the solute molecule (i.e., the whole system $ClCH_3Cl^-$ with any internuclear distances) with solvent molecules along the direction of motion on the PES. To solve the former problem, the authors assumed that the nucleophile, a carbon atom in the methyl group, and the leaving chlorine atom are always located on a straight line. Relatively simple stereochemical mechanism and reaction path allowed the PES to be

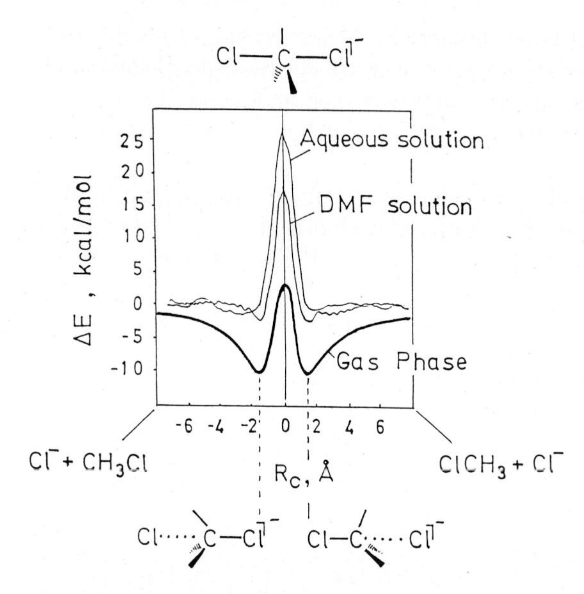

Fig. 5.3. Energy profile of the degenerate reaction $Cl^- + CH_3Cl^- \rightarrow ClCH_3 + Cl^-$ in the gas phase, in water and DMF; for the gas phase the change in internal energy E was calculated by the ab initio method 6–31G, for solutions the mean force potential W was found using Monte Carlo simulation. (Adapted from Ref. [55])

approximated by the analytical function of only one variable, the distance C—Cl (r_{CCl}). The latter problem can be dealt with in a standard fashion. 250 molecules have been taken into account. Figure 5.3 shows the dependence of the internal energy in the gas phase and of the moderate force potentials in water and DMF on the distance r_{CCl}. The greater changes, with respect to the gas phase, occur in aqueous solution. The gas phase minima of internal energy approximately 5 kcal/mol deep are to a considerable degree smoothed out in water solution and should not affect reaction kinetics. Thus, in water the prereaction complex vanishes and the substitution reaction becomes concerted. At the same time, in DMF the ion-dipole complexes are, apparently, retained—indirect experimental evidence in favor of this is discussed in Ref. [56]. This warrants the assumption that in DMF and similar solvents the S_N2 reactions proceed unconcertedly as a two-step process analogously to the gas-phase reactions. Thus, the experimentalist is confronted with the problem of detecting the ion-dipole complexes in those solvents. The calculated free activation energies in water (26.2 ± 0.5 kcal/mol) and DMF (19.3 ± 0.5 kcal/mol) excellently agree with the experimental data of 26.5 and 22.7 kcal/mol, respectively.

Of special interest is the elucidation of the causes leading to an increase in activation barriers when going from gas to solvent. Let us break up the total activation barrier, according to Refs. [55, 56], into two contributions: the internal barrier of the solute molecule and the barrier arising from different hydrations of the ion-dipole complex and the transition state. Calculations show that the former contribution equals roughly 14 kcal/mol, and the latter is somewhat lower 11.7 kcal/mol. Hence both components must be taken into account.

Experimentalists often hold the view that the desolvation of the nucleophile is in the initial stage so high that the activation barrier is practically determined by it. According to the MC calculations [55, 56], this claim appears to be erroneous, at least as regards the reaction under consideration. Equally erroneous seems to be the conclusion, drawn by Dewar and co-workers from MNDO [17] and AM1 [57] calculations and an analysis of experimental data, to the effect that the free energies of solvation of the ion-dipole complex and the transition state are roughly equal.

Thus, the first results of the Monte Carlo modelling of reactions in solution call for cautious attitude towards the conclusions based on calculations in supermolecular approximation.

5.1.1.5 *Reactions with Retention of Configuration of the Carbon Atom*

Retention of the stereochemical configuration of the central carbon atom in the S_N2-type reactions has not been so far verified experimentally. Reports to the contrary, such as Ref. [58], have not withstood careful testing [59, 60], at the same time provoking active theoretical studies of this problem.

The principal question is how to achieve, by varying the nucleophilic reactant Y and the leaving group X, stabilization of the alternative to II [Eq. (5.2)] pentacoordinate transition state C_s structures of the type XXI which form through frontal approach of the nucleophilic reactant IX:

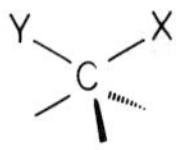

XXI 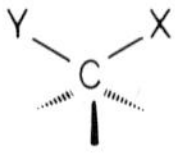 XXIa 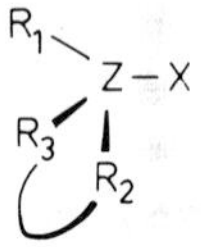 XXII

This problem may conveniently be examined in terms of the orbital interactions. Decisive is in this case the interaction between the frontier orbitals, namely, the lone pair orbital n_Y and the σ^* orbital of the C—X bond. The larger the overlap integral $S_{n\sigma^*}$, the larger will, evidently, be the energy of two-electron stabilization of Eq. (4.8) associated with this interaction. Anh and Minot [61, 62] have analyzed the role of various factors that control relative magnitudes of the integrals $S_{n\sigma^*}$ for the frontal (retention of configuration) VIII and the rear-side VII (inversion of configuration) approaches. By inspecting the orbital overlap schemes shown in formulae VII and VIII one may easily understand which factors of the electronic structure of the nucleophilic reactant, the leaving group and the central atom favor the retention of configuration.

1. *The nucleophile.* To be able to reduce the negative overlap component during the frontal approach, the nucleophilic reactant must have a contracted orbital of the electron lone pair. Then the integral $S_{n\sigma^*}$ may, in the frontal approach, exceed the value of $S_{n\sigma^*}$ in the rear-side approach, which will result in retention of configuration of the tetrahedral carbon atom. Clearly, the contracted frontier orbitals are characteristic of nucleophilic reactants with electronegative and low polarizable central atoms, i.e., those which, according to Pearson's [63] classification, belong to hard bases such as F^-, RO^- anions and H_2O^-, NR_3-like neutral nucleophilic reactants.
2. *The leaving group.* A drop in the contribution from the orbital of the leaving group X to the σ^*MO of the C—X bond leads, upon the front-side approach, to diminution of the negative overlap component. Such a trend can take place when the energy of the orbital X is considerably smaller than that of the partner, which is achieved through an increase in the electronegativity of the leaving group H_2O^+, F.
3. *The central atom.* An analogous effect may result from an increase in the energy level of the hybrid orbital of the central atom linked to the leaving group X. A lowering of the electronegativity of the central atom will lead to a decrease in the contribution from the orbital X to the MO σ^*_{ZX} ($Z = Si$, Sn, Ge), as compared to σ^*_{CX}, and thereby to an inrease of the integral $S_n\sigma^*$ in the case of the frontside approach VIII. Of importance is also the type of the hybrid orbital of the central atom, which it uses to maintain a link with the leaving group X. The more pronounced is the s-character of this orbital, the more it is extended towards X and the larger is the integral $S_{n\sigma^*}$ for the front-side approach. The s-character of this orbital is enhanced with the increase in the angles R—Z—X at the site XXII, which occurs upon inclusion of the fragments R into the strained three or four-membered cycle.

Indeed, in the case of cyclopropane derivatives (a three-membered cycle) the calculations [64] using the extended Hückel method point to the preferability of the reaction path providing for retention of configuration. In actual practice, the S_N2-type reactions for the series of cyclopropane derivatives proceed at very low rates and no reliable confirmation of the retention of configuration at the carbon being attacked at the stage of kinetical control has so far been produced [60]. For the cyclobutane derivatives (a four-membered cycle), ab initio (STO-3G) calculations [65] of alternative reaction paths bear witness to the preferability of a rear-side approach with inversion of configuration.

On the other hand, in the case of the derivatives of four-coordinate silicon ($Z = Si$) the retention of configuration of tetrahedral silicon atom is a well-established and frequently-registered fact, see the review under Ref. [66]. According to predictions of the orbital model, "hard" nucleophiles and "hard" leaving groups facilitate the retention of configuration of the central Si atom—a statement known as the Corriu rule [66].

Theoretically one may think of one more mechanism of the S_N2 reactions that would lead to retention of configuration of the central atom. It has to do with the formation of an intermediate pentacoordinate structure capable of polytopal rearrangement. The basic mechanistic scheme of such reactions will be examined in Sect. 5.5.6.

5.1.2 The S_N1 Reactions

The reactions which proceed by the S_N1 mechanism with dissociation of the C—X bond in compounds of type I have been studied theoretically much less thoroughly than the S_N2 reactions. Two reasons may explain this. First, the one-determinant MO method, because of its shortcomings noted earlier (see Sect. 2.2.4), cannot adequately describe the bond-breaking processes. For example, the value of the energy of the C—F bond dissociation calculated by the CNDO/2 method exceeds nearly four times the experimental value (about 120 kcal/mol), moreover, this calculation gives an incorrect idea as to the charge separation ($CH_3^{+0.4}F^{-0.4}$) in a completely dissociated system. Secondly, in order to describe a dissociation reaction in solution, the solvation effect has to be directly accounted for.

The following S_N1 reactions were calculated more rigorously than others, with the solvent (S) effect taken into account using the supermolecular approximation [67–70]:

$$CH_3F(S)_n \rightarrow CH_3^+(S)_j + F^-(S)_k \quad S = H_2O, HF, CH_4, H_2 \tag{5.7}$$

$$CH_3N_2^+(S)_n \rightarrow CH_3^+(S)_j + N_2(S)_k \quad S = H_2O, HF \tag{5.8}$$

The first solvation shell of the methyl cation in water and HF can satisfactorily be represented by five and that of the fluorine anion—by six solvent molecules, i.e., the total number of the solvent molecules included in the supermolecular approximation is 11. The C—X distance was taken as the reaction coordinate and all other geometry parameters, includingg those of the solute environment, were optimized. The most important result of the calculations of Eq. (5.7) by the CNDO/2 and ab initio (STO-3G basis set) methods is the detection of three minima along the MERP. The first of these characterized by the distance $r_{CF} = 1.388$ Å corresponds to the hydrated undissociated molecule CH_3F. The second minimum corresponds to $r_{CF} = 3.480$ Å. There are no solvent molecules between the ions CH_3^+ and F^-, hence they are located in one cage and may be structurally described as a contact ion pair of type IV. The third minimum corresponds to a completely dissociated system ($r_{CF} = 5.463$ Å), i.e., the solvent-separated ion pair V with each ion surrounded by its own solvation shell with $n = 11$, $j = 5$, and $k = 6$ in Eq. (5.7).

The above-presented picture does not, on the whole, change when HF, another polar solvent, is used. The structure of the solvent-separated pair CH_3F of type V in this solvent is given by the formula XXIII ($r_{CF} = 5.665$ Å) [69].

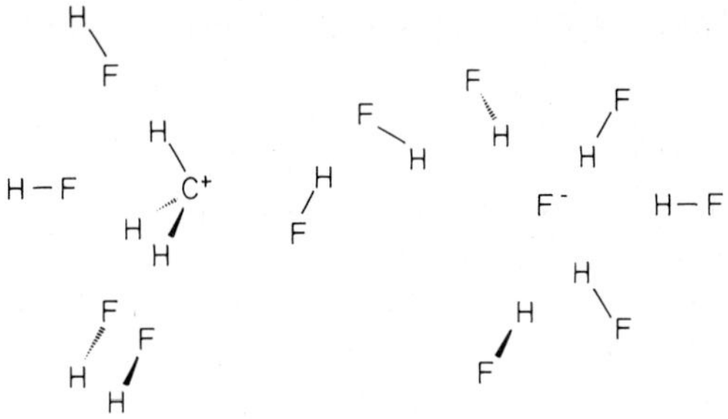

XXIII

Upon going to solution in methane ($n = 9$)—a model of nonpolar and nonpolarizable solvent—the minima on the PES, which are associated with the ion pairs, vanish. The effect of the solvent is felt only in some lowering of the calculated dissociation energy. So as to assess the role of a nonpolar but readily polarizable solvent, the authors of Ref. [67] chose H_2($n = 11$). The calculation revealed, apart from solvated undissocated CH_3F, only one additional minimum at $r_{CF} = 2.92$ Å, $j = 5$, $k = 6$. In this structure, one H_2 molecule is incorporated into the cavity between the ions, and this sytem may be represented as a solvent-separated ion pair in which one solvent molecule is shared by two solvation shells formed upon ionic dissociation.

No intermediate structures were found by the CNDO/2 calculations [69, 70] in the dissociation reaction of the methyl diazonium cation of Eq. (5.8). This finding is in accord with experiment [71]. It is interesting to observe the change in the equilibrium structure $CH_3N_2^+$ when passing from gas phase to solution ($n = 11$), and with the increase in the dielectric permittivity:

	r_{CN}, Å	HCN, deg
Gas	1.39	108.5
HF	1.44	109.2
H_2O	1.53	90

The calculations, discussed in this Section, of dissociation reactions in solution of the type CH_3—X compounds, may be regarded as the beginning of a serious theoretical assessment of the structural aspect of the ion pair concept which has proved quite valuable in the analysis of the mechanisms of organic reactions but has, so far, been very rarely invoked in regard to specific geometry types.

5.2 Electrophilic Substitution of the Tetrahedral Carbon Atom

In electrophilic substitution reactions, the electron pair of the breaking bond C—X stays with the carbon atom, while the departing electropositive group X leaves the molecule in the form of a cationic particle. Primarily the organometallic compounds (X = M) undergo these reactions whose C—M bond is polarized in such a way that a considerable negative charge is accumulated at the carbon atom.

The spectrum of mechanisms of the electrophilic substitution at a saturated carbon atom is very broad [1, 72–74]. The bimolecular S_E2 exchange reactions in the series of organo-mercury compounds have been studied in the most detailed manner. For these reactions as well as for the S_E2 reactions of other organometallic compounds, the derivatives of Li, Mg, B, Sn, Ge, a four-center transition state of XXIV type is assumed:

$$R_1R_2R_3C{-}MX + RMY \rightleftharpoons [R_1R_2R_3C \cdots M(X) \cdots Y \cdots M(R)] \rightleftharpoons R_1R_2R_3C{-}MR + MXY \quad (5.9)$$

No gas-phase reactions of electrophilic substitution have been known so far, while the formation of the transition state structure of Eq. (5.9) in solution is greatly (often decisively) affected by medium factors, the specific solvation and the nucleophilic catalysis. Even though the true structure of a transition state of the S_E2 reaction is much more complicated than XXIV, this structure correctly reflects the stereochemistry of substitution at the tetrahedral carbon atom, namely, the retention of configuration of the carbon atom bonds observed experimentally in most S_E2 reactions (the S_E2 rule, Ref. [1]).

A theoretical interpretation of this important rule was for a long time based on the assessment of relative stability of various structures of the pentacoordinate carbon atom which carry, unlike II in Eq. (5.2), a positive charge. The simplest model is represented by the methonium ion CH_5^+, and the simplest S_E2 reaction is an attack by a proton upon the methane molecule: $CH_4 + H^+$.

Ab initio calculations [75] with an extended basis set and inclusion of correlation corrections have shown that the proton approaches the methane molecule along the C—H bond and follows this reaction path up to the distance C—H^+ (1.85 Å):

$$\geq C{-}H \cdots\cdots H^+ \longrightarrow \geq C^+ \langle^{H_1}_{H_2}$$

XXV, C_s

After this the reaction complex rearranges into a structure of C_s-symmetry (compare approach 2 on scheme IX, Sect. 5.1.1.1)—analogous to XXIV in the configuration of the bonds of the central carbon atom.

Another possible C_s-structure of XXVa type (M = H) is energetically nearly equivalent to XXV, whereas the structure of the methonium ion with D_{3h}-symmetry is unfavored (Table 5.3).

XXVa, C_s XXVb, C_s XXVc, D_{3h}

The bonds C—$H_{(1)}$, C—$H_{(2)}$ in XXV are considerably loosened as compared to the bonds C—H which constitute the pyramidal basis. Their length calculated with the 4–31G basis set amounts to 1.25 Å, i.e., it is longer by 0.2 Å than the C—H bonds of the pyramidal form XXVc (M = H). This fits quite well with the model of the transition state XXIV.

It would, however, be incorrect to regard the preferability of the C_s structure of the methonium ion over the bipyramidal D_{3h} structure as a theoretical confirmation of the S_E2 rule, as it is sometimes done, for example, in Ref. [2]. First, the energy difference between these structures is considerably smaller than in the case of the corresponding anions XII. Secondly, ab inito calculations [80], carried out recently, of more realistic models of the transition states of the S_E2 reactions XXVb, c which contain two metal atoms attached to he carbon, have shown the relative stability of the C_s and D_{3h} structures to depend strongly on the nature of the entering and the leaving group.

Table 5.3 lists the results of these and some other nonempirical calculations of relative stability for various geometrical structures of the pentacoordinate carbon atom which carry a positive charge.

Table 5.3. Relative energies (in kcal/mol) of cation $CH_3M_2^+$ structures XXV, according to ab initio calculations

M	Basis set	XXVa, C_s	XXVb, C_s	XXVc, D_{3h}	Ref.
	4–31G	0	0	7.2	[75, 78]
H	6–311G*	0	0.1	6.4	[81]
	6–311G*(MP4)	0	0.1	11.7	[79, 81]
Li	4–31G	5.0	5.5	0	[80]
	6–31G*	2.4	—	0	[80]
BeH	4–31G	7.9	7.9	0	[80]
Na	STO-3G*	0	—	8.8	[80]
MgH	STO-3G*	0	—	0.1	[80]

There are quite a few examples, discussed in the literature [72, 73, 80], of inversion of configuration in the S_E2 reactions. In particular, an inversion of the bond configuration was registered in prochiral carbon atoms of the methylene groups of alkyl lithium tetramers $(ROH_2Li)_4$ in solution [82]. This result is consistent with the data of Table 5.3 which show preferability of the D_{3h} form of $CH_3Li_2^+$.

Generally, one may state that the PES of the reaction of electrophilic substitution in the region of a transition state is much simplified as compared with the nucleophilic substitution. Therefore, the effects of solvation and small amounts of catalysts, particularly those coordinating the metal centers in XXIV, may considerably exceed the magnitude of the above-examined structural effects. One may appropriately point to the calculations on the hydrated methonium ion performed by the CNDO/2 method with a special parametrization in supermolecular approximation, and with the surrounding of 5 and 10 water molecules taken into account. They led to the conclusion that the pyramidal C_{4v} structure, rather than the structures C_s or D_{3h}, is in solution energetically the most favored [83]. The structure of the first hydrated shell is shown by the formula XXVI:

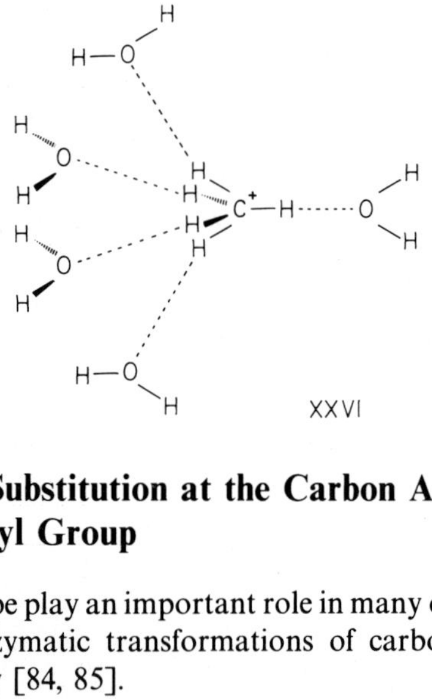

XXVI

5.3 Nucleophilic Substitution at the Carbon Atom of the Carbonyl Group

The reactions of this type play an important role in many organic and bioorganic processes since all enzymatic transformations of carboxylic acid derivatives belong to this category [84, 85].

5.3.1 The Stoichiometric Mechanism

As has been shown by Bender's elegant experiments [85] using the isotope labelling technique, the reaction of nucleophilic substitution at the carbonyl carbon proceeds not as a concerted one-step transfer of the acyl group, but rather as a two-step associative process by the addition-elimination (AdE) mechanism $B_{Ac}2$ with the formation of the tetrahedral intermediate XXVII:

$$R-C(=O)X + Y^- \rightleftharpoons R-C(O^-)(X)-Y \rightleftharpoons R-C(=O)Y + X^- \quad (5.10)$$

XXYII $Y = H^-, OH^-, H_2O, Hal^-, CN^-, NH_3, \ldots$
$X = H, Hal, OR, NH_2$

The XXVII type structures or their derivatives have in some cases been fixed experimentally in solution [86]. Adducts formed by anions with carbonyl compounds were also registered by the ICR method in a number of gas-phase ion-molecular reactions of the type of Eq. (5.10): in the reaction of $CF_3CO_2^-$ with trifluoroacetic anhydride [87], Cl^- with acetyl chloride [88], HO^- with $HCOOCH_3$ [89]. In the first two cases, the authors interpreted the experimental data in favor of a tetrahedral intermediate structure of these adducts XXVIII, XXIX. However, this conclusion requires, apparently, further confirmation. Ab initio (4–31G + polarization functions) calculations [90] of the gas-phase nucleophilic displacement in acyl chlorides did not support the formation of the tetrahedral intermediate in this reaction as a general rule, though confirming the formation of a 1:1 adduct of Cl^- with CH_3COCl which was classified as a loose charge-transfer complex. The fact that no ^{18}O exchange is observed between HO^- and $HC(^{18}O)(OCH_3)$ in an ICR experiment led to the conclusion that the tetrahedral species of the XXVII type is likely to represent a transition state rather than an intermediate structure [41, 89]:

$$F_3C-C(O^-)(OCOCF_3)-OCOCF_3 \qquad H_3C-C(O^-)(Cl)-Cl$$

XXVIII XXIX

The tetrahedral intermediate XXVII is, as a rule, more favored thermodynamically than the starting reactants. Most of the calculations made so far support this claim as well as the conclusions drawn in Refs. [91, 92] as to the formation of XXVII in the ion-molecular reaction of Eq. (5.10) without a barrier. The energy profile depicted in Fig. 5.4a is compatible with this character of the reaction. Table 5.4 presents some ab initio and semiempirical calculation data on the energetics of gas-phase reactions of the type of Eq. (5.10).

The shape of the energy profile of the ion-molecular reaction shown in Fig. 5.4a does not fit the interpretation by Brauman [100] of the data on reaction kinetics obtained with use of the ICR technique. Even though the reactions described by Eq. (5.10) are in the gas-phase extremely fast, their velocity is still 3–10 times lower than the collision frequencies of ions and molecules. As in the case of the S_N2 reactions of Eq. (5.2), this behavior may be explained by the emergence of an internal energy barrier for the nucleophilic substitution reaction. According to the scheme suggested in Ref. [100], which rests on the calculations

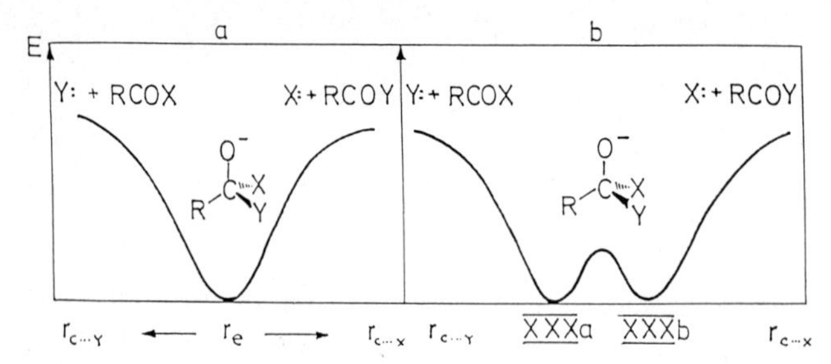

Fig. 5.4a. Typical potential curve of the nucleophilic substitution reaction proceeding by the AdE mechanism obtained in theoretical calculations; **b** potential energy curve of Eq. (5.10) that originates from the interpretation [100] of the ICR experiments

based on the RRKM theory, such a barrier arises at the stage of the formation of the structure with a tetrahedral carbon atom from the initially formed ion-molecular adduct XXXa (Fig. 5.4b):

$$Y^- + RCOX \rightleftharpoons \underset{XXXa}{(Y\cdots RCOX)^-} \rightleftharpoons \underset{XXXb}{(X\cdots RCOY)^-} \rightleftharpoons RCOY + X^-$$

In the general case, according to the microscopic reversibility requirements, the formation of a second adduct XXXb should be assumed. Although the energy of the tetrahedral structure lies, as may be seen from Fig. 5.4b, lower on the energy axis than the reactants' energy, it none the less corresponds to the transition state of the interconversion reaction of the adducts XXXa, XXXb rather than to the minimum on the PES, as in Fig. 5.4a.

XXXa, XXXb are thought [101] to be quite “loose” complexes in which the structure of the carbonyl reactant changes insignificantly. The nature of the forces that form such complexes should have electrostatic character. The calculations [93, 94, 96–99] do indeed reveal appreciable attraction of the negative ion towards the electrophilic carbon atom of the carbonyl group being attacked at various angles of approach to the plane of that group. At large distances, favored is a collinear disposition of the nucleophilic reactants and the C=O bond (see below Fig. 5.7) dictated by the electrostatic forces. In ab initio (6–31 + G*) calculations of the gas-phase reaction between hydroxide anion and formaldehyde [96], it was found that very little change in the reactants' structures occurs when they approach each other in a linear fashion until they are 2.74 Å apart. However, the interaction energy at this distance rises to – 19.1 kcal/mol.

Madura and Jorgensen [96] carried out sophisticated ab initio and Monte Carlo calculations for the nucleophilic addition reaction $HO^- + HCHO$ in aqueous solution. In this solution, the free energy gradually grows with the diminishing distance between the reactants from infinity to 3.0 Å. In this region, an increase in the solvation energy is offset by the increase in the ion-dipole

Table 5.4. Calculated energies of tetrahedral intermediates or transition states of XXVII type relative to noninteracting reactants[a]

Reaction	Calculation method basis set	r_{CY}, Å	ΔE, kcal/mol	Ref.
	7s 3p/4s	1.12	−48	[93]
$HCHO + H^-$	4–31G	1.362	−65	[94]
	MINDO/3	1.12	−60	[95]
	6–31 + G*	1.47	−35	[96]
$HCHO + HO^-$(XXXIIc)	MINDO/2	1.50	−70	[97]
	4–31G	1.516	−98	[94]
$HCOF + F^-$	4–31G	1.44	−61	[98]
	PRDDO	1.44	−187	[98]
$HCONH_2 + HO^-$	MP2/4–31G*	1.75	−40	[99]
$HCONH_2 + CH_3O^-$	4–31G	1.55	−38	[98]
	PRDDO	1.55	−241	[98]
$HCHOH^+ + H_2O$(XXXIIb)	4–31G	1.67	−130	[94]
	MP3/6–31G*	1.624	34	[100]
$HCHO + NH_3$	STO-3G	1.50	259	[94]
	PRDDO	1.59	221	[98]
$HCHO + CH_3NH_2$	3–21G	1.603	29	[100]
$HCHO + NHF_2$	3–21G	1.959	45	[100]
$HCHO + H_2O$(XXXIIa)	4–31G	1.44	55	[94]
	MINDO/2	1.5	15	[97]

[a] Negative sign before ΔE refers to a stable tetrahedral intermediate, positive sign—to a transition state

attraction. In the region between 2.0 and 3.0 Å, the gas-phase energy hardly changes and there is no compensation for desolvation effects, which leads to a rapid growth of the free energy (Fig. 8.3) up to the transition state with $r_{O \cdots C} = 2.05$ Å. For the gas-phase reaction, a sharp drop in energy starts at this point due to the formation of the chemical bond HO···C. This drop is also reflected in the curve profile of the free energy in water. It should be stressed that, according to the Monte Carlo calculations, the term desolvation describes lessening of the strength of intermolecular hydrogen bonds with retention of their total number when moving from noninteracting reagents to the transition state. The force of the hydrogen bonds, formed by OH^-, decreases by about 40%. Thus, unlike the S_N2 reactions, the activation barrier of addition of OH^- to formaldehyde in water is fully determined by the solvent. The reaction profile calculated in Ref. [96] coincides with that obtained by Weiner et al. for the hydrolysis of formamide with a hydroxide ion [99].

Ab initio 4–31G calculations [34] of the cyanide ion nucleophilic attack upon formaldehyde have reproduced quite well the initial portion of the potential curve shown in Fig. 5.4b. As the reactants draw together, first a minimum is revealed on the PES corresponding to the ion-dipole adduct XXXIa and then a maximum associated with the transition state XXXIb which separates XXXIa and the tetrahedral intermediate XXXIc:

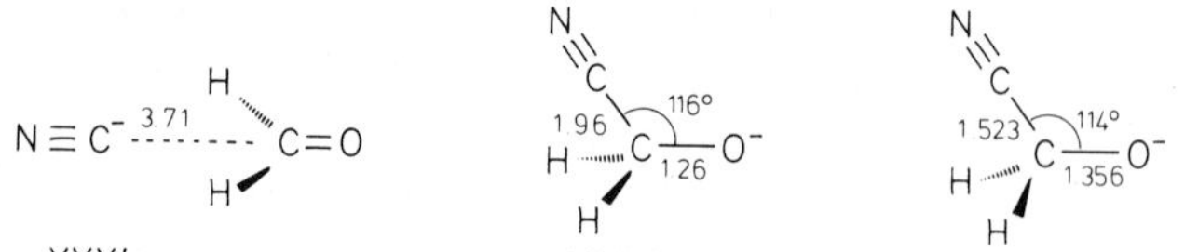

XXXIa	XXXIb	XXXIc
$N\equiv C^{-}\cdots(3.71)\cdots CH_2=O$	$N\equiv C\cdots(1.96)\cdots CH_2-O^-$ (116°, 1.26)	$N\equiv C-(1.523)-CH_2-O^-$ (114°, 1.356)
ΔE, kcal/mol -16	-8	-14

The intermediate XXXIc corresponds to a minimum on the PES (in contrast to the scheme in Fig. 5.4b) which lies a mere 14.1 kcal/mol below the starting reactants. However, formation of XXXIc could not be registered in the ICR experiments [92], owing, probably, to low exothermicity of the reaction (cf. the data of Table 5.4).

5.3.2 Homogeneous Catalysis

As may be seen from the data of Table 5.4, the interaction of neutral nucleophiles (such as NH_3, H_2O) with the carbonyl reactants corresponds to the repulsion potential. Detailed ab initio (STO-3G and 4–31G) calculations [94, 102–104] showed that as the water and the formaldehyde molecules draw nearer along the reaction path (of the type depicted in Fig. 5.7), the repulsion between them increases and that the dipolar structures XXXIIa, in contrast to the ions XXXII XXXIIc, do not belong to the minima regions on the PES and do not conform to the bound structures:

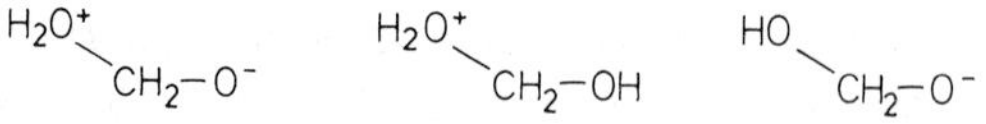

XXXIIa	XXXIIb	XXXIIc
$H_2O^+-CH_2-O^-$	$H_2O^+-CH_2-OH$	$HO-CH_2-O^-$

Stabilization of the ionic tetrahedral intermediates manifests the specific acid (XXXIIb) and base (XXXIIc) catalysis of the nucleophilic substitution reactions. Energetically favorable proves to be a two-step formation of the tetrahedral intermediate with prior protonation of the substrate (specific acid catalysis) or deprotonation of the nucleophilic reactant (specific base catalysis). The chief factor which helps to overcome the repulsion potential and provides for the exothermicity of the formation of the intermediates XXXIIb, XXXIIc is the drawing together of the levels of the frontier orbitals of reactants and the effective mechanism of charge transfer (Fig. 5.5).

The calculations point to a possibility of a theoretical explanation of the effectiveness of the homogeneous catalysis. However, it should be emphasized

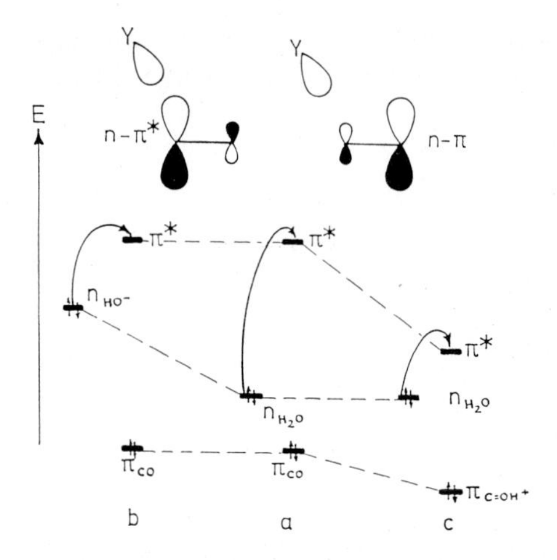

Fig. 5a–c. General pattern of the orbital interactions that determine the formation of the O—C—O bound in the nucleophilic addition of Y to the carbonyl group: **a** interaction of a water molecule with the carbonyl compound; **b** base catalysis; **e** acid catalysis

that the predominant mechanism of acceleration of a nucleophilic reaction at the carbonyl carbon atom is not the specific but rather the general acid-base catalysis [84, 105]. In the simplest manner, this type of catalysis can theoretically be modelled by the bimolecular reaction $CH_2O + H_2O$ with the reactants approaching each other as in XXXIII:

$$H_2C{=}O + H_2O \rightarrow [\text{XXXIII}]^{\ddagger} \rightarrow CH_2(OH)_2 \quad (\text{not via XXXIIa}) \qquad (5.11)$$

The calculations [94, 102–104] indicate that the repulsion potential is overcome when the separating distance of $r_{O\cdots C} = 1.674$ Å is reached and the structure XXXIII corresponds to the genuine transition state with a single negative force constant ($\nu^{\ddagger} = i$ 1997 cm^{-1}, 4–31G). The 42 kcal/mol energy barrier of the reaction is, however, much higher than its experimental values for the formaldehyde hydration reaction. Since a stepwise route via the dipolar ion XXXIIa is ruled out because it does not correspond to a minimum on the PES, the reaction runs concertedly. According to Jencks [106], such reactions should be classified as the reactions with enforced concertedness.

A sharp lowering of the energy barrier is brought about when the reaction of Eq. (5.11) is catalized with an additional water molecule. A necessary condition for the development of the catalytic effect is the drawing together of the reactants in a configuration of the XXXIV type which provides for the possibility of a bifunctional catalysis. The reaction channels characterized by the structure XXXV (model of general acid catalysis) or XXXVI (model of general base catalysis) are ineffective merely leading to some lowering of the repulsion potential relative to XXXIIa without any basic changes in the PES pattern (absence of the bound states).

XXXIV XXXV XXXVI

The calculations [102–104] bear witness to a concerted development of the reaction of methanediol formation from the adduct XXXIV where the process of making bonds with the nucleophilic center and of proton transfer are synchronized. In a search for the transition state, a new effective modification of the coordinate regime was used characterized by independent variation of not one but rather two geometry parameters and an optimization of all others for each fixed value of the reaction coordinate parameters [107]. Figure 5.6 features the calculated structures of the initial complex $CH_2O{\cdot}(H_2O)_2$ XXXIV, the reaction product, hydrated methanediol, and the transition state that corresponds to a saddle point on the 24-dimensional PES of the reaction. The energy barrier amounts, according to calculations, to a mere 6.7 kcal/mol relative to the reactants, CH_2O and the water dimer $(H_2O)_2$, i.e., is reduced by 35 kcal/mol, once a second water molecule enters into reaction.

According to the calculations [108], inclusion of four water molecules into the calculation of the formaldehyde hydration reaction is sufficient for the appearance of a local minimum whose structure corresponds to the dipolar form. When number of the solvent molecules reaches six, the minimum becomes rather deep (over 5 kcal/mol). Thus, dipolar intermediates may, in principle, exist in solution.

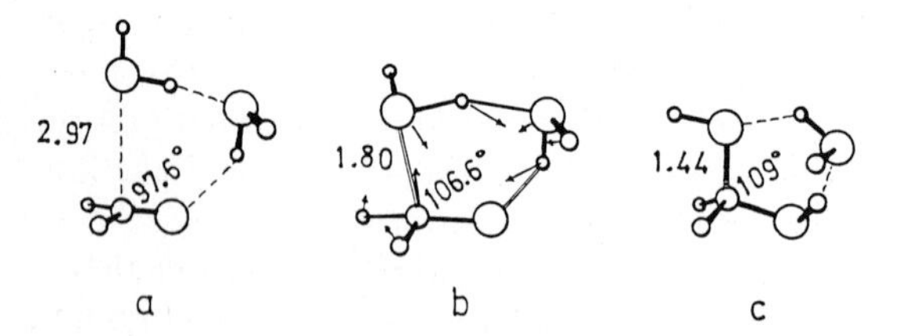

Fig. 5.6a–c. Pathway of formaldehyde hydration reaction $CH_2O + 2H_2O$, ab initio (STO-3G) calculated [102]: **a** hydrate complex $CH_2O{\cdot}(H_2O)_2$; **b** transition state of the reaction (components of the transition vector corresponding to imaginary vibration frequency $\nu^{\ddagger} = i$ 547 cm^{-1} are shown); **c** hydrated methanediol $CH_2(OH)_2{\cdot}H_2O$ is the reaction product. (Adapted from Ref. [102] with permission from the American Chemical Society)

This result, is apparently, of a more or less general significance for the reactions of nucleophilic substitution at the carbonyl carbon. Thus, quantum chemical calculations predict appreciable acceleration of ammonolysis reactions of esters [93] and carboxylic acids [108] when an additional ammonia molecule takes part in their catalysis. The structure, found by a nonempirical calculation [109], of a transition state in the concerted reaction of formation of formaldehyde XXXVII in the reaction:

$$NH_3 + HCOOH + NH_3 \longrightarrow H_2NCHO + H_2O + NH_3$$

is fairly close to XXXIV.

The calculations [109] carried out at a quite high level of approximation MP2/6–31G* have demonstrated excellent agreement of the calculated values of $\Delta G = -0.24$ kcal/mol and $T\Delta S = 0.4$ kcal/mol with the experimental values for the gas phase at 298 K. The energy barrier associated with the transition state XXXVII 13.9 kcal/mol is significantly lower than that for the noncatalyzed bimolecular reaction $HCOOH + NH_3$ (39 kcal/mol):

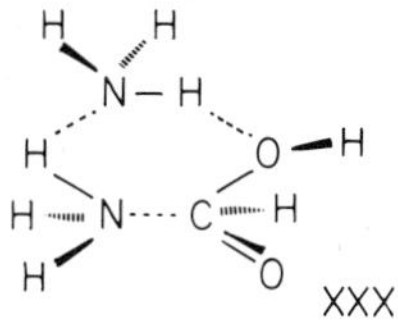

H H N–H H O–H H–N····C–H H O XXXVII

The above-considered calculational data point to high effectiveness of the bifunctional catalysis in hydrolytic reactions and the reactions related to these, due to involvement of molecular chains of water and ammonia, as well as to the preferability in these reactions of a concerted mechanism. This conclusion is fairly general and is corroborated by calculations on other types of nucleophilic reactions, such as hydrolysis of methyl fluoride, tautomerization of pyridine in aqueous solution etc. [110]. An advisable piece of work would apparently, be an analysis, in the light of the conclusions discussed, of mechanisms of the catalytic act in enzymic hydrolysis reactions of the ester and peptide bonds. In the most advanced up-to-date models for, e.g., the reactions with participation of α-chymotrypsin (see Ref. [111]), the steps of the base and the acid catalysis are separated. The latter is commonly thought [84, 111] to be operative at the stage of enzymic decomposition of the tetrahedral intermediate. However, taking into account the possibility of realization of the conformationally excited states of the active enzymic center, it would not be hard to think of some realistic schemes of concerted mechanisms, the more so that the fast growing body of calculational material continuously supplies fresh evidence in favor of such mechanisms.

5.3.3 Stereochemistry of the Reaction

The stereochemical outcome of the reactions in question is determined by: 1) the direction of approach of the attacking nucleophile to the carbonyl plane at the stage of formation of the tetrahedral intermediate and 2) the relation of the character of dissociation of this intermediate to its conformation.

5.3.3.1 The Direction of Nucleophilic Attack and Orbital Steering

The reaction path for addition of the simplest nucleophilic reactant, the hydride ion, calculated by an ab initio and the MINDO/3 method is depicted in Fig. 5.7. At large distances between the hydride ion and the electrophilic center, predominant are the electrostatic forces of interaction between the reactants, and the nucleophile occupies a position collinear with the C=O bond, as in XXXIa. At the distances less than 3 Å, the role of the orbital interaction, i.e., the exchange repulsion and charge transfer effects, becomes decisive. The attacking nucleophile leaves the plane of the carbonyl group and approaches the carbonyl fragment, in the zone of formation of the new bond, at an angle of 109.5°.

This picture closely agrees with predictions of the structural correlation method (Sect. 1.3.5.3, Fig. 1.18) and conforms well to the predictions founded on a study of the regions of charge concentration and charge depletion, as defined by the Laplacian of electron density, to identify the sites of nucleophilic attack of formaldehyde and acetaldehyde [112]. It may easily be interpreted by considering the interactions between the frontier orbitals of reactants. The main

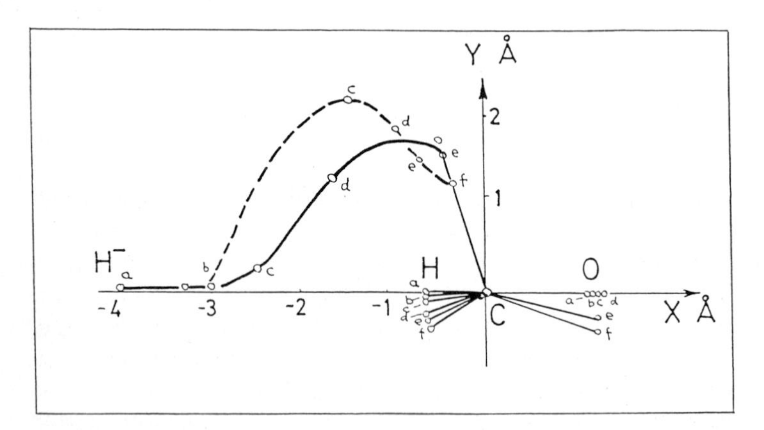

Fig. 5.7. Minimal energy path of attack upon the carbon atom of the formaldehyde molecule by the hydride-anion according to ab initio [93] (*dashed line*) and MINDO/3 [95] (*solid line*) calculations. The circles a, b, c, d, e, f denote the positions of the atoms H, C, O of the molecule being attacked and H^-, OH^-, with the distances between the nucleophile and the carbon atom being, respectively, 4.0, 3.0, 2.5, 1.5, and 1.12 Å. In the initial state the formaldehyde molecule lies in the XOZ plane

interaction consists in charge transfer from the lone pair orbital n_Y of the nucleophilic reactant to the lower antibonding MO of the π^* type of the carbonyl group resulting in two-electron stabilization. The destabilization depends on exchange repulsion between the electron pairs of the n_Y and π orbitals (Fig. 5.5). According to Eqs. (4.8), (4.9), the former interaction grows stronger with the narrowing of the energy gap between the n_Y and the π^* orbitals, while the latter, the unfavorable one, gets weaker with the widening of the gap between the n_Y and the π orbitals.

Figure 5.5 featuring the displacement of these orbitals along the energy axis visualizes the reason for the strongly pronounced effects of the base and the acid catalysis (see Sect. 5.3.2). Furthermore, considering the role of the overlap integral in the interactions discussed, one may arrive at the conclusion that the angle of attack by the nucleophilic reactant should somewhat exceed 90° in order that the destabilizing effect of the $n_Y - \pi$ interaction be diminished [113–116].

It is commonly assumed that one of the most important functions of an enzyme in the bioorganic reactions, particularly, in reactions of transfer of the acyl groups by the nucleophilic substitution mechanism is manifested in the drawing together of the interacting centers and their mutual orientation that should correspond to the optimal reaction pathway. Storm and Koshland [117] studied the kinetics of an extensive series of lactonization reactions, i.e., intramolecular nucleophilic addition of hydroxyl to a carboxylic group. Upon passing from the γ-hydroxybutyric acid XXXVIIIa to the norbornane hydroxy acid XXXVIIIc, the cyclization rate increases by four orders of magnitude:

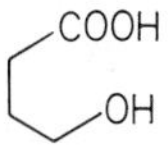

COOH OH CH$_2$OH COOH OH COOH

XXXVIIIa $K_{rel} = 1$ XXXVIIIb $K_{rel} = 80$ XXXVIIIc $K_{rel} = 13000$

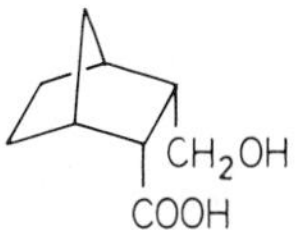

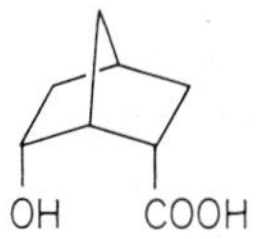

Having compared these data with the geometry of the molecules calculated by the molecular mechanics method, the same authors concluded that in XXXVIIIc optimal orientation is achieved of the nucleo philic group OH with respect to carbonyl, namely, the angle of approach of 98°, while even insignificant deviations ($\pm$ 10°) from it appreciably inhibit the reaction. The requirement for such refined adjustment of the reaction site became known as the orbital steering concept provoking lively discussion [118, 119].

Theoretical checking of this concept led to a more precise definition of its scope. Using as an example the ion-molecular reaction of addition of the methoxide anion to formaldehyde (see Table 5.4), which served as a model for the reaction in complexes of the enzyme trypsin with its substrates, Scheiner and Lipscomb [98] have shown that the region of the attractive potential on the reaction PES,

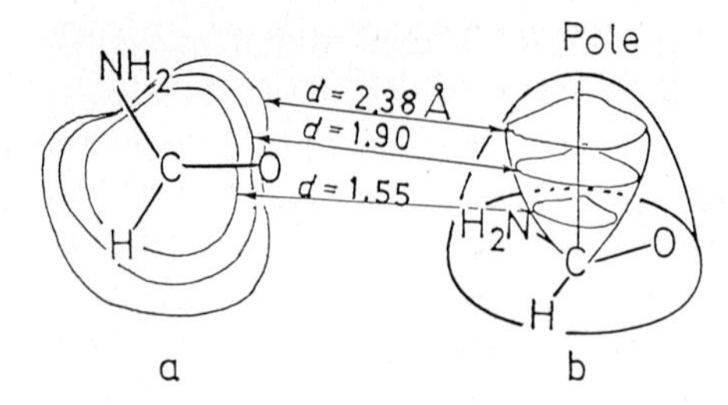

Fig. 5.8a, b. Region of the attractive potential (inside the "reaction funnel") of the system $CH_3O^- + HCONH_2$ constructed from calculation data [98] by the PRDDO method for three distances O=C....O (d_1, d_2, d_3 in Å units): **a** projections of the zero isoenergetic curves onto the equatorial plane; **b** the "reaction funnel" outside which repulsion of the reactants is observed. (Adapted from Ref. [98], with permission from the American Chemical Society)

in which the methoxyde anion is attracted to formamide without any barrier to form the addition product XXVII, is in reality quite wide.

Using the distance O=C···O as the reaction coordinate and optimizing, parallel with the variation of this distance, the bond lengths C=O and C—N, the angle of approach of the nucleophilic reactant to the C=O bond and other angular parameters, the authors of Ref. [98] found that the region of attractive potential—the "reaction funnel"—takes up about 20% of the whole half-sphere with the center at the carbonyl carbon (Fig. 5.8). Thus, although the optimal angle of approach amounted, as had been expected on the basis of the orbital arguments, to 110°, the reaction channel was on the whole rather broad and even sizeable deviations from the optimal route of approach could not inhibit the nucleophilic addition reaction.

Also the more recent calculations [119] based on a modified molecular mechanics model parametrized according to results of a complementary ab initio (3–21G) study of relevant transition state structures, did lend support to the orbital steering concept in its original formulation. No significant changes in angularity parameters of the reaction site had been found in this model for hydroxy acids XXXVIII. The drastic differences in their reactivity were assigned to the differences in the steric strain arising upon distortion of the hydroxy acid carbon frameworks to fit the geometric demands of proper transition state structures. As calculations [119] have shown, in this case a fairly broad range of angularities is possible in the PES region that preceds the formation of the HO···C=O bond. Moreover, since in these reactions the rate determining step is believed to be the breakdown of the tetrahedral intermediate [120], the angularity of the nucleophilic attack leading to its formation is unlikely to be a controlling factor.

5.3.3.2 *Stereochemical Control of the Breakdown of the Tetrahedral Adduct*

The second step of the AdE mechanism of nucleophilic substitution i.e., the breakdown of the tetrahedral intermediate XXVII, is stereoselective and depends on the conformation of XXVII.

As a result of an analysis of numerous data on the direction of hydrolysis reactions of ethers and amides which proceed via formation of the type XXVII intermediates, a simple phenomenological theory was evolved of stereoelectronic control of fragmentation of tetrahedral intermediates [120, 121]. This theory rests on the classical notion of the hybrid orbitals of electron lone pairs (ELP) of heteroatoms.

Let in Eq. (5.10) $X = OR'$ and the attacking nucleophile $Y = OCH_3$. The tetrahedral intermediate XXXIX may undergo decomposition in two directions, as shown in Eqs. (5.10) and (5.12), and two esters XL and XLI that are formed in this case may exist in two conformations each.

According to Ref. [121], the direction of elimination of an alkoxide ion from the intermediate XXXIX depends on its conformation and is controlled by the orientation of the ELP orbitals relative to the bonds being split. The basic principle operative in this process is that those bonds C—O (C—Y, C—X) are broken which are antiperiplanar relative to the two ELP geminal heteroatoms.

This principle will be exemplified by the cleavage of XXXIX. Since usually the gauche conformations are favored with the rotations about the carbon-heteroatom bond (the gauche effect, Ref. [122]), XXXIX may be represented as containing a set of nine *gauche* conformations formed by rotation about the bonds C—OR′, C—OCH_3. Three of these XXXIXa–c, are given below arising from rotation of the methoxyl group relative to one of the fixed conformations of the group OR′. It is these conformations of the tetrahedral intermediate XXXIX which are formed at the kinetic control stage as a result of addition of the methoxide ion to ester XL in its *E* conformation.

$$\underset{\text{XL,E}}{R{-}C({=}O){-}O{-}R_1} \rightleftharpoons \underset{\text{XXXIX}}{R{-}C(OR_1)(O^-){-}OCH_3} \rightleftharpoons \underset{\text{XLI,E}}{R{-}C({=}O){-}O{-}CH_3}$$

$$\underset{\text{XL,Z}}{R{-}C({=}O){-}O{-}R_1} \rightleftharpoons \text{XXXIX} \rightleftharpoons \underset{\text{XLI,Z}}{R{-}C({=}O){-}O{-}CH_3} \qquad (5.12)$$

Upon its formation, the conformer XXXIXa can undergo decomposition in reverse direction only without displacing the leaving group $R'O^-$ as only one ELP of the C—O^- bond is antiperiplanar with respect to the bond C—O in this conformation. The conformers XXXIXb and XXXIXc have each two anti-periplanar ELP's relative to both the C—O and C—OCH_3 bond, and give, upon substitution, ester XLI in the conformations *E* and *Z*, respectively. When the conformational composition of XXXIX is known, one may predict the conformations of the substitution products.

XXXIX a

CH_3O^- + R—C(OR$_1$)=O ⇌ XXXIX b ⇌ XLI, E + $^-OR_1$

XL, E

XXXIX c ⇌ XLI, Z + $^-OR_1$

The interactions between the orbitals of geminal bonds in the tetrahedral fragments CXYZ (X, Y, Z=N, O, F...)—see Refs. [93, 120–122]—serve as the theoretical basis for the above rule of breakdown of the type XXXIX structures. Crucial is the n→σ* interaction, associated with a charge transfer from the ELP of the heteroatom to the antibonding σ-orbital of the geminal bond. This interaction is maximal in the case of the antiperiplanar orientation of the ELP of the heteroatom X and of the bond C—Y as shown in XLII:

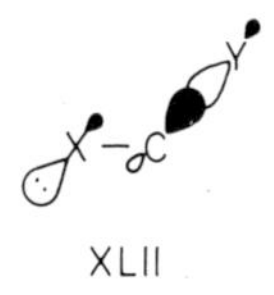

XLII

As is evident from the scheme XLII, the orbital interaction of this type leads to an increase in electron population, the strengthening and shortening of the bond C—X and, conversely, to a lowering in population, the loosening and lengthening of the bond C—Y. The latter will, therefore, be cleaved more readily in the decomposition process of the tetrahedral intermediates of type XXXIX. Table 5.5 contains selective data of nonempirical calculations for some conformations of aminodihydroxy-methane XLIII which may be regarded as model of the intermediate compound in the biologically important reaction of base catalyzed hydrolysis of the imidoesters.

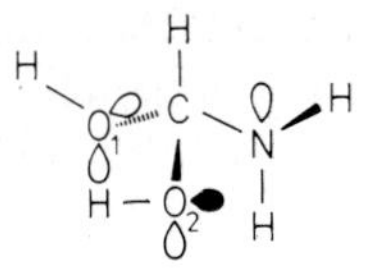

XLIIIa 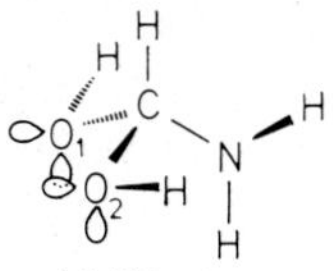 XLIIIb

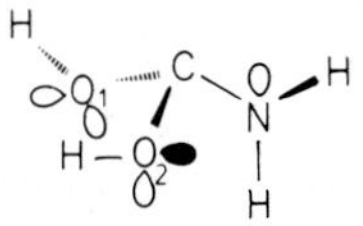

XLIIIc 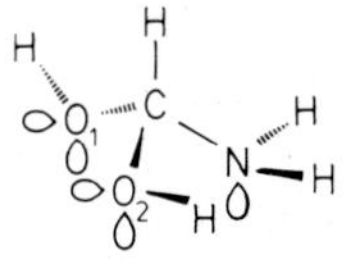 XLIIId

The calculation data are in complete accord with the predictions of the simple orbital model of the $n \rightarrow \sigma^*$ interaction thus confirming the assumption as to the stereoselectivity of the breakdown reaction of the intermediates XXVII and XXXIX under kinetic control.

It should be noted that, in virtue of the microscopic reversibility principle, the above-described rule also regulates the type of approach of the attacking nucleophile to the carbonyl group, in other words, the exact conformation of the axially nonsymmetrical nucleophile at the stage of the formation of the tetrahedral intermediate.

5.4 Aromatic Electrophilic Substitution Reactions

Electrophilic aromatic substitution reactions are among the most important reactions of organic chemistry. A canonical textbook example of the best understood stoichiometric mechanism is given by the nitration reaction where,

Table 5.5. Relative energies (ΔE), electron populations (P) and lengths (l) of the C—O, C—N bonds in various conformations of aminodioxymethane ab initio calculated with a DZ basis set [123]

Conformation	ΔE kcal/mol	Bond	Number of ELP	P_{CX}	l, Å
XLIIIa	5.0	C—O_1	1	0.387	1.427
		C—O_2	2	0.313	1.447
		C—N	0	0.490	1.430
XLIIIb	7.2	C—O_1	0	0.427	1.404
		C—O_2	1	0.363	1.425
		C—N	2	0.368	1.461
XLIIIc	0	C—O_1	1	0.371	1.415
		C—O_2	1	0.372	1.417
		C—N	1	0.430	1.440
XLIIId	17.4	C—O_1	0	0.393	1.406
		C—O_2	0	0.393	1.406
		C—N	2	0.314	1.485

as has reliably seen ascertained, the active electrophilic reagent is the nitronium cation NO_2^+ [1, 124, 125]. Similarly to many other substitution reactions, the aromatic electrophiliic substitution is believed to occur by the addition-elimination (AdE) mechanism:

$$ArH + X^+ \rightleftharpoons [ArH \rightarrow X^+] \rightleftharpoons \text{(XLV)} \rightleftharpoons [ArX \rightarrow \overset{+}{H}] \rightleftharpoons ArX + H^+ \quad (5.13)$$

XLIV XLV XLIVa

In the initial stage, the electrophile X^+ forms with the aromatic compound ArH an intermediate complex XLIV. It is commonly assumed that for the ArH compounds with an aromatic ring, activated by donor substituents or through annelation, the XLIV structure corresponds to the π-complex in which X^+ is located over the ring plane. For the compounds of this type, the reaction of formation of XLIV is characterized by very high rates which are limited only by the collision frequency of reactants of up to $10^{10}\,s^{-1}$ [126].

The rate-determining step for the overall process of Eq. (5.13) is, as a rule, the formation of an adduct, i.e., the substituted cyclohexadienyl cation XLV or, much less frequently (e.g., in a sulfonation reaction), the decomposition of XLV. The XLV type cations were termed the σ-complexes or the Wheland complexes, although, as has recently been noted [127], the first to have pointed, in a well-founded manner, to the formation of ion intermediates in the electrophilic substitution reactions were, back in 1928, Pfeiffer and Wizinger [128].

At present, there are numerous experimental proofs of the formation of σ-complexes XLV during electrophilic substitution reactions in solution. The salts of these cations are quite stable and not rarely can be preparatively isolated [124, 127, 129]. However, the question whether the cations XLV and, more generally, any ion adduct can be formed in the gas-phase ion-molecular reactions of electrophilic substitution provoked active discussion [120, 131]. Ultimately it was found that in case a gas-phase reaction in conditions of a mass-spectrometric, ICR [91, 132–134] or radiochemical [135] experiment proceeds under a fairly high vacuum (<13 Pa) when the emerging ion adducts (X = Alk, Allyl, NO_2, NO, OCOR) can scatter the redundant energy through collisions, these adducts acquire considerable stability (exothermicity of the reaction exceeds 100 kcal/mol). This fact is regarded as a serious evidence for the conclusion that the structure of the ion adducts corresponds to the σ-complex of the XLV type. Indeed, ab initio 4–31G calculations have shown that the structure XLVI corresponds to the global minimum on the PES of $C_6H_7^+$ [136]. The hydrogen bridged structure is a transition state of the reaction of the 1,2-hydrogen shift in XLVI. Owing to incomplete geometry optimization in the calculations on XLVII, its energy might have been overestimated against XLVI: the experimental energy barrier (solution $HF—SbF_5—SO_2FCl—SO_2F_2$) is 10 kcal/mol (see Ref. [129]).

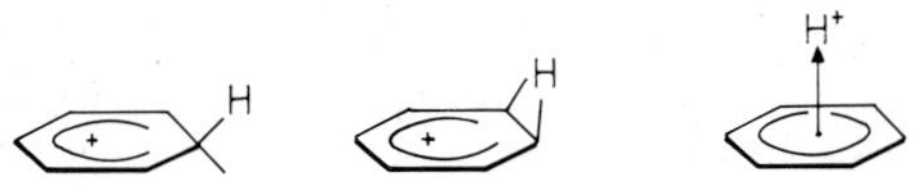

XLVI XLVII XLVIII

As regards the structure of the π-complex XLVIII, the calculations [136] did not detect a minimum on its PES. An analysis [98] of the results of a number of other nonempirical and semiempirical calculations on ion adducts in the reactions of electrophilic aromatic substitution ($X = H, F, CH_3$) has led to the conclusion that generally the symmetric type XLVIII structures of π-complexes are not stable. This conclusion has been supported by semiempirical and ab initio calculations of the mechanisms governing the gas-phase reactions of nitrosation [137, 138] and nitration [139–141] of benzene.

A thoroughgoing experimental study of this reaction [133] had demonstrated that its primary stable product both in the gas phase and in solution is a π-complex which, owing to equivalence of the ring protons observed in the ^1H NMR spectrum (FSO_3H, $-78°C$), was assigned the symmetrical structure XLIX. The π-complex is transformed, allegedly via the σ-complex stage L, into the end product, protonated nitrosobenzene:

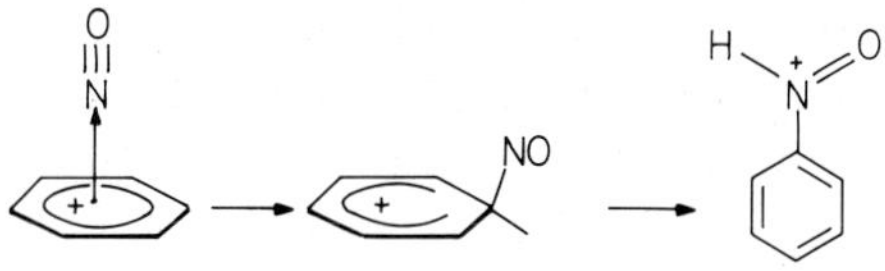

XLIX → L → LI (5.14)

The MINDO/3 calculations [137] are in accord with basic outline of the proposed scheme showing at the same time that the conventional views on the structure of the π-complexes as the high-symmetry systems should be modified. The structure XLIX corresponds, similarly to XLVIII, to a point at the top of a flat hill on the PES of $C_6H_6NO^+$. On the other hand, the calculations reveal on the PES of Eq. (5.14) two local minima corresponding to the π-complexes XLIX and XLIXb. Even though their true symmetry is much lower than the effective C_{6v} symmetry registered by the NMR spectra, it is, in fact, a seeming contradiction. The calculations reveal low-energy barrier rearrangements associated with either the "gliding" of the nitroso group over the ring plane in XLIXa or the switching of the C—N bonds in XLIXb, which must lead to the total averaging of the ring protons on the NMR time scale.

The conclusion as to the preferability of the type XLIXa π-complex structure and its fluxionality, at which the calculations [137] arrived, has recently found a strong experimental support. X-ray structural data [142] have been obtained on the durene and hexamethylbenzene π-complexes, with nitrosonium cation counter-ioned by the tetrachloroaluminate-anion, showing close similarity to the gas-phase structure XLIXa. Both solution and solid-phase ^{13}C-NMR spectra

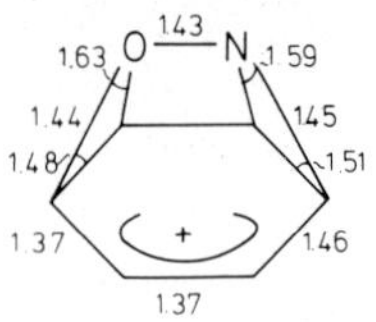

XLIXa
ΔH_f = 229 kca/mol
ΔH_f (exp.) = 211 ± 5 kcal/mol

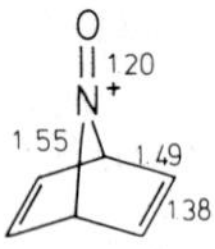

XLIXb, C_{2v}
ΔH_f = 243 kcal/mol

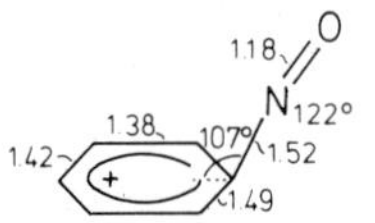

La, C_s
ΔH_f = 224 kcal/mol

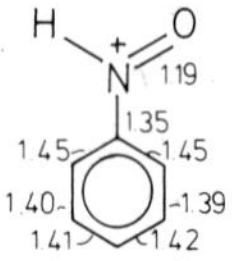

LI
ΔH_f = 192 kcal/mol
ΔH_f (exp.) = 203 ± 7 kcal/mol

of these complexes manifested rapid intramolecular processes which were assigned to a gliding of the nitroso moiety over the arene plane.

Although the calculations have revealed, in addition to the π-complexes, the presence of a minimum on the PES in the region of the σ-complex structure La, the PES is in this zone flattened and the solutions obtained are triplet unstable (see Sect. 2.2.5). The π-complex XLIXa transforms via a slightly distorted structure XLIXb or La into the stable structure LIa of *N*-protonated nitrosobenzene.

This form in its planar C_s-conformation LI was also found to be the most stable conformation of *N*-, *O*-, and *C*-protonated nitrosobenzenes by ab initio calculations [138] at the best (MP2/6–31G + D) level achieved in that study. It was also found that the binding energy of the σ-complex L (*ipso* C-protonated nitrosobenzene) was only about 6 kcal/mol and, in the light of a comparison with the more thoroughly investigated model ethylene + NO^+ system, it hardly belongs to a minimum on the $C_6H_6NO^+$ PES. Interestingly, the σ-complex structure (XLV, R = H, X = NO_2) generally expected to be the intermediate of the reaction of nitration of benzene was also shown by ab initio (3–21G) and accompanying MINDO/3 calculations to belong to the saddle point on the $C_6H_6NO_2^+$ PES [141].

An important feature of the electronic structure of the π-complex XLIXa is that at the stage of its formation there occurs the complete electron transfer from the aromatic molecule to the electrophile. The total charge of the nitroso group in XLIXa equals—0.03e (MINDO/3), i.e., this structure corresponds to the pair of benzene cation radical-nitrogen oxide.

According to Olah [143], the formation of the π-complexes with the aromatic compounds activated by donor substituents explains the paradox, known for this series, of the loss of substrate selectivity with simultaneous retention of the positional (regio) selectivity in the nitration and alkylation reactions. It is argued that the formation of the π-complex is the rate-limiting step and since the substituents in the aromatic nucleus affect but slightly its π-donating properties, the reactivity of various molecules under the conditions of competition for the π-bonding of the electrophile is roughly equal (in the case of the strongest electrophiles $k_{toluene}/k_{benzene} = 1.6$). By contrast, the positional selectivity is regulated by the formation of a σ-complex at the XLIV→XLV stage where relative reactivity of the *o*-, *m*-, *p*-positions plays an important role.

Such an explanation of the selectivity paradox is compatible with the concept of electron transfer mechanisms in an electrophilic aromatic substitution (for a general review of this concept see Refs. [144, 145]). The electron transfer occurs at the stages of formation of the intermediate complex XLIV and the σ-complex XLV. In 1959 Brown put forward a hypothesis that there should exist a charge transfer stage responsible for the formation of the π-complex XLIV in a nitration reaction. A similar view is held by some other authors [146–148], even if the adduct XLIV is not necessarily regarded as a π-complex. In the nitration and nitrosation reactions, the electron transfer from an activated aromatic nucleus onto the lower-lying vacant MO's of the ion acceptor is thermodynamically quite advantageous (20–111 kcal/mol), therefore the detailed representation of the reaction scheme of Eq. (5.13):

$$ArH + X^{+} \longrightarrow [\overset{+}{ArH^{\cdot}} \cdots\cdots X^{\cdot}] \longrightarrow \overset{+}{Ar}\begin{matrix} H \\ X \end{matrix} \tag{5.13a}$$

which assumes that the σ-complex XLV forms from the biradical pair XLIVb [140, 144, 145, 149, 150] appears realistic enough. Direct experimental evidence for the reactions between aromatic radical cations and NO_2 affording the formation of σ-bonded intermediates was reported in Ref. [151]. The positional selectivity must in this case be regulated by the distribution of spin density in the nucleus and such a correlation does indeed take place [147, 152]; thus, the retention of the positional selectivity with the simultaneous loss of the substrate selectivity has been accounted for in a reasonable manner.

These calculations have been supported by the analysis of the mechanism of electrophilic substitution reactions based on studies of stability of the HF solutions. In terms of the RHF approximation, the detection of triplet instability at certain portions of the reaction path indicates that the energy levels of the reactants' frontier orbitals draw closer together and that a biradical electron configuration makes a strong contribution which means that the electron transfer mechanism is operative. In order to determine such portions of the reaction path, it is sufficient to analyze the character of the least eigenvalue of the ${}^{t}\lambda_{+}$ matrix of triplet instability [Eq. (2.24)]—see Ref. [153]. The solution is unstable for ${}^{t}\lambda_{+} < 0$. In the corresponding regions of the configurational space, one should pass to the UHF approximation where the electron transfer state is characterized by the appearance of spin density (see Fig. 2.1), i.e., by the so-called spin density waves in systems with the even number of electrons.

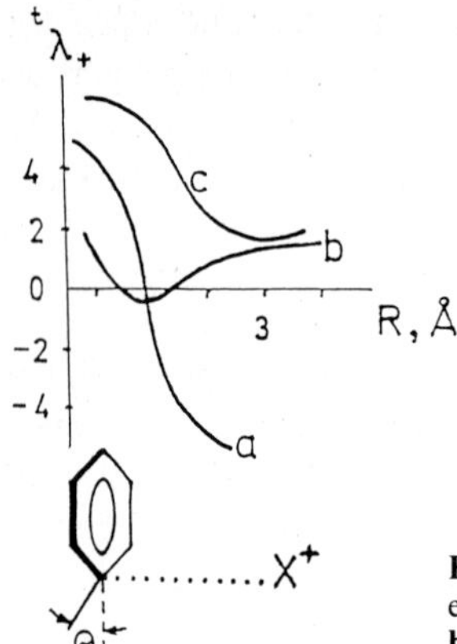

Fig. 5.9a, c. Dependence of $^t\lambda_+$ values on the distance from the electrophile X^+ to the plane of the benzene molecule: **a** $X = H^+$; **b** $X = NO_2^+$; **c** $X = CH_3^+$. Calculated by the CNDO/2 method [153]

As may be seen from Fig. 5.9, in the case of such strong electrophiles as H^+ or NO_2 the electron transfer states are found already at a long distance from the aromatic molecule being attacked and it is to be noted that when a nitronium cation is present, the biradical structure is realized at the formation stage of both the π- and σ-complex. On the other hand, for the system $C_6H_6 \cdots CH_3^+$, the RHF solutions are stable along the whole reaction path, which points to the absence of an electron transfer stage in alkylation reactions. The data presented in Fig. 5.9 refer to gas-phase reactions, however, the conclusions made on this level of approximation are fully valid also in the case of solutions. This has been evidenced by calculations [154] in which the solvation in a polar solvent was taken into account by means of the point charges method (see Sect. 3.3.2). The order of the electrophiles $F^+ > H^+ > Cl^+ > Br^+ > CH_3^+$ reflecting the sizes of the electron transfer zone parallels the decreasing order of their electronegativity. A special position in this sequence of electrophiles belongs to NO_2^+ for which the electron transfer region, with the nitronium ion approaching benzene, extends to the σ-complex structure.

Irrespective of how the σ-complex is formed from the adducts XLIV or XLIVb, it is not clear whether the anomalous selectivity phenomenon can be fully understood on the basis of the Olah hypothesis. Ab initio STO-3G calculations [155] on the structures LII and LIII, which model, respectively, the "late" and the "early" transition states of electrophilic substitution, i.e., of the σ- and π-complexes in the vicinity of those structures, have shown that both approximations lead to identical predictions in regard to positional selectivity. The structure LII is based on optimized geometry of the cyclohexadienyl cation, and in the structure LIII the hydroxonium cation has been selected as an electrophilic particle:

Table 5.6 lists the data on relative energies of the transition states for the substitution at various positions in the toluene benzene ring. These data show that if the structure LIII reproduces correctly the transition state of the

LII LIII

π-complex formation, then Olah's assumption of low-positional selectivity of the substituted aromatic nucleus with respect to the electrophile at this stage is not correct and, consequently, it does not explain the selectivity paradox.

The loss of the substrate selectivity of electrophilic reagents in the nitration and alkylation reactions of aromatic compounds represents a particular case of the phenomenon detected for a number of reactions between strong electrophiles and nucleophiles, namely, that for these reactions relative reactivity of the electrophile (selectivity) does not depend on the type of nucleophile [156]. Whereas Olah's hypothesis [143] explains the reason for such uncommon behavior as lying in a displacement of limiting reaction stage, other more recent suggestions take into account the solvation effects of the electrophilic components of the reaction. According to Ref. [156], no desolvation of the electrophile occurs in the transition state, while a higher reactivity of active electrophiles is offset by strong screening of the reactive center by a stable solvation shell. By contrast, another explanation [157] admits such desolvation which is claimed to be the stronger, the weaker the electrophile. This dampens relative activity of electrophilic reagents and leads to a loss in selectivity.

The Pross hypothesis [157] is supported by theoretical calculations (MINDO/3) of an attack on the benzene molecule by the hydrated electrophiles H^+ [158] and CH_3^+ [110, 159]. In the case of the strong electrophile H^+, the dehydration process does not practically occur in the transition state at the step of the σ-complex formation. Different is the development of the reaction with CH_3^+ and $CH_3OH_2^+$. Having taken as a reaction coordinate the distance benzene—CH_3^+ (as in Fig. 5.9) and optimized all other geometry parameters in each point, the authors of Ref. [159] obtained for the gas-phase reaction an energy profile which contained an intermediate at $r = 3.3$ Å and a transition

Table 5.6. Relative energies of type LII, LIII transition states according to ab initio STO-3G calculations [125] in kcal/mol

Type of transition state	LII "late" $X^+ = H^+$	LIII "early" $X^+ = H_3O^+$
Benzene + X^+	0	0
Toluene + X^+ (*para*)	−9.5	−2.9
Toluene + X^+ (*meta*)	−1.3	−1.1
Toluene + X^+ (*ortho*)	−6.3	−2.2
Toluene + X^+ (*ipso*)	+1.4	+4.2

state at $r = 3.2$ Å lying lower by, respectively, 9.1 and 2.6 kcal/mol than the reactants. When r attains the value of 1.5 Å, a σ-complex is formed stabilized at 88.7 kcal/mol relative to the reactants, which is in good agreement with experimental data [160]. With hydrated CH_3^+ the situation radically changes. The transition state of the σ-complex formation reaction grows tighter ($r = 1.98$ Å); its energy level lies higher by 30 kcal/mol than that of the reactants. The relative stability of the σ-complex is considerably lowered (to 20 kcal/mol relative to the reactants). Of particular significance is the fact that for CH_3^+, as a weaker electrophile than H^+, the process of partial desolvation is clearly registered in the transition state; the distance $CH_3—OH_2^+$ increases by 0.2 Å, while the electronic population of that bond which controls its strength is halved.

These results support the view that a very important part of the overall reaction coordinate is due to the reorganization of the solvent which should not be ignored when examining the intrinsic mechanism of a reaction.

5.5 Nucleophilic Substitution at the Nitrogen, Phosphorus, and Sulfur Centers

The theoretical studies of these reactions are relatively scarce being chiefly represented by the data of semiempirical calculations; even so the results obtained provide some valuable information on the stereochemical route of nucleophilic attack upon heteroatomic centers and bear witness to the general character of the AdE scheme in nucleophilic substitution reactions.

5.5.1 Substitution at the Nitrogen Atom of Nitroso and Nitro Groups

The nucleophilic attack on the nitrogen atom of a nitroso group is the limiting stage of condensation reactions of nitroso compounds with amines, methylene-active, and organomagnesium compounds, of Ehrlich-Sachs, Fischer-Hepp, Burton, and other reaction [161, 162].

So as to determine the optimum path of approach of a nucleophile to the nitroso group, model reactions of the associative nucleophilic substitution in the molecules of nitroxyl and nitrosyl flouride were studied by the MINDO/3 method [163]:

$$Y^- + \underset{\underset{O}{\|}}{N}{-}X \rightleftharpoons \left[Y{-}\underset{\underset{O}{|}}{N}{-}X\right]^- \rightleftharpoons Y{-}\underset{\underset{O}{\|}}{N} + X^- \qquad (5.15)$$

X,Y = H,F

LIV

Figure 5.10 presents the data of a calculation of the MERP of Eq. (5.15) where Y = F and X = H. Similarly to the reactions of addition to the carbonyl group (see Fig. 5.7), the nucleophilic attack takes place in the π-bond plane and its angularity is controlled by the overlap of the orbitals n_Y and π^*_{NO}. An appreciable charge transfer from the nucleophilic reagent to substrate occurs when it comes sufficiently close to the nitrogen atom being attacked (~ 2.0 Å) and a considerable overlapping of those orbitals takes place. In that region, the attacking nucleophile leaves the YOX plane nearing the N=O bond at an obtuse angle whose magnitude depends on the nature of Y and X: 119° (Y = X = H); 109.2° (Y = F, X = H); 154.4° (Y = H, X = F). A dependence of the stereoelectronic demands of nucleophilic substitution reactions on the type of the nucleophile (its hardness or softness) was also found in the case of the substitution at the carbonyl carbon atom [97, 98, 114]. The elimination of the leaving group X starts developing only after the barrierless formation of the intermediate ion LIV.

The same mechanism is predicted by the calculations [163] also for the substitution reactions at the nitrogen atom of a nitro group. They are less common than Eq. (5.15) and proceed readily only in very few cases. Thus, the nitro-containing compounds do not practically exchange their oxygen atoms when treated with water in acidic or alkaline medium [167]. As a model for studying the MERP peculiarities, a degenerate transformation was taken:

$$\mathrm{OH^- + O_2N{-}OH \rightleftharpoons \underset{LV}{[HO{-}N(O)_2{-}OH]^-} \rightleftharpoons HO{-}NO_2 + OH^-} \qquad (5.16)$$

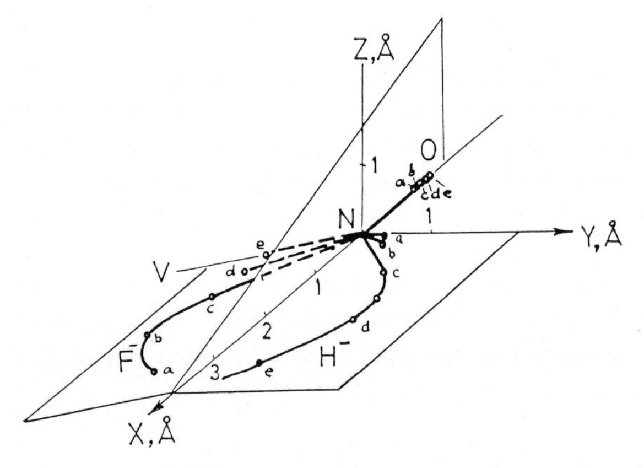

Fig. 5.10. Minimal energy path of attack upon a nitrogen atom of the HNO molecule by the fluoride ion and the trajectory of the hydride ion abstracted from the intermediate $HNOF^-$, calculated by the MINDO/3 method [163]. The points a, b, c denote the positions of the atoms H, N, O, F^- with the distance between the nucleophile F^- and the nitrogen atom being, respectively, 3.0, 2.0 and 1.5 Å (relative energies are 30.6 and 27.0 kcal/mol). The points a–e denote the positions of the atoms F, N, O, H^- with the distance between the abstracted H^- and the nitrogen atom equalling 2.0 and 3.0 Å, respectively, (relative energies are 13.6 and 87.5 kcal/mol). In the initial state the HNO molecule lies in the XOY plane, while the FNO molecule lies in the plane VOX. The energy of the intermediate $HNOF^-$ is taken as the zero point (point b, structure LIV)

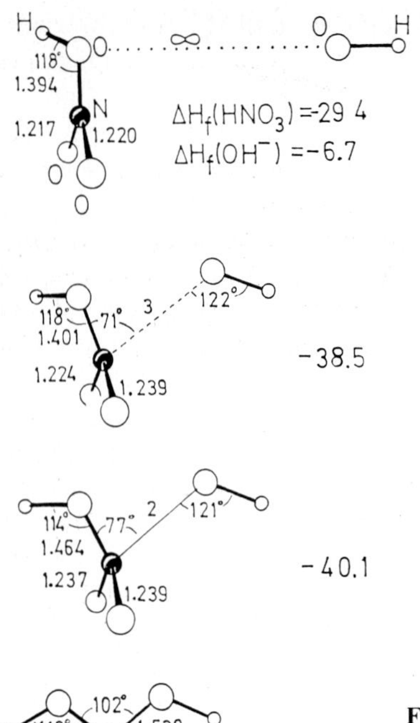

Fig. 5.11. Snap-shot pattern of the MERP for various distances between the nucleophile OH^- and the nitrogen atom being attacked of the HNO_3 molecule [163]. Bond lengths are given in Å, ΔH_f in kcal/mol

Figure 5.11 depicts a snap-shot pattern of the MERP of Eq. (5.16) where the distance $H_1O^- \cdots N$ was taken as the reaction coordinate. Similarly to Eq. (5.15), the addition stage is barrierless: drawing together of the reactants lowers monotonically the energy of the supermolecule up to the point when the adduct LV is formed. The nucleophilic attack occurs strictly in the bisecting plane of the nitro group, with the angle between the direction of attack and the plane of the group equalling in the reaction zone approximately 130°.

The ion LV, whose calculated structure is shown in Fig. 5.11, is an anion of the orthonitric acid H_3NO_4 the existence of which in sufficiently concentrated solution of HNO_3 has been proven.

5.5.2 Substitution at the Dicoordinate Sulfur Atom

Reactions of this type are numerous and important for biological systems [164–166]. An ion-molecular reaction of mixed sulfur dihalogenide with the fluoride ion was studied by the CNDO/2 method in order to determine the optimal trajectory for an approach of the nucleophile to the dicoordinate sulfur atom [167]:

$$\mathrm{F^- + \underset{\displaystyle Cl}{\underset{|}{S}}{-}F \rightleftharpoons \underset{\displaystyle \underset{LVI}{Cl}}{\left[F{-}\underset{|}{S}{-}F\right]^-} \rightleftharpoons F{-}\underset{\displaystyle Cl}{\underset{|}{S}} + F^-} \tag{5.17}$$

Figure 5.12 shows the sections of the PES of Eq. (5.17) for several distances $F^- \cdots S$ which serve as reaction coordinates in three different directions. As may be seen, the orientation of the attacking ion along the axial direction relative to the more electronegative ligands F^- in the plane of the molecule being attacked (route 1 in Fig. 5.12) takes place already at a quite large distance (4.0 Å) from the sulfur atom. This reaction path is more favored by about 5 kcal/mol than the path **3** in which the attacking fluoride ion is oriented axially with respect to the less electronegative leaving group Cl. No reaction valley exists for a symmetrical approach of the nucleophile **2** to the model sulfide. These conclusions are in complete agreement with the predictions made by the structural correlation method (see Sect. 1.3.5.3). These predictions are based on an analysis of a large number of crystalline structures of dicoordinate sulfur compounds with additional intermolecular nucleophilic coordination in crystal [168].

The T-shaped structure LVI is strongly stabilized with respect to the separate reactants and the reaction leading to its formation along the paths **1** and **3** proceeds without the activation energy (see Fig. 5.12). Although the CNDO/2 method overestimates the stability of anion complexes (see Sect. 2.3.2.1), there are reasons to believe that the type LVI T-structures are genuine intermediates rather than transition states in the reactions of the S_N2 type. The same conclusion follows also from the results of ab initio (DZ basis set) calculations [169] on the reaction of nucleophilic cleavage of the disulfide bond:

$$\mathrm{HS-SH + Y^- \rightarrow HS-Y + SH^-}$$

$$\mathrm{Y = H, F, Cl, OH, SH}$$

which predict existence of stable intermediate anions with the T-shaped structure. Stability of these structures is corroborated by, for example, a recently

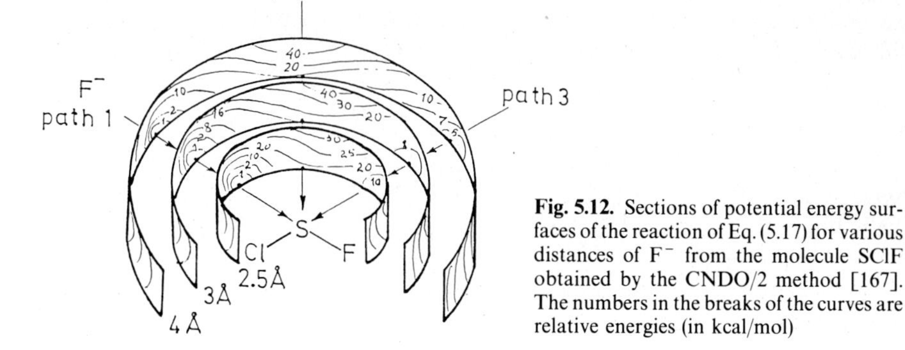

Fig. 5.12. Sections of potential energy surfaces of the reaction of Eq. (5.17) for various distances of F^- from the molecule SClF obtained by the CNDO/2 method [167]. The numbers in the breaks of the curves are relative energies (in kcal/mol)

performed synthesis of the salts of the tricoordinate sulfuranide anion [170, 171]:

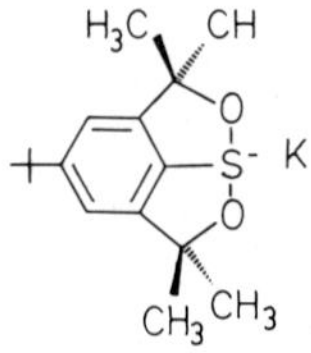

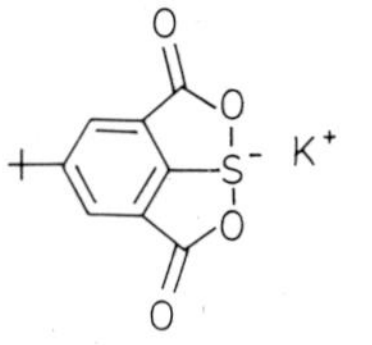

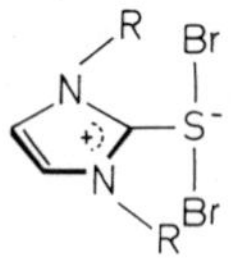

LVII LVIII LIX

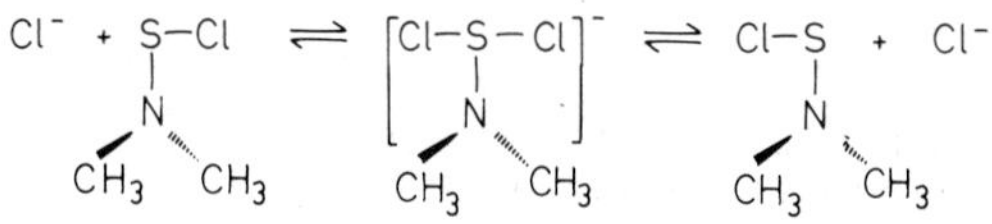

$$Cl^- + S(Cl)N(CH_3)_2 \rightleftharpoons [Cl{-}S(N(CH_3)_2){-}Cl]^- \rightleftharpoons Cl{-}S(N(CH_3)_2) + Cl^- \tag{5.18}$$

LXa LXb

The calculated reaction path **1** and the type of the intermediate structures LVI correspond to an inversion of configuration at the dicoordinate sulfur atom (in the two-dimensional space—Ref. [172]). In actual fact, this result can be observed in reactions of this type of the compounds with axial chirality, namely, the derivatives of amino sulfenyl chlorides LX.

Reactions of degenerate racemization of compounds LX associated with rotation about the S—N bond are considerably accelerated upon addition of the chloride ion. This acceleration can be accounted for by the switching on of a supplementary mechanism of configuration inversion at the sulfur atom, i.e., the reaction of nucleophilic substitution in Eq. (5.18)—see Ref. [173].

5.5.3 Substitution at Tricoordinate Sulfur and Phosphorus Centers

Simple derivatives of sulfoxide LXI and phosphine LXII with substituents considerably differing in electronegativity were taken as convenient models for studying stereochemistry of these reactions.

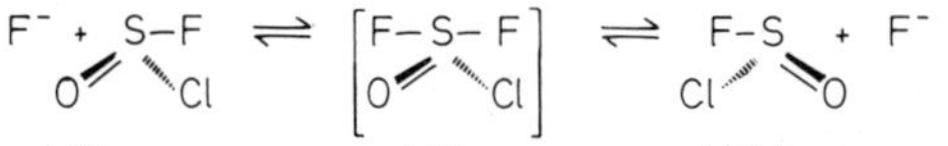

$$F^- + S(F)(O)(Cl) \rightleftharpoons [F{-}S(O)(Cl){-}F] \rightleftharpoons F{-}S(Cl)(O) + F^- \tag{5.19}$$

LXIa LXIc LXIb

$$F^- + P(F)(Cl)(H) \rightleftharpoons [F{-}P(Cl)(H){-}F]^- \rightleftharpoons F{-}P(H)(Cl) + F^- \tag{5.20}$$

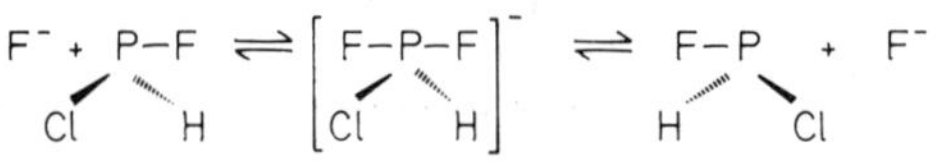

LXIIa LXIIc LXIIb

CNDO/2 calculation results [174] are in principle very close to those described in the preceding section. As may be seen from Fig. 5.13, featuring the

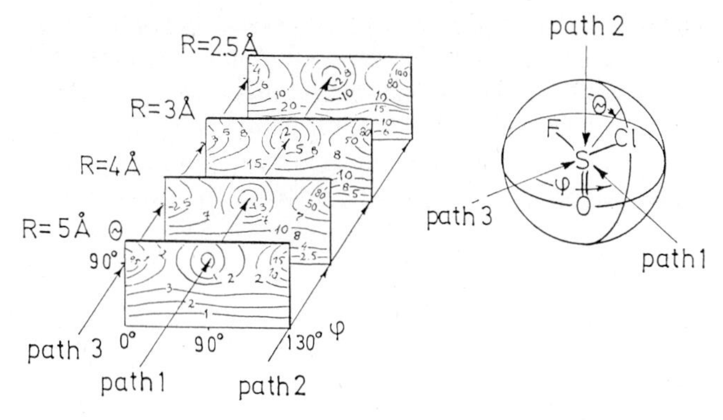

Fig. 5.13. Sections of the PES's for several distances R of a fluoride ion from the central atom in the reacting system of Eq. (5.19)—Ref. [176]. The numbers in the breaks of the curves are relative energies of the levels in kcal/mol. Orientation of the molecule SOFCl and spheric coordinates are shown on the right-hand scheme. Path 2 can be realized from every angle

calculation data for sections of Eq. (5.19) PES, out of three possible routes for the nucleophilic attack leading to the formation of an intermediate anion LXIC in a stable bisphenoid configuration, only for two there are corresponding reaction valleys. The lines along the bottoms of these valleys are strictly collinear with one of the bonds of the initial pyramidal molecule. Energetically advantageous are the reaction paths associated with a nucleophilic approach from the side opposite to the most polar bonds S—F and P—F. This behavior is dictated by the polarity rule for bisphenoid structures LXIc and LXIIc which predicts stability of the diaxial configuration of the most electronegative substituents and drastic loosening of axial bonds [175, 176]. The orientation of the attacking fluoride ion in the axial direction along the bond S—F in the compound LXI and the bond P—F in LXII (reaction path **1** in Fig. 5.13) takes place already at relatively large distances (4.0–4.5 Å) between the ion and the molecule being attacked. No reorientation occurs when they draw further together.

The calculated minimum energy paths of Eqs. (5.19) and (5.20) and the intermediate structures LXIc and LXIIc expected to be formed on these paths are compatible with the stereochemical outcome of the substitution reaction associated with the inversion of the pyramidal bond configuration at tricoordinate sulfur and phosphorus centers. Such a stereochemical outcome was obtained experimentally in reactions of various sulfinyl [177] and phosphinyl [178] derivatives.

5.5.4 Substitution at Tetracoordinate Phosphorus

Of particular interest among numerous transformations of this type are the hydrolysis reactions of the phosphoric acid esters which occupy an important place in the series of enzymic processes [179]. Belonging to the second row elements, phosphorus forms hypervalent bonds much more readily than carbon and nitrogen. This fact was one of the reasons for the assumption, first advanced by Westheimer [180], to the effect that the trigonal-bipyramidal structure LXIII emerging in the reactions of nucleophilic substitution at the tetracoordinate phosphorus represents not a transition state, as in the S_N2 reactions of derivatives of tetrahedral carbon, but rather an intermediate. Indeed, in the series of reactions:

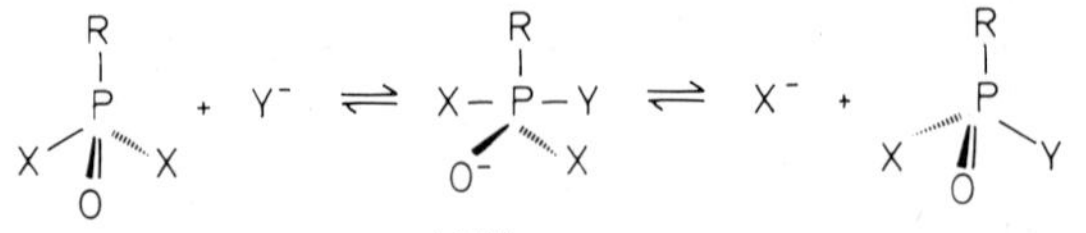

$$R-P(=O)X_2 + Y^- \rightleftharpoons X-P(R)(O^-)(X)-Y \rightleftharpoons X^- + R-P(=O)(X)Y \qquad (5.21)$$

LXIII

the cyclic [181] and also the acyclic, such as ($R = CH_3$, $X = F$, $Y = NH_2$)—see Ref. [182]—phosphoranes LXIII were observed in the NMR spectra and even isolated preparatively. Their formation in such reactions was also predicted from quantum chemical calculations in both the CNDO/2 and the STO-3G ab initio approximation [183, 184]. Among the transformations of Eq. (5.21), the reaction of basic hydrolysis of dimethyl phosphate, where $X = OCH_3$, $R = O^-$, $Y = OH^-$, was studied in the most detailed fashion:

$$(CH_3O)PO_2^- + OH^- \rightleftharpoons [(CH_3O)_2PO_3H]^2$$

LXIV

$$\rightleftharpoons CH_3OPO_3H^- + CH_3O^-$$

At the stages of the formation and decomposition of the pentacoordinate intermediate LXIV, calculations reveal the transition states which link the reactants and products with the trigonalbipyramidal adduct LXIV and resemble it in structure (their chief difference being in the lengthening by about 0.1 Å of the forming and breaking bonds). The attacking nucleophile takes up the axial position and the leaving group is represented by the axial methoxyl group, which is in accord with the polarity rule.

The calculations produced quite an interesting result, namely, a sizeable stereoelectronic effect consisting in the dependence of transition state energies at both reaction stages on the conformation of the methoxyl group in LXIV. Whereas the initial *gauche-gauche-* and *gauche-trans-*conformations of the dimethyl phosphate anion are energetically nearly equal, the energies of the transition states that correspond to the following three conformations of intermediate phosphorane LXIV differ appreciably:

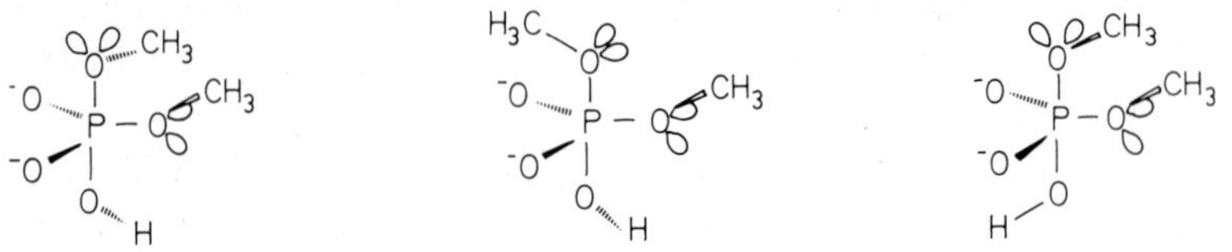

LXIVa (g,g,-g) LXIVb (t,g,-g) LXIVc (g,t,t)

At the stage of elimination of the leaving methoxyl group from the intermediate structure LXIV, the most significant effect for the transition state manifests itself: the *gauche*-conformation of the equatorial methoxyl group in LXIVa and LXIVb stabilizes this state at approximately 11 kcal/mol and greatly facilitates the departure of the axial methoxyl group. It is not hard to see a complete analogy between this stereoelctronic effect and that examined in Sect. 5.3.3.2 for the decomposition of tetrahedral adducts in the reactions of carbonyl compounds. The equatorial methoxyl group has in the *gauche*-conformation one electron pair which is antiperiplanar relative to the bond C—O, breaking upon departure of the axial methoxyl group, and loosens this bond in accordance with the orbital interaction mechanism XLII. In the case of a *trans*-conformation of the equatorial methoxyl group in LXIVc, there are no lone pairs antiperiplanar with respect to this bond. It may be assumed that the considerable acceleration of hydrolysis reactions of cyclic phosphates that involve cleavage of the axial bonds C—O in the five-membered cycle where the structural conditions LXIV are realized could be accounted for by the stereo electronic effect just described.

5.5.5 Substitution at Pentacoordinate Phosphorus

Even though the nucleophilic subtitution reactions in the phosphoranes series have been studied experimentally fairly well [181, 185], practically no data on their stereochemistry were reported. This question was addressed in the semiempirical CNDO/2 and MINDO/3 calculations [186] using as model examples a series of simple phosphoranes attacked by a hydride ion:

$$H^- + PH_3R_{(1)}R_{(2)} \rightleftharpoons \underset{\text{LXV}}{[PH_4R_{(1)}R_{(2)}]^-} \rightleftharpoons \underset{R_{(1)},R_{(2)}=CH_3,H}{H^- + PH_3R_{(1)}R_{(2)}} \tag{5.22}$$

Figure 5.14 presents a snapshot sequence of the approach of the hydride ion along the MERP to the phosphorane PH_5. As the reaction coordinate, the distance was taken between the attacking nucleophile and the phosphorus atom. The rest of the geometry parameters were optimized. As is evident from Fig. 5.14, the hydride ion, already at quite large distances from the center being attacked, is oriented in the equatorial plane of the trigonal bipyramidal phosphorane molecule. Drawing together of the interacting reagents proceeds strictly along the bisecting line of the angle between the equatorial bonds. The reaction runs without any barrier with gradual lowering of the total energy of the supermolecule and leads to the formation of the octahedral intermediate ion PH_6^-. The decisive geometrical restructuring of the system starts in the reaction zone ($R \approx 2.0$–2.2 Å). At earlier reaction stages, a week distortion begins showing up of D_{3h} structure of the PH_5 molecule toward C_{2v}. It is consistent

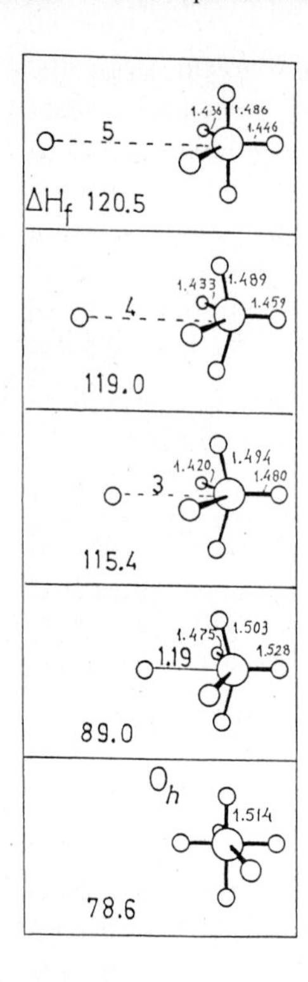

Fig. 5.14. MERP of Eq. (7.22), $R_1 = R_2 = H$. Heats of formation ΔH_f of the supermolecule are given in kcal/mol for several distances R in Å units between the hydride-ion and PH_5 molecule being attacked [186]

with the low-frequency deformational vibration e′ characteristic of the trigonal-bipyramidal structures.

The calculated MERP of nucleophilic addition may be explained when examining the structure of the lowest-vacant MO of phosphorane PH_5. With D_{3h} symmetry of the molecule, this orbital is fully localized in the equatorial plane of LXVI, which agrees with the results of ab initio calculations on PH_5 [187]. Clearly, the optimal overlapping of the occupied orbital of the nucleophile H^- and the vacant *p*-orbital of phosphorus in the lowest vacant MO of PH_5 will be achieved precisely on the approach route shown in Fig. 5.14. Increased overlapping is aided by the $D_{3h} \rightarrow C_{2v}$ distortion of the PH_5 molecule near the reaction zone for LXVII:

LXVI, e′ LXVII, a_1 LXVIII, e′

The calculations [186] have shown that in the case of substituted phosphoranes favored would be an approach of the nucleophile from the inside of the angle $R_{(1)}PR_{(2)}$ made by the most electropositive groups $R_{(1)}$ and $R_{(2)}$ of the initial molecule. This result will be understood after examining the form of the lowest-vacant MO of phosphorane $PH_3(CH_3)_2$ LXVIII that has the greatest amplitude precisely in the realized (Fig. 5.14) direction of approach of the hydride ion.

The general conclusions on the mechanism of associative nucleophilic substitution at pentacoordinate phosphorus come down to two points

1) The first step of the substitution reaction consists in adding the nucleophile to form a stable hexacoordinate phosphorus ion; owing to the equatorial approach of the nucleophile to the central atom, a *cis*-position of the most electronegative groups is realized in the intermediate octahedral ion.
2) Departure of the leaving electronegative group takes place in the *cis*-configuration. In the foregoing examples of Eq. (5.22), these groups were the hydrogen atoms compared with which the methyl groups are electropositive. By extrapolating these statements to a general case, a comprehensive scheme may be set up.

The most electronegative ligand X occupies an apical position. The rest of the ligands, less apicophilic than X or $X_{(1)}$, are numbered with 1, 2 designating the most electropositive groups. The symbols for the stereoisomers (pairs of apical ligands are indicated) are given in accordance with the nomenclature for the trigonal-bipyramidal systems [181]:

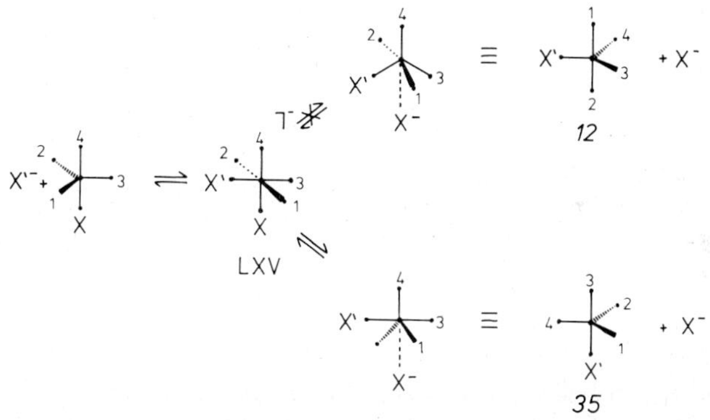

(5.23)

Once the octahedral intermediate LXV with a *cis*-configuration of the leaving and entering groups is formed, two variants of a *cis*-departure of X are possible which are stereochemically not equivalent. Only one of these leads to the formation of the stable structure [35] with the electronegative group $X_{(1)}$ in apical position. Between this structure and the structure [12], a polytopal rearrangement is possible due to a single Berry pseudorotation (4 is the pivot ligand). The energy barrier of this pseudorotation may, at sufficiently high apicophility of $X_{(1)}$, reach 15–20 kcal/mol, even in the absence of steric hindrances. Consequently, this secondary process cannot determine the stereochemical outcome of Eq. (5.23).

Thus, stereochemistry of elimination of a nucleophilic group from the octahedral intermediate is governed by energy factors which determine relative stability of various topomers of the trigonal-bipyramidal structures and can be described by the polarity rule. As concerns the initial octahedral intermediate, the *cis*-elimination of the leaving group may be regarded as a manifestation of the known *cis*-effect of the ligands in octahedral structures of the nontransition elements [188]. The following scheme explains quite well high mobility of the equilibrium which is shifted in solution to the left and in crystalline state to the right.

LXIX ⇆ LXIXa

That this equilibrium associated with a six-fold degenerate rearrangement is established, is the result of an orientation of the nucleophilic center highly favorable for nucleophilic attack upon the pentacoordinate phosphorus atom. The molecular geometry of spirophosphorane LXIX fixes the anionic center with respect to pentacoordinate phosphorus immediately on the MERP of the nucleophilic addition (see Fig. 5.14). The reaction is also facilitated by the fact that the oxygen center of the monocoordinate bidentate ligand is less electronegative than that of the dicoordinate ligand [189].

5.5.6 Inclusion of the Polytopal Rearrangements of Intermediates in the Overall Reaction Scheme

The calculation results presented in Sects. 5.5.1–5.5.5 bear witness to the formation of stable intermediates in reactions of associative nucleophilic substitution. In case these intermediates contain as a central atom the second and lower row elements, their structures are mainly the derivatives of a trigonal bipyramid structure and are capable of low-energy barrier polytopal rearrangements. Earlier we discussed the detailed physical mechanism of such rearrangements that lead to various permutations of the ligands linked with the central atom for the case of the tricoordinate T-shaped and tetracoordinate bisphenoidal structures (see Sect. 1.3.3.2) and for the pentacoordinate trigonal-bipyramidal structures (Sect. 1.3.5.3) The stages of polytopal rearrangements of intermediates are particularly important for a characterization of the stereochemical course of reactions. In cyclic system, owing to steric restraints, the requirements of the principle of microscopic reversibility, i.e., the necessity of axial approach of the entering and axial departure of the leaving group, cannot be satisfied without inclusion of the stage of polytopal rearrangement. A generalized formulation of this principle [180] states that the necessity to meet the above requirements can contain also the condition that they be satisfied at different stages of reaction.

Such reactions, referred to as the reactions with the AdRE mechanism (addition—rearrangement—elimination), are particularly characteristic of intramolecular rearrangements of both the reversible [190, 191] and the irreversible [180, 181] type. The most important stereochemical consequence of the reactions with this mechanism consists in the retention of configuration of the central atom at which there occurs a substitution rather than an inversion of configuration as would be the case with the concerted mechanism. A particularly elegant confirmation of this conclusion is given by the following rearrangement [192]:

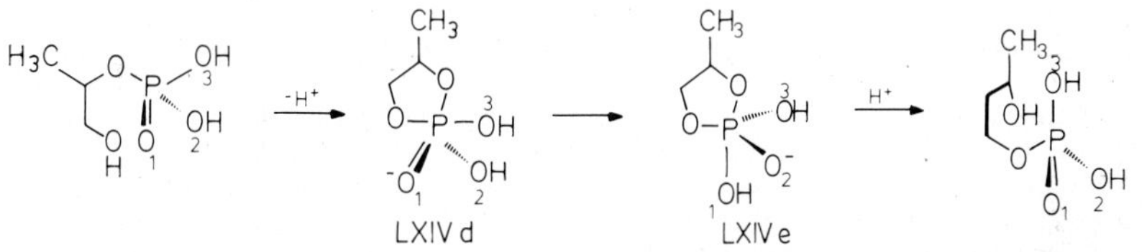

$-H^+$ → LXIV d → LXIV e $\xrightarrow{H^+}$ (5.24)

The oxygen atoms $O_{(1)}$–$O_{(3)}$ in the starting optically active compound were labelled with ^{16}O, ^{17}O, ^{18}O isotopes. No inclusion of the isotope ^{18}O from the solvent $H_2{}^{18}O$ occurred in the course of the reaction. Complete retention of stereochemical configuration of the phosphorus atom was observed after the rearrangement of Eq. (5.24).

References

1. Ingold CK (1969) Structure and mechanism in organic chemistry. Cornell Univ Press, Ithaca, New York
2. Lowry TH, Richardson KS (1981) Mechanism and theory in organic chemistry. Harper & Row, New York
3. Breslow R (1965) Organic reaction mechanisms. Benjamin, New York
4. Dneprovskii AS, Temnikova TI (1979) Theoretical foundations of organic chemistry (in Russian). Khimia, Leningrad
5. Woodward RB, Hoffmann R (1970) The conservation of orbital symmetry. VCH, Weinheim
6. Guthrie RD (1975) J Org Chem 40:402
7. Roberts DC (1978) J Org Chem 43:1473
8. Dewar MJS, Dougherty RC (1975) The PMO theory of organic chemistry. Plenum Press, New York
9. Yamaguchi K (1982) Intern J Quant Chem 22:459
10. Bunton CA (1963) Nucleophilic substitution at a saturated carbon atom. Elsevier, New York
11. Streitwieser A (1962) Solvolytic displacement reactions. McGraw-Hill, New York
12. Beletskaya IP (1975) Usp Khim (Russ Chem Rev) 44:2205
13. Walden P (1985) Ber 28:1287, 2766; (1987) Ber 30:3146
14. Cremaschi P, Gamba A, Simonetta M (1972) Theor Chim Acta 25:237
15. Dannenberg JJ (1976) Angew Chem Intern Ed Engl 15:519; (1976) J ACS 98:6261
16. Minkin VI, Kletskii ME, Minyaev RM, Simkin BY, Pichko VA (1983) Zh Org Khim 19:9
17. Carrion F, Dewar MJS (1984) J ACS 106:3533

18. Viers JW, Schug JC, Stovall MD, Seeman JI (1984) J Comput Chem 6:598
19. Ishida K, Morokuma K, Komornicki A (1977) J Chem Phys 66:2153
20. Bantle S, Ahlrichs R (1978) Chem Phys Lett 53:148
21. Dedieu A, Veillard A (1972) J ACS 94:6730
22. Keil F, Ahlrichs R (1976) J ACS 98:4787
23. Alagona G, Ghio C, Tomasi J (1981) Theor Chim Acta 60:79
24. Schlegel HB, Mislow K, Bernardi F, Bottoni A (1977) Theor Chim Acta 44:245
25. Bader RFW, Duke AJ, Messer RR (1973) J ACS 95:7715
26. Wolfe S, Mitchell DJ, Schlegel HB (1981) J ACS 103:7692
27. Müller K, Brown LD (1979) Theor Chim Acta 53:75
28. Ritchie CD, Chappell GA (1970) J ACS 92:1819
29. Dedieu A, Veillard A (1979) in: Quantum theory of chemical reactions. Daudel R, Pullman A, Salem L, Veillard A (eds) Reidel, Dordecht, 1:69
30. Basilevsky MV, Koldobsky SG, Tikhomirov VA (1986) Uspekhi Khimii (Russ Chem Rev) 55:1667
31. Westheimer FH, Haake PC (1961) J ACS 83:1101
32. Müller K (1980) Angew Chem Intern Ed Engl 19:1
33. Bürgi H-B (1975) Angew Chem Intern Ed Engl 14:460
34. Mitchel DJ (1981) Theoretical aspects of S_N2 reactions. PhD Thesis. Kingston, Queens Univ, p 167
35. Ugi I, Marquarding D, Klusacek H, Gillespie P (1971) Acc Chem Res 4:288
36. Basilevsky MV, Koldobsky SG, Tikhomirov VA (1982) Zh Org Khim 18:917
37. Dewar MJS (1978) in: Further perspectives in organic chemistry. Elsevier, Amsterdam, p 107
38. Yamabe S, Ihira N, Hirao K (1982) Chem Phys Lett 92–172
39. Riveros JM, Breda AC, Blair LK (1973) J ACS 95–4066
40. Dougherty R, Dalton J. Roberts JD (1974) Org Mass Spectrom 8:77, 81
41. Riveros JM, Jose SM, Takashima K (1986) Adv Phys Org Chem 21–197
42. Lieder CA, Brauman JI (1974) J ACS 96:4028.
43. Brauman JI, Olmstead WN, Lieder CA (1974) J ACS 96:4030
44. Olmstead WN, Brauman JI (1977) J ACS 99:4219
45. Pellerite MJ, Brauman JI (1980) J ACS 102:5993; (1983) J ACS 105:2672
46. Karlström G, Engström S, Jönsson B (1978) Chem Phys Lett 57:390
47. Bohm DK, Mackay GI Payzant JD (1974) J ACS 96:4027
48. Marcus RA (1968) J Phys Chem 72:891; (1964) Ann Rev Phys Chem 15:155
49. Albery WJ (1979) Pure Appl Chem 51:949
50. Shaik SS, Schlegel HB, Wolfe S (1988) JCS Chem Commun:1322
51. Reichardt C (1969) Lösungsmittel-Effekte in der organischen Chemie. VCH Weinheim
52. Simkin BY, Sheikhet II (1989) Quantum chemical and statistical theories of solutions. Computational methods and their applications (in Russian) Khimia Moscow
53. Morokuma K (1982) J ACS 104:3732
54. Bohme DK, Mackay GI (1981) J ACS 103:978
55. Chandrasekhar J, Smith SF, Jorgensen WL (1985) J ACS 107:154
56. Chandrasekhar J, Jorgensen WL (1985) J ACS 107:2974
57. Dewar MJS, Storch DM (1985) JCS Chem Commun:94
58. El Gomati T, Lenoir D, Ugi I (1975) Angew Chem Intern Ed Engl 14:59
59. Maryanoff GA, Ogura F, Mislow K (1975) Tetrahedron Lett:4095
60. Vergnani T, Karpf M, Hoesch L, Dreiding AS (1975) Helv Chim Acta 58:2524
61. Anh NT, Minot C (1980) J ACS 102:103
62. Minot C (1981) Nouv J Chim 5:319
63. Pearson J (1971) Usp Khim (Russ Chem Rev) 40:1259
64. Stohrer W-D (1974) Chem Ber 107:1795; (1976) 109:285
65. Cremaschi P, Simonetta M (1976) Chem Phys Lett 44:70
66. Corriu RJP, Guerin C (1980) J Organometal Chem 198:231
67. Demontis P, Fois E, Gamba A (1983) THEOCHEM 93:231
68. Cremaschi P, Gamba A, Simonetta M (1977) J Chem Soc Perkin Trans II:162
69. Demontis P, Fois E, Gamba A, Manunza B, Suffritti GB, Simonetta M (1982) J Chem Soc Perkin Trans II:783
70. Simonetta M (1984) Chem Soc Rev 13:1

71. McGarrity JF, Smyth T (1980) J ACS 102:7303
72. Reutov OA, Beletskaya IP, Sokolov VI (1972) Reaction mechanisms of organometallic compounds. Khimia Moscow
73. Beletskaya IP (1976) in: Mechanisms of heterolytic reactions Nauka, Moscow, p 5
74. Reutov OA (1975) J Organometal Chem 100:219 (1978) Tetrahedron 34:2827
75. Dyczmons V, Kutzelnigg W (1974) Theor Chim Acta 33:239
76. Minkin VI, Minyaev RM, Zhdanov YA (1987) Nonclassical structures of organic compounds. Chap 6, Mir Publ, Moscow
77. Olah GA, Surya Prakash GK, Williams RE, Field LD, Wade K (1987) Hypercarbon chemistry. Chap 5, Wiley, New York
78. Pople JA (1982) Ber Bunsenges Phys Chem 86:806
79. Hehre WJ, Radom L, Schleyer PR, Pople JA (1986) Ab initio molecular orbital theory. Chap 7, Wiley, New York
80. Jemmis ED, Chandrasekhar J, Schleyer PR (1979) J ACS 101:527
81. Schleyer PR, Apeloig Y, Arad D, Luke BT, Pople JA (1983) Chem Phys Lett 95:477
82. Clark T, Schleyer PR, Pople JA (1978) Chem Commun:137
83. Cremashi P, Simonetta M (1975) Theor Chim Acta 37:341
84. Jenks WP (1969) Catalysis in chemistry and enzymology. McGraw-Hill, New York
85. Bruice TC, Benkovic SJ (1966) Bioorganic mechanisms. Benjamin, New York
86. Capon B, Ghosh AK, Grieva DMA (1981) Acc Chem Res 14:306
87. Bowie JH, Williams BD (1974) Australian J Chem 30:795
88. Asubiojo OI, Blair LK, Brauman JI (1975) J ACS 97:6685
89. Takashima K, José SM, do Amaral AT, Riveros JM (1987) JCS Chem Commun:1255
90. Yamabe S, Minato T (1983) J Org Chem 48:2972
91. Bowie JH (1980) Acc Chem Res 13:76
92. Bohme DK, Mackay GI, Tanner SD (1980) J ACS 102:407
93. Bürgi HB, Dunitz JD, Lehn JM, Wipff G (1974) Tetrahedron 30:1563
94. Williams IH, Maggiora GM, Schowen RL (1980) J ACS 27:7831
95. Kletskii ME, Minyaev RM, Minkin VI (1980) Zh Org Khim 16:686
96. Madura JD, Jorgensen WL (1986) J ACS 108:2517
97. Brustein KY, Khurgin YuI (1977) Izvest Akad Nauk SSSR (ser khim) :1490
98. Scheiner S, Lipskomb WN, Kleier DA (1976) J ACS 98:4770
99. Weiner SJ, Singh UC, Kollman PA (1985) J ACS 107:2219
100. Yamataka H, Nagase S, Ando T, Hanafusa T (1986) J ACS 108:601
101. Asubiojo OI, Brauman JI (1979) J ACS 101:3715
102. Williams IH, Spangler D, Femec DA, Maggiora GM, Schowen RJ (1983) J ACS 105:31; (1980) J ACS 102:6619
103. Williams IH (1987) J ACS 109:6299
104. Williams IH, Spangler D, Maggiora GM, Schowen RL (1985) J ACS 107:7717
105. Litvinenko LM, Oleynik NM (1981) Organic catalysts and homogeneous catalysis (in Russian) Naukova Dumka, Kiev
106. Jencks WP (1976) Acc Chem Res 9:425; (1980) 13:161
107. Williams IH, Maggiora GM (1982) THEOCHEM 89:365
108. Burstein KY, Isaev AI (1985) Izvest Akad Nauk SSSR (ser khim) :1066; (1985) Zh Strukt Khim 26:16
109. Oie T, Loew GH, Burt SK, Binkley JS, MacElroy RD (1982) Int J Quant Chem: Quantum Biol Symp 9:224; (1982) J ACS 104:6169
110. Bertran J (1983) J Mol Struct, THEOCHEM 93:129
111. Volkenstein MV, Golovanov IB, Sobolev VI (1982) Molecular Orbitals in enzymology. Nauka, Moscow
112. Bader RFW, MacDougall PJ, Lan CD (1984) J ACS 106:1594
113. Anh NT, Eisenstein O (1977) Nouv J Chim 1:61
114. Liotta CL, Burgess EM, Eberhardt WH (1984) J ACS 106:4849
115. Bachrach SM, Streitwieser A (1986) J ACS 108:3946
116. Houk KN, Paddon-Row MN, Rondan NG, Wu YD, Brown FK, Spellweyer DC, Metz JT, Li Y, Loncharich (1986) Science 231:1108
117. Storm DR, Koshland DE (1972) J ACS 94:5805, 5814
118. Menger FM (1983) Tetrahedron 39:1013; (1985) Acc Chem Res 18:128

119. Dorigo AE, Houk KN (1987) J ACS 109:3698
120. Kirby AJ (1983) The anomeric effect and related stereoelectronic effects at oxygen. Springer-Verlag, Berlin-Heidelberg-New York
121. Deslongchamps P (1983) Stereoelectronic effects in organic chemistry. Pergamon Press, New York
122. Wolfe S (1972) Acc Chem Res 5:102
123. Lehn JM, Wipff G (1978) Helv Chim Acta 61:1274
124. Koptyug VA (1976) Zh Vsesoyzn Khim Obszh im DI Mendeleeva 21:247
125. Schofield K (1980) Aromatic nitration. Cambridge Univ Press, London
126. Coombes RG, Moodie RB, Schofield K (1968) J Chem Soc Sec B:800
127. Farcasiu D (1982) Acc Chem Res 15:46
128. Pfeiffer P, Wizinger R (1928) Liebigs Ann 461:132
129. Koptyug VA (1984) Topics Current Chem v 122
130. Morrison DJ, Stanney K, Tedder JM (1981) J Chem Soc Perkin Trans 11:838, 967
131. Cacace F (1982) J Chem Soc Perkin Trans II :1129
132. Miller DL, Lay JO, Gross ML (1982) Chem Commun:970
133. Reents WD, Freiser BS (1980) J ACS 102:271
134. Lehman TA, Bursey MM (1976) Ion cyclotron resonance spectrometry. Wiley, New York
135. Akulov GP (1976) Usp Khim (Russ Chem Rev) 45:1970
136. Hehre WJ (1977) in: Schaefer H (ed) Applications of electronic structure theory. Plenum Press, New York, 4:277
137. Minkin VI, Minyaev RM, Yudilevich IA, Kletskii ME (1985) Zh Org Khim 21:926
138. Radhavachari K, Reents WD, Haddon RC (1986) J Comput Chem 7:265
139. Politzer P, Jayasuriya K, Sjoberg P, Laurence PR (1985) J ACS 107:1174
140. Feng J, Zheng X, Zerner MC (1986) J Org Chem 51:4531
141. Gleghorn JT, Torossian G (1987) J Chem Soc Perkin Trans II :14303
142. Borodkhin GI, Nagy SM, Gatilov YV, Mamatyuk VI, Mudrakovskii IL, Shubin VG (1986) Doklady Akad Nauk SSSR 288:1364
143. Olah GA (1971) Acc Chem Res 4:240
144. Eberson L, Radner F (1987) Acc Chem Res 20:53
145. Todres ZV (1978) Uspekhi Khim (Russ Chem Rev) 47:260; (1985) Tetrahedron 41:2771
146. Nagakura S, Tanaka J (1959) Bull Chem Soc Jpn 32:734
147. Pedersen EB, Petersen TE, Torssell K, Lawesson SO (1973) Tetrahedron 29:579
148. Eberson L, Nyberg K (1976) Adv Phys Org Chem 12:1
149. Perrin CL (1977) J ACS 99:5516
150. Clemens AH, Ridd JH, Sandall JPB (1984) J Chem Soc Perkin Trans II :1659, 1667; (1985) J Chem Soc Perkin Trans II :1225
151. Schmitt RJ, Ross DS, Buttrill SE (1981) J ACS 103:5265; (1984) J ACS 106:926
152. Epiotis ND (1983) Pure Appl Chem 55:229
153. Takabe T, Takenaka K, Yamaguchi K (1976) Chem Phys Lett 44:65
154. Abronin IA, Gorb LG, Litvinov VP (1983) in: Advances in chemistry of nitrogen-containing heterocycles. Rostov-on-Don
155. Santiago C, Houk KN, Perrin CL (1979) J ACS 101:1337
156. Ritchie CD (1972) Acc Chem Res 5:348; (1978) Pure Appl Chem 50:1281
157. Pross A (1976) J ACS 98:776
158. Sordo T, Arumi M, Bertran J (1981) J Chem Soc Perkin Trans II:708
159. Sordo T (1982) Chem Phys Lett 85:225
160. Cacace F, Giacomello P (1977) J ACS 99:5477
161. Patai S (ed) (1982) The chemistry of amino, nitroso and nitro compounds and their derivatives. Suppl F, Wiley, New York
162. Akiba K, Inamoto N (1977) Heterocycles 7:1131
163. Kletskii ME (1981) PhD Thesis. Rostov-on-Don
164. Block E (1981) Reactions of organosulfur compounds. Academic Press, New York
165. Allison WS (1976) Acc Chem Res 9:293
166. Zefirov NS, Makhonkov DI (1982) Chem Rev 82:615
167. Minkin VI, Minyaev RM (1977) Zh Org Khim 13:1129
168. Rosenfield RE, Parthasarathy R, Dunitz JD (1977) J ACS 99:4860
169. Pappas JA (1977) J ACS 99:2926

170. Lau PHW, Martin JC (1978) J ACS 100:7077
171. Arduengo AJ, Burgess EM (1977) J ACS 99:2376
172. Prelog W (1969) Usp Khim (Russ Chem Rev) 38:952
173. Raban M, Noyd DA, Berman L (1975) J Org Chem 40:752
174. Minyaev RM, Minkin VI, Kletskii ME (1978) Zh Org Khim 14:449
175. Minkin VI, Minyaev RM (1975) Zh Org Khim 11:1993; (1977) Zh Strukt Khim 18:274
176. Chen MML, Hoffmann R (1976) J ACS 98:1647
177. Tillet JG (1976) Chem Rev 76:747
178. Kyba EP (1976) J ACS 98:4805
179. Frey PA (1982) Tetrahedron 38:1541
180. Westheimer FH (1968) Acc Chem Res 1:70
181. Luckenbach R (1973) Dynamic stereochemistry of pentacoordinated phosphorus and related elements. Thieme Stuttgart
182. Granoth I, Segall Y, Waysbort D, Shurin E, Leader H (1980) J ACS 102:4523
183. Gorenstein DG, Luxon BA, Findlay JB, Momii R (1977) J ACS 99:4170
184. Gorenstein DG, Luxon BA, Findlay JB (1979) J ACS 101:5869
185. Koenig M, Munoz A, Wolf R (1976) J Chem Soc Perkin Trans II:955
186. Minyaev RM, Minkin VI (1979) Zh Strukt Khim 20:842
187. Altmann JA, Yates K, Csizmadia IG (1983) J ACS 98:1450
188. Nefedov VI, Gofman MM (1978) Ligand interactions in inorganic compounds. Itogy nauki i teckniki, 6, Moscow
189. Pickering M, Jurado B, Springer CS (1976) J ACS 98:4503
190. Minkin VI, Olekhnovich LP, Zhdanov YA (1981) Acc Chem Res 14:210
191. Minkin VI (1989) Pure Appl Chem 61:661
192. Buchwald SL, Pliura DH, Knowles JR (1982) J ACS 104:845

Chapter 6

Addition Reactions

6.1 Electrophilic Additions to Multiple Bonds

The spectrum of mechanisms governing the reactions of electrophilic addition to multiple bonds is quite broad. The solvent, the polarity of electrophilic agent, the type and conformation of substituents at a multiple bond and its polarity all have a substantial effect on kinetics and stereochemistry of the reactions in question [1–3]. The most general and important features of various mechanisms are summarized in the following scheme:

$$\mathrm{C{=}C + XY \rightarrow \underset{I}{[C{=}C\cdot XY]} \xrightarrow{(1)} \underset{IIa}{\overset{X^{+}}{C{-}C}\;Y^-} \rightleftharpoons \underset{IIb}{XC{-}C^{+}\;Y^-} \rightarrow XC{-}CY;\quad I \xrightarrow{(2)} XC{-}CY} \qquad (6.1)$$

Two principal mechanisms govern the transformation of the initially formed adduct I into the final product: one is polar or ionic [1], it includes the slow step of formation of a bridged or open carbocation II, the other is molecular [2]. The formation of the cations II in solution and gas phase has been detected experimentally and theoretical studies of the ionic mechanism of electrophilic addition have primarily been directed towards determining the structure and relative stability of their isomeric forms. Some results of calculations are given in Table 6.1.

The data of Table 6.1 show that the inclusion of electron correlation leads to considerable stabilization of the bridged structure IIa as compared to IIb. The same effect is achieved when polarization functions are included in the AO basis set [4]. Such a trend is quite typical of the nonclassic structures, to which many π-complexes IIa belong [7], and must be taken into account for correct assessment of their relative stability. Both isomeric structures of the cations

Table 6.1. The relative energies (in kcal/mol) of isomeric cations $C_2H_4X^+$ of types IIa, IIb. Positive values indicate the more stable bridged form IIa

X	MINDO/3 [4]	MNDO [5]	4–31G [6]	CEPA [4]
H	7.9	−15.1	−7.4[b]	7.7
CH_3	6.0	−15.8	−0.5[b]	11.0 (7.9)[a]
F	31.1	−18.6	−7.4[c]	1.5
Cl	29.2	5.5	11.7	30.1[d]
NH_2	–	31.8	46.6	(43.1)[a]
OH	24.1	5.0	9.1	22.0 (20.1)[a]
SH	40.2	16.7 [4]	20.1	48.3 (5.3)[a]

[a] Experimental data are given in parentheses;
[b] Data compiled from Ref. [5];
[c] 6–31G orbital basis set [6];
[d] IIa is a stable form

correspond to the minima on the PES separated by rather low energy barriers. Figure 6.1 shows calculation data for the PES of the isomerization of the 2-fluoroethyl cation obtained by the MNDO method with localization of the transition states through mimimization of the gradient norm. The calculations have revealed an interesting feature of the MERP, namely the fact that it connects the bridged and the open forms of the cation II (X = F) via two successive transition states[1].

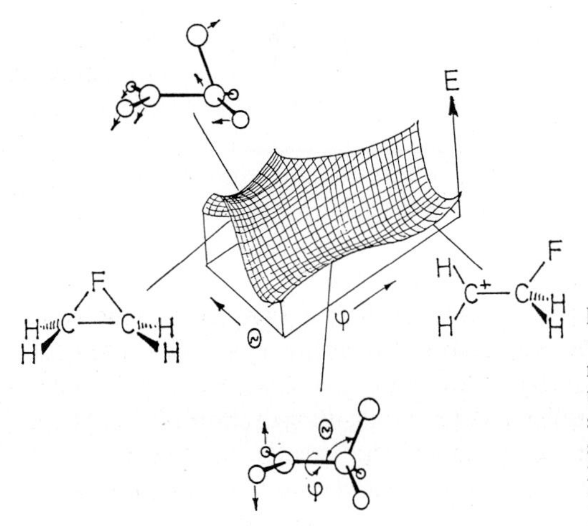

Fig. 6.1. PES of the isomerization IIa → IIb (X = F) calculated by the MNDO method and represented as a function of two angles. (Adapted from Ref. [5]) Transition vectors are shown of two transition state structures that link the isomers of the 2-fluoroethyl cation ($\Delta E = 22.5$ and 7.4 kcal/mol relative to IIb)

[1] According to the principle of structure stability established by Fernandez and Sinanoglu [8, 9] a maximum or a minimum must exist on a PES between any two saddle points linked by a gradient line. It was, however, stressed that many real PES's, whose form does not follow, for example, the condition of nondegeneracy of the second derivatives matrix, do not necessarily obey this principle [10]. For another example of two consecutive transition state structures see the case of enantiotopomerization of the [4.4.4.4] fenestrane anion [11]

From the data of Table 6.1 a conclusion may be drawn that the stabilization of the bridged cyclic form of the intermediate cation IIa is facilitated by the groups X with the central atom of the second row in the Periodic Table, and within the same row by the less electronegative groups. This fact can be explained in terms of the orbital interactions (see Sect. 4.2) by examining the role of the orbital overlap (electronegative X have contracted orbitals affording a low overlap with the π-MO of the multiple bond) and the difference between the energy of the frontier orbitals X^+ and that of the multiple bond (effect of the direct and the reverse donating)—see Refs. [5, 7].

In the molecular mechanism of addition, the step of ionization of the primary complex I is absent and Eq. (6.1) occurring as a bimolecular one leads to *cis* (*syn*) addition. The data of ab initio calculations predict the formation of stable adducts of ethylene with HF [12, 13] and HCl [14]. The structure of those adducts corresponds to the π-complex of type III.

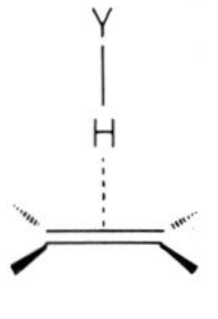

III
Y = F, Cl

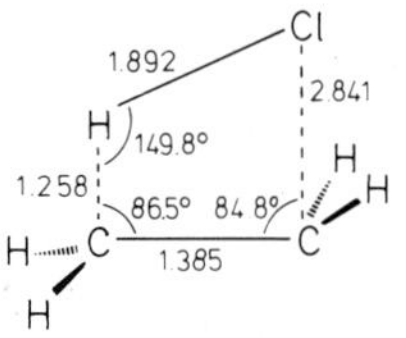

IV
ΔE = 36 kcal/mol (3-21G)

The calculated energy of stabilization of such an adduct (DZ-type basis set) is about 5 kcal/mol. Structure IV corresponds to the transition state of bimolecular addition of HCl to ethylene localized by the gradient norm minimization method.

A similar structure of the four-centered late transition state V was found in ab initio calculations [15], with the electron correlation accounted for by the CI method, on the reaction of addition of HCl to the triple bond of nitriles RCN (R = H, CH_3). The primary adduct having linear structure acquires in the course of reaction the planar L-shaped form and then, moving along the MERP strictly in a common plane (in accordance with the symmetry rules, see Sect. 1.3.4.3), transforms to V.

Unlike the four-centered transition states IV, V, the reaction of addition of HCl to silaethene proceeds, as predicted by the calculations [14], via the two-centered early transition state VI which is similar to the initial adduct.

The calculated activation energy is in good agreement with the experimental value for the gas-phase reaction (2.4 ± 0.2 kcal/mol).

It should be noted that conclusions on the structure of intermediate complexes, formed in molecular addition reactions of the type HY reactants to olefines, ought not to be directly extended to the case of reactions with halogens. Thus, the detailed ab initio 3–21G calculations [16] with a fairly complete geometry optimization show that the adduct VIIa, analogous to III, corresponds

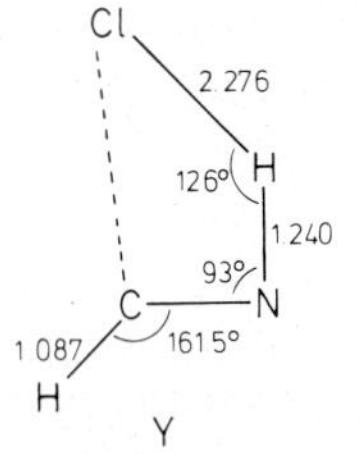

Y
$\Delta E = 59$ kcal/mol (4-31G)

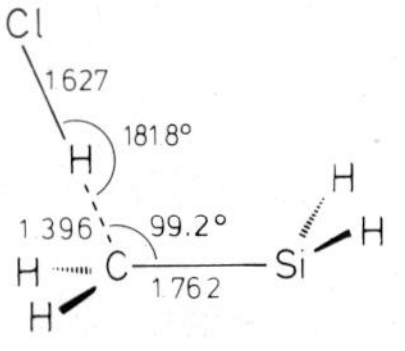

YI
$\Delta E = 3.6$ kcal/mol (3-21G)

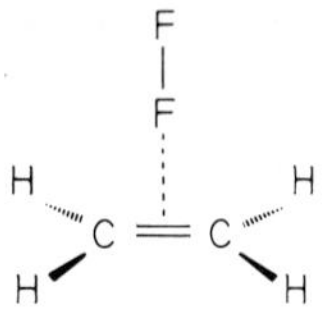

YIIa
$\Delta E = -0.5$ kcal/mol

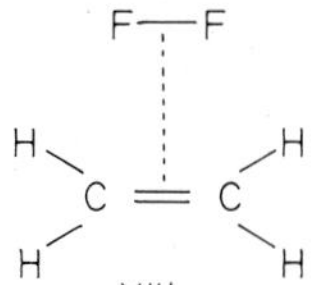

YIIb
$\Delta E = -2.0$ kcal/mol

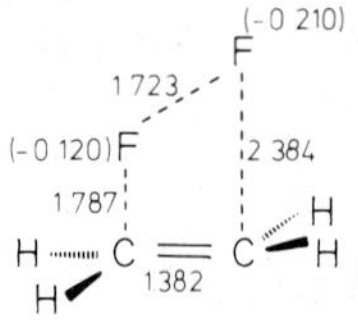

YIIc
$\Delta E = 50.1$ kcal/mol (3-21G)

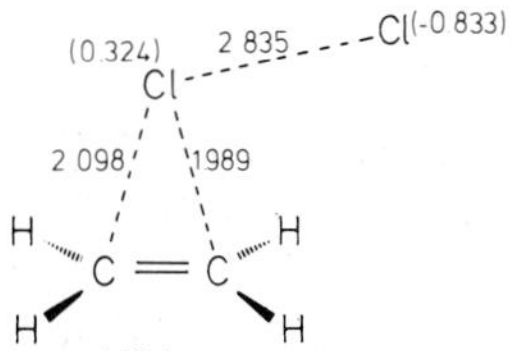

YIId
$\Delta E = 50$ kcal/mol (3-21G)

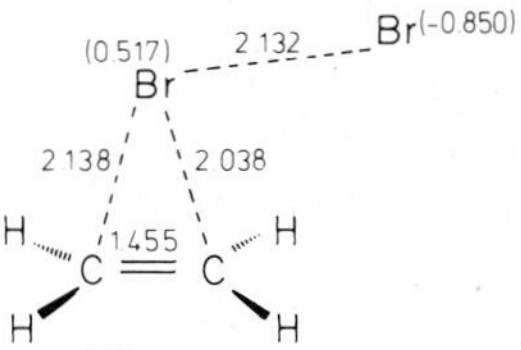

YIIe
$\Delta E = 61.4$ kcal/mol (3-21G)

to the local minimum only. The structure VIIb is the most stabilized with respect to the starting components, in it the halogen molecule lies in the plane of the ethylene molecule.

The transition state structure for the addition of molecular fluorine to ethylene VIIc was found by the 3–21G calculations [17] to closely resemble that for the addition of HCl (IV). Of special interest are the computational results obtained on transition state structures VIId, VIIe for gas phase electrophilic addition reactions of molecular chlorine and bromine to ethylene. Whereas the four-centered structure VIIc for the fluorination of ethylene indicates concerted *cis*-addition, which is in accord with experimental finding for this reaction [18], the transition state geometries for chlorination and bromination may be regarded as cyclic halonium ions backed by halogenide counter-ions. Noteworthy is that the calculations [17] predict the heterolysis to occur intrinsically without any assistance from polar solvents. The three-centered structures VIId, VIIe help to clarify the reason for *trans*-stereoselectivity of the chlorination and bromination reactions of ethylene [1, 19].

The data obtained are useful for the understanding of various details of the intrinsic methanism governing the reactions of electrophilic addition, but, on the whole, theoretical studies of the mechanism of this important class of reactions are still in their early stage. Considering the complexities of the kinetic scheme of addition reactions (inclusion of the solvent, trimolecularity) and the sometimes determinative role of the medium, one may recognize all the difficulties of such studies. The key problems confronting the researcher are the interpretation of the complex stereochemistry of the reaction whose general theoretical study has not actually begun yet, and the effect the substituents at the double bond have on the nature of an elementary act of the limiting step. For example, when electron-donating substituents at the carbon atoms of the double bond are present already at an early stage of the formation of the type II intermediate cation, the emergence of a biradical state (electron transfer to an electrophile) is possible emphasizing instability of the corresponding RHF solutions [20].

6.2 Nucleophilic Addition to Alkenes

In the nucleophilic addition, the double bond of alkene is in the limiting step attacked by the nucleophilic component Y of the reactant XY being added, rather than by the electrophilic component X, as in Eq. (6.1). For this mechanism to be sufficiently effective, Y must possess a high-lying bonding MO that can provide adequate overlap with the π^*-MO of the double bond. The free anions Y^- satisfy this condition best. The addition of Y^- to the double bond of alkene is usually the rate-determining stage of the addition-elimination (AdE) mechanism of substitution reactions at the sp^2-carbon atom [21, 22]:

(6.2)

The trajectory of nucleophilic attack upon the double bond of VIII is quite close to that of the addition of nucleophiles to the carbonyl bond. For the addition reaction of the hydride-ion (Y = H) to ethylene (R = Z = H), and propene ($R = CH_3$) the angle θ is 123° (according to the ab initio 3–21G calculations in Ref. [23]. The driving force of the reaction is the charge transfer from the electron lone pair orbital of the nucleophile Y to the π^*-orbital of olefine. The latter has, unlike the orbital π^*_{CO} (see Fig. 4.1), no loop on the α-atom of carbon and is delocalized, which diminishes the overlap integral $S_{n\pi^*}$ and thereby the energy of interaction ΔE^2 from Eq. (4.10) of the nucleophile with alkene as compared to the energy of interaction with the carbonyl compound in Eq. (5.10). In order to promote the nucleophile addition in Eq. (6.2), it is necessary to polarize the π^*-orbital of an alkene so as to increase the magnitude of the $n_Y\pi^*_{C=C}$ overlap.

The special role of this effect becomes apparent in the reactions of nucleophilic addition to the double bond of alkenes coordinated with the organometallic groups formed by transition metals. Many ethylene complexes of this type XII readily react with various nucleophiles, while ethylene VIII itself is in these reactions inert.

The groups ML_n formed by transition metals have a MO of b_2-symmetry whose energy level is very close to that of the ethylene π^*-MO (Fig. 4.4), which gives rise to the so-called back donation effect stabilizing the structure XII:

Y Y

ML_n ML_n

XII XIII $ML_n = [CpFe(CO)_2]^+$, $[CpPdPPh_3]^+$ $Fe(CO)_4$

As seen from the orbital interaction diagram in Fig. 6.2, this effect leads to widening of the energy gap between the n_Y and π^* orbitals. Moreover, admixture of the orbital b_2 diminishes the overlap between the above orbitals.

The origin of activation of nucleophilic addition in XII was explained [24] as being due to a shift of the organometallic group ML_n, in the course of the reaction or even in the initial structure, from the center of the double bond (η^2-coordination) to its end (η^1-coordination). The orbital $(\pi^* - \lambda b_2)$ in the

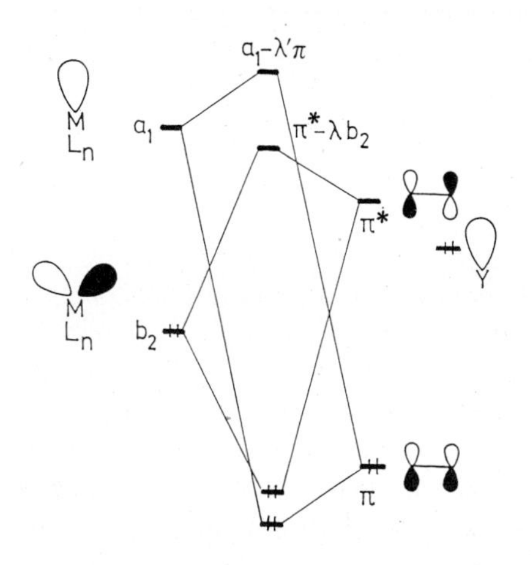

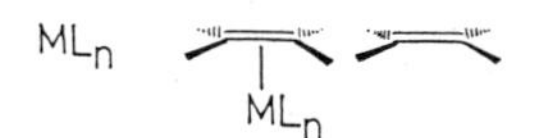

Fig. 6.2. Frontier orbitals of the η^2 complex XII according to Ref. [24]

η^1-complex XIII is polarized in such a manner that the overlap with the orbital n_Y sharply grows (XIV). For the characterization of nucleophilic reactivity an index was proposed corresponding to the value of the overlap integral $S_{n\pi^*}$ between the s-orbital of the nucleophile H^- and the π^*-orbital of the multiple bond at the distance of 2.0 Å with the angle of approach shown by XV optimized:

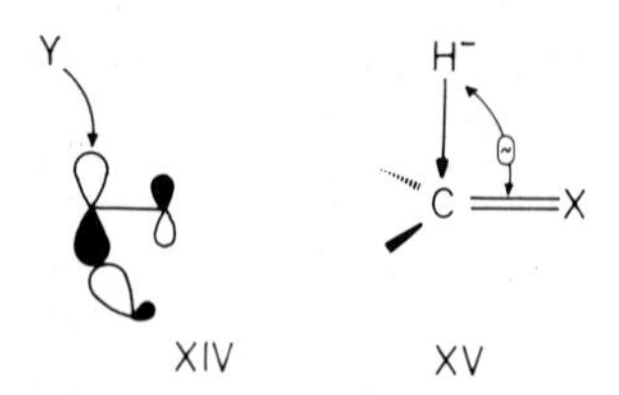

In the case of formaldehyde (X = 0), the maximal value of $S_{n\pi^*}$ is achieved at $\theta = 100°$ (cf. Fig. 5.7), for ethylene and the complexes XII and XIII this angle is 105–112°, being smaller for the η^2-complexes than for unsubstituted ethylene. Only with quite considerable shifts of the ML_n group from the center of the double bond, the value of $S_{n\pi^*}$ in XIII exceeds that for noncoordinated ethylene. The enhanced nucleophilic reactivity of the η^2-complexes XIII can be explained by considering the isolobal analogy between the organometallic and the organic groups. The group $CpFe(CO)_2^+$ is isolobal to $Fe(CO)_5^{2+}$ and CH_3^+, Hal^+, while the η^1-complex XIII is isolobal to the cation IIb. Thus, for reactions of the compounds XIII with nucleophiles favorable electronic conditions are imitated for electrophilic addition [Eq. (6.1)].

The above-considered activation mechanism for π-complexes of the XII type is, apparently, of fairly general significance. The rearrangements XII ⇄ XIII, commonly referred to as the π, σ-rearrangements, have been verified experimentally for many complexes of the transition metals ($X = ML_n$), they lie at the root of a number of processes of homogeneous catalysis [25].

The values of $S_{n\pi^*}$ are crucial for determining the regioselectivity of nucleophilic addition to the double bonds of the α-enones XVI that possess two reaction centers C_4 and C_2:

XVI XVIa XVIb

The soft nucleophiles preferably attack the C_4 atom, which can be well accounted for by greater localization of the π^*-orbital at this center in comparison with C_2. The shapes of molecular orbitals of the species XVI obtained by means of STO-3G ab initio calculations [26] help clarify the reason for which the complexation with metal cations and the protonation results in the change of the reaction site. On going to lithium complex, the localization of the π^*-MO at the carbonyl carbon increases, so that the electrophilic assistance leads to the shift of the most active reaction center from C_4 to C_2.

In case the groups R or Z in VIII are chiral, the double bond plane gets diastereotopic and the approaches of Y from one and the other side of the bond become inequivalent (asymmetrical induction)—the choice of the approach will depend on the conformation of R. For the reactions of nucleophilic addition to alkenes and carbonyl compounds the sterical models represented by the Newman projection formulas XVII are usually invoked for the rationalization of stereoselectivities:

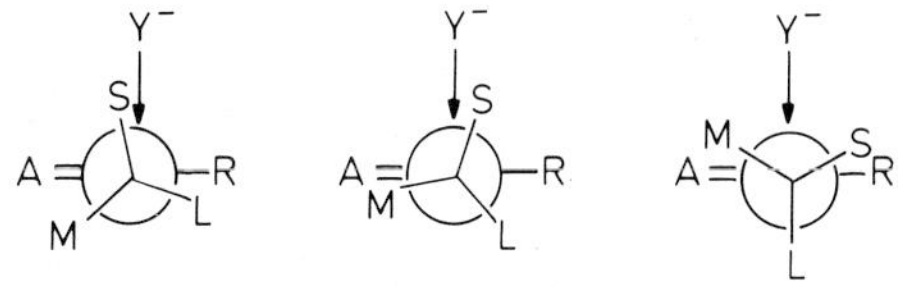

XYIIa Cram model | XYIIb Karabatsos model | XYIIc Felkin model

$A = CR_1R_2; O$

On these formulas Z = CLMS where L, M, S are the substituents of analogous type diminishing in size in the order shown (with L, M, S standing for Large, Medium, Small).

Calculations carried out using the minimal (STO-3G) [27] and the 3–21G [23, 28] basis sets point to preferability of the so-called Felkin model. Felkin [29, 30] was the first to deduce that the transition states of nucleophilic additions would have the staggered arrangement with respect to the attacking nucleophile. The reason for this stereoelectronic effect lies in the favorability of the anti-periplanar orientation of the C_2—L and of the forming C_1—Y bonds. This can be explained by purely steric factors which in electronic terms are the minimal exchange repulsion of electrons in σ-orbitals of those bonds and more favorable conditions for the overlapping of the $\pi^*_{C=A} - \sigma_{CL}$ orbitals resulting in the lowering of the $\pi^*_{C=A}$ level and thereby in the enhancement of the effect of two-electron $n_Y - \pi^*_{C=A}$ stabilization.

6.3 Nucleophilic Addition to a Triple Bond

The double CC bond of alkenes is rather inert in regard to the nucleophiles. As opposed to it, the triple bonds of alkynes readily react with water, alcohols, amines, with *trans*-addition being the favored process [31]:

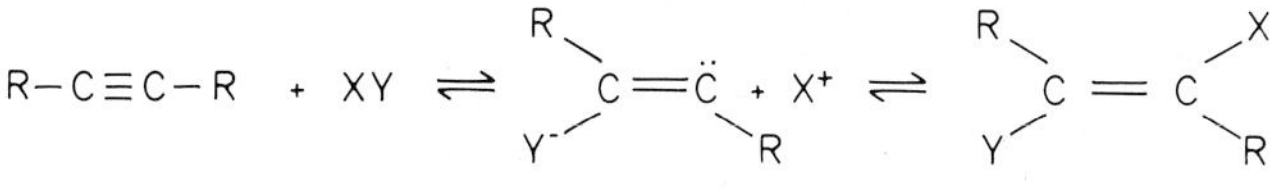

$$R{-}C{\equiv}C{-}R + XY \rightleftharpoons (R)(Y^-)C{=}\ddot{C}(R) + X^+ \rightleftharpoons (R)(Y)C{=}C(X)(R) \qquad (6.3)$$

In the generalized scheme of Eq. (6.3), the first stage, i.e., the attack by the nucleophilic center Y(Y^-) is the key step determining the stereochemistry of an addition reaction. This stage has the crucial effect on the general direction of reactions of nucleophilic substitution at the carbon atom of a triple bond. An analysis of electronic and steric demands of Eq. (6.3) may conveniently be made on the simplest model.

Several ab initio calculations on Eq. (6.4) indicate that the hydride-ion attacks the triple bond at an obtuse angle of 110–120° [32], 127–128° [33, 34]. The planar structure of the transition state XVIII [34], calculated with the 4–31G

$$H^- + H{-}C{\equiv}C{-}H \longrightarrow \left[\text{XVIII}\right]^{\ddagger} \longrightarrow \begin{matrix} H \\ H \end{matrix}\!\!>C{=}C\!\!<\begin{matrix} H \\ - \end{matrix} \qquad (6.4)$$

XVIII: H···C 2.133, ∠127°; C–C 1.224; C–H 1.046, ∠156°; C–H 1.056, ∠148°

basis set and with correlation corrections accounted for (CI 3×3), belongs to a transition state of the "early" type. An analogous transition state structure is realized when the anion F^- is the attacking nucleophile [32].

The character of deformations of the transition state structurally may readily be explained by the preferability of *trans* addition. Indeed, for a *cis* addition the stage of topomerization of the vinyl anion XIX is needed, which requires overcoming the energy berrier of the planar inversion amounting for the vinyl anion to 35 kcal/mol. The *trans* distortion of the C—H bonds exhibited by the triple bond when it is approached by a nucleophile is one more reason for the remarkable ease with which the nucleophilic addition reactions of this type proceed.

Figure 6.3 displays the changes in the shape and the energy levels of the frontier orbitals of *trans*-distorted acetylene. The geometry deformation of the XVIII type not only provides for the localization of the π^*-orbital in the direction

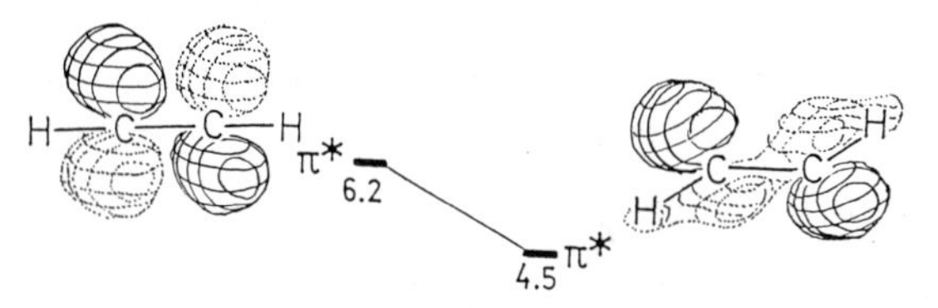

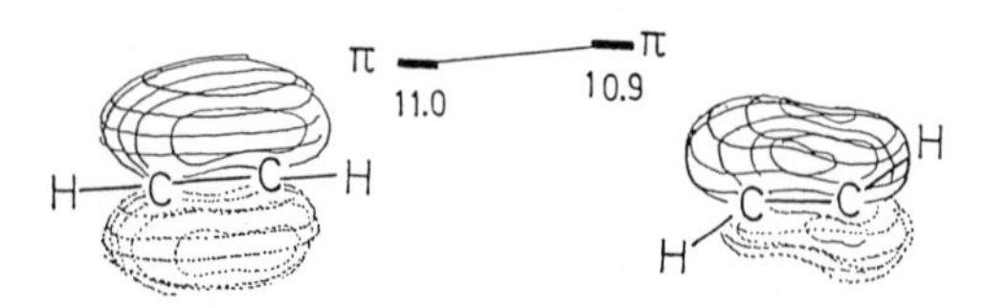

Fig. 6.3. Variation in the shape of MO's and in the energy levels (in eV) of the frontier π-orbitals on the acetylene molecule that lie in the same plane as the approaching nucleophile due to the bonds C—H being *trans*-bent as in XVIII. The data of an ab initio calculation (4–31G) for acetylene in equilibrium geometry and in the geometry XVIII [34]

of the approaching nucleophile thereby enhancing the overlap $S_{n\pi^*}$, but also effects a sharp drop in the level of the π^*-orbital bringing it down to the level of the n orbital of the nucleophile. Both effects rapidly grow with the increase in the angle of unbending of the C—H bonds [35]. At the same time, the unfavorable effect of an increase in the energy level of the occupied π-orbital of acetylene, which leads to stronger four-electron $n \rightarrow \pi^*$-repulsion, is relatively weak.

The pattern of adaptation of the orbital and energy structure of the alkyne molecule to the demands of the reaction of addition to a triple bond, shown in Fig. 6.3, is, in principle, retained also in addition reactions to heteronuclear triple bonds. This conclusion follows, for example, from the structure of the transition state V in the reaction of addition of hydrogen chloride to nitriles, which at the stage of the final product formation is characterized by the making of the bond C—Cl. The angle of nucleophilic attack is in the region of the transition structure 126°.

A deformation of the bonds of the digonal carbon atom, analogous to V and XVIII, has also been found in the reaction of nucleophilic addition of a hydride-ion to iron pentacarbonyl which gives the formyl complex XXI:

$$Fe(CO)_5 + H^- \longrightarrow \text{XX} \longrightarrow Fe(CO)_4CHO^- \quad \text{(XXI)} \qquad (6.5)$$

XX: O, C, 106°, H^{-}]‡

The ab initio calculations [36] carried out with an extended basis set point to particular importance of the two-electron $n \rightarrow \pi^*_{CO}$ interaction that stabilizes the adduct. As the energy level of the π^* orbital of the axial group CO lies lower than that of the equatorial group and the positive charge is at the axial carbon greater, the nucleophilic attack is directed against the axial center XX.

The group $Fe(CO)_4$ being isolobal to CH_2 (see Sect. 4.4), iron pentacarbonyl may be regarded as an organometallic analog of ketene $CH_2{=}C{=}O$ and Eq. (6.5) as comparable to nucleophilic addition of allene-type carbon to the digonal atom.

References

1. De la Mare PBD, Bolton R (1982) Electrophilic addition to unsaturated systems. Elsevier, Amsterdam
2. Smit V (1977) Zh Vsesoyuzn Khim Obszh im DI Mendeleeva 22:30
3. Billups WE, Houk KN, Stevens RV (1986) in: Bernasconi C (ed) Investigations of rates and mechanisms of reactions. V 6, Part I, 4/e Chap X, Wiley, New York, p 701
4. Lischka H, Köhler HJ (1979) J ACS 101:5297; (1979) Chem Phys Lett 63:326
5. Dewar MJS, Ford GP (1979) J ACS 101:783
6. Hopkinson AL, Lien MH, Csizmadia IG, Yates K (1980) Theor Chim Acta 55:1

7. Minkin VI, Minyaev RM, Zhdanov YA (1987) Nonclassical structures of organic compounds. Mir Publ Moscow
8. Fernandez A, Sinanoglu O (1984) Theor Chim Acta 66:147
9. Fernandez A (1985) Theor Chim Acta 67:229
10. Basilevsky MV, Ryaboy VM (1987) in: Veselov MG (ed) Current problems of quantum chemistry. The quantum chemical methods. The theory of intermolecular interactions and solid state. Khimia, Moscow (in Russian)
11. Minyaev RM, Minkin VI (1985) Zh Org Khim 21:1361
12. Kollman PA, Allen LC (1972) Chem Rev 72:283
13. Volkmann D, Zurawski B, Heidrich D (1982) Intern J Quant Chem 22:631
14. Nagase S, Kudo T (1983) Chem Commun :363
15. Alagona G, Tomasi J (1983) THEOCHEM 91:263
16. Toyonago B, Peterson MR, Schmid GH, Csizmadia I (1983) THEOCHEM 94:363
17. Yamabe S, Minato T, Inagaki S (1988) Chem Commun:532
18. Rozen S, Brand M (1986) J Org Chem 51:3607
19. Freeman F (1975) Chem Rev 75:439
20. Yamaguchi K (1982) Intern J Quant Chem 22:459
21. Rybinskaya MI (1967) Zh Vsesoyuzn Khim Obszh im DI Mendeleeva 12:11
22. Modena G (1971) Acc Chem Res 4:73
23. Paddon-Row MN, Rondon NG, Houk KN (1982) J ACS 104:7162
24. Eisenstein O, Hoffmann R (1980) J ACS 102:6148; (1981) J ACS 103:4308
25. Nakamura A, Tsutsui M (1980) Principles and applications of homogenous catalysis. Wiley, New York
26. Lefour JM, Loupy A (1978) Tetrahedron 34:2597
27. Anh NT, Eisenstein O (1977) Nouv J Chim 1:61
28. Houk KN, Wu Y (1987) in: Bartmann A, Sharpless KB (eds) Stereochemistry of organic and bioorganic transformations. VCH, Weinheim, p 247
29. Cherest M, Felkin H, Prudent N (1968) Tetrahedron Lett :2199
30. Cherest M, Felkin H (1968) Tetrahedron Lett :2205
31. Viehe HG (ed) (1969) Chemistry of acetylenes. Marcel Dekker, New York
32. Eisenstein O, Proctor G, Dunitz JD (1978) Helv Chim Acta 61:1538
33. Dykstra CE, Arduengo AJ, Fukunaga T (1978) J ACS 100:6007
34. Strozier RW, Caramella P, Houk KN (1979) J ACS 101:1340
35. Hoffmann DM, Hoffmann R, Fisel CR (1982) J ACS 104:3858
36. Nakamura S, Dedieu A (1982) Theor Chim Acta 61:587

CHAPTER 7

Low-Energy Barrier Reactions. Structural Modelling

7.1 The Principle of Correspondence Between Structures of the Initial and the Transition State of Reaction

Each type of the addition and substitution reactions dealt with in the preceding chapters is characterized by a distinct type of the inherent transition state structure as well as the optimal trajectory for drawing together of the reactants in the zone of its formation. Usual structural changes, such as introduction of substituents, do not substantially alter the configuration of interacting centers in the neighborhood of a transition state, provided that the general type of the reaction mechanism stays unchanged. This principle holds also for intramolecular reactions. Since in this case interacting fragments are contained within one and the same molecule, steric hindrances may arise for certain orientations of these even in spite of the possible closeness of reacting centers. Whenever conformations which topologically correspond to transition state structures fall into such a sterically inaccessible region, the reaction cannot proceed intramolecularly. First this was stated by Eschenmoser et al. [1] when studying the 1,5-methyl rearrangements:

SO_2, O, CH_3, CH^-, SO_2Ar → SO_3^-, CH_3, CH, SO_2Ar

(7.1)

SO_2, O, CH_3, S, CH_3 → SO_3^-, CH_3, $\overset{+}{S}$, CH_3

Despite close proximity of all the centers of the reactive site, trigonal-bipyramidal configuration of type II (see Chap. 5) with diaxial position of the entering and the leaving groups cannot be achieved in the six-membered transition-state structure. Equation (7.1) proceed exclusively by intermolecular route.

Taking into account topological structures of transition states and characteristics of the reaction trajectories related to the MERP's, a general principle may be formulated of selection of the low-energy barrier intramolecular reactions according to the geometry and fluxionality of the initial structure [2–4]. In order to be capable of fast intramolecular rearrangements, the molecule, in its ground state, must be rigidly fixed in a structure similar in its geometrical characteristics to those of the transition state of the reaction or display a high conformational (polytopal) flexibility permitting it to take on such a structure without overcoming significant energy barriers. In case these conditions are met, the rearrangement can be strongly affected by a proper selection of the substituents attached to the reacting fragments.

On this basis, structural relationships have been established for several types of reaction mechanisms, which allow one to predict whether a rearrangement can occur in an intramolecular way and to assess the ease with which a reaction may develop depending on the initial structure of the compound being rearranged. Let us now consider two important examples, which manifest a deep connection between these relationships and the conclusions drawn from a theoretical analysis of the stereochemistry of organic reactions.

7.2 Nucleophilic Rearrangements and Tautomerizations

The rearrangements of Eq. (7.2) belong to the transformations of this type. By their mechanism, they are classified under the reactions of nucleophilic substitution at the central atom of a migrating group

$$\mathrm{X(MR_n)\!-\!Z_m\!=\!Y}\ (\mathrm{Ia}) \rightleftharpoons \mathrm{X\cdots(M^{-}R_n)\cdots Y,\ Z_m^{+}}\ (\mathrm{Ib}) \rightleftharpoons \mathrm{X\!=\!Z_m\!-\!Y(MR_n)}\ (\mathrm{Ic}) \qquad (7.2)$$

X,Y = O, S, NR

-Z- = CR, N, conjugated chains

MR_n = $CR_1R_2R_3$, COR, Ar, NO, PR_1R_2, POR_1R_2, SR, SOR, SO_2R

As has been shown earlier, practically all reactions of nucleophilic substitution at the main group element center follow the addition-elimination (AdE) mechanism. Its crucial stage is the formation of an intermediate compound in which the central atom of the group M in Ic supplements its electron shell with the electron pair of the attacking nucleophile (Y in Ia). It has been demonstrated [2] that the relative orientation of the forming and the breaking bonds M—Y and M—X in Ic can be reliably predicted on the basis of the simple rules of the valence state electron pair repulsion theory advanced by Gillespie and Nyholm [5]. According to this model, all structures of the intermediates Ic are divided into two groups depending on relative orientation of the forming and breaking bonds M—Y and M—X.

The first group comprises intermediates II characterized by angular arrangement of the bonds which is realized in tricoordinate planar and pyramidal as well as in tetracoordinate tetrahedral structures. The second group consists of intermediates Ic with trigonalbipyramidal, bisphenoidal, and T-shaped structures III of the reaction site. These structures exhibit characteristic lengthening of the axial bonds with the most electronegative centers X and Y invariably occupying the axial positions.

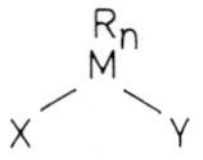

II $\varphi(XMY) = 100 \div 120°$

$l_{XY} = 2.3\text{-}2.8$ Å

$MR_n = MgR, BR_1R_2,$ COR, Ar, NO, NO_2

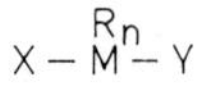

III $\varphi = 180°$

$l_{XY} = l_{MX} + l_{MY} + 0.6 = 3.5\text{-}4.5$ Å

$MR_n = CR_1R_2R_3, PR_1R_2, \overset{+}{P}R_1R_2R_3, POR_1R_2,$ SR, SOR, SO_2R

The topology of the reactive site (the parameters φ, l_{XY}) imposes certain constraints upon the length and configuration of the chain that links X and Y. It is not difficult to lay down the corresponding rules of "selection". Thus, in the case of conjugated chains, intramolecular l, j-shifts of the migrants of type II can occur in the systems of type IV–VI:

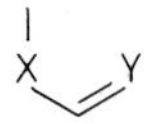

IV, j=3 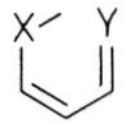 V, j=5 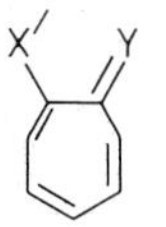 VI, j=9

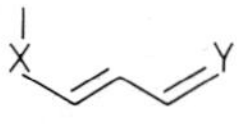

VII, j=5  VIII, j=5 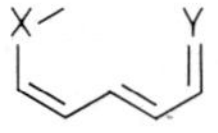 IX, j=7

In the systems of type VII the reactive centers X and Y are located at a distance too large for the 1,5-σ-bond shift. By contrast, in the system VIII, in spite of a fairly close promixity of the reactive sites, the reaction is inhibited due to the unfavorable angular parameter φ.

The simplest chain Z, which provides conditions for migration of the type III groups, possesses a *cis-trans*-configuration and corresponds to the case of a 1,7-sigmatropic migration via an intermediate eight-membered ring compound.

The conclusions based on the model proposed agree quite well with the experimental data and quantum chemical calculations [4]. Consider, as an illustration, a MINDO/3 calculation of the reaction path for the degenerate 1,3-acetyl rearrangement of amidine X which corresponds to system IV:

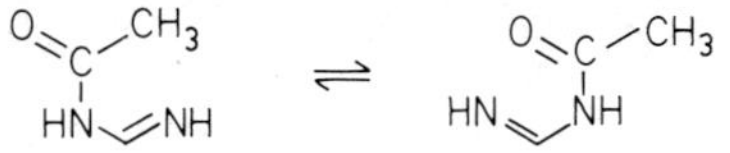

Xa ⇌ Xb

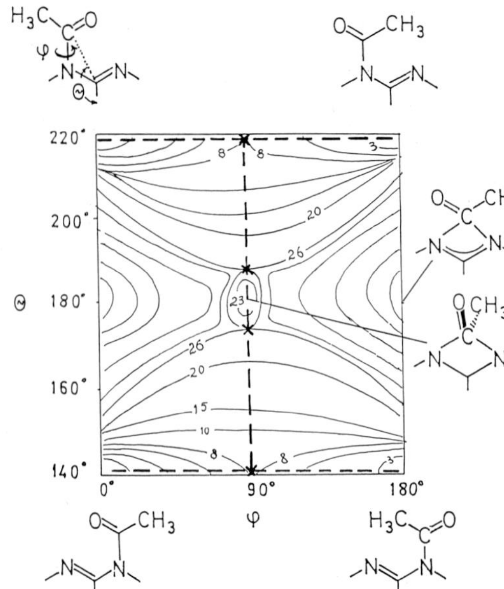

Fig. 7.1. PES of the acetyl group 1,3-shift in the amidine system in the coordinates θ and φ according to MINDO/3 calculations [2]. Dotted lines show the MERP's of the 1,3 shift and rotation about the C(O)—bond

As may be seen from Fig. 7.1, the N→N transfer is preceded by rotation of the acetyl group into a plane orthogonal to the amidine fragment. Such a rotation provides for an orientation of the nucleophile Y and the π-bond of the carbonyl group required by intrinsic mechanism of the reaction (see Fig. 5.7).

Calculations show that the angle of nucleophilic attack θ in amidines of type X amounts in the transition state structure zone to approximately 125°, which appreciably exceeds optimal values for the reactions of this type (see Sect. 5.3.3.1). Accordingly, activation energies, found experimentally, of acetyl migrations in amidines are fairly high. Once the angle θ is optimized, the barriers of acyl migrations in the derivatives of enoles of 1, 3-diketones XI and tropolones XII are sharply lowered thus falling within the activation scale of tautomeric reactions.

In the case of derivatives of pyrazole XIII, which correspond to the structures of type VIII, intramolecular acyl transfers are unfeasible and reaction can proceed at high temperatures only by the intermolecular route.

Also the prediction has proved correct of a molecular configuration, in which intramolecular alkyl shifts are realized, i.e., an intramolecular nucleophilic substitution at the tetrahedral carbon atom. It has been found [6] that a fast ($10^2\ s^{-1}$ at 25°C) bond-switching process associated with the rupture-formation of the C—S bonds of the anchored C_{sp_3} center does occur in the degenerate rearrangement of the 1, 8-*bis*-(arylthio)-anthraćene-9-carbinyl cation XIV in solution:

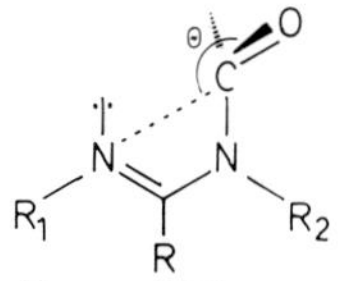

X, $\Theta = 125°$
$\Delta G^{\ddagger}_{298} = 22 \div 26$ kcal/mol

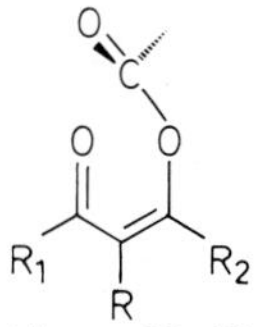

XI, $\Theta = 80 \div 85$
$\Delta G^{\ddagger}_{298} = 15 \div 17$ kcal/mol

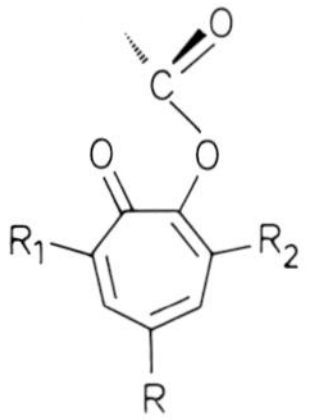

XII, $\Theta = 100°$
$\Delta G^{\ddagger}_{298} = 8 \div 12$ kcal/mol

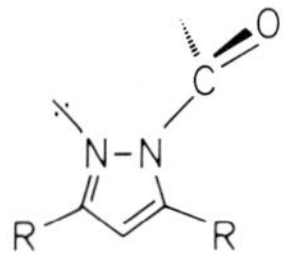

XIII, $\Theta = 150°$
$\Delta G^{\ddagger}_{298} > 25$ kcal/mol

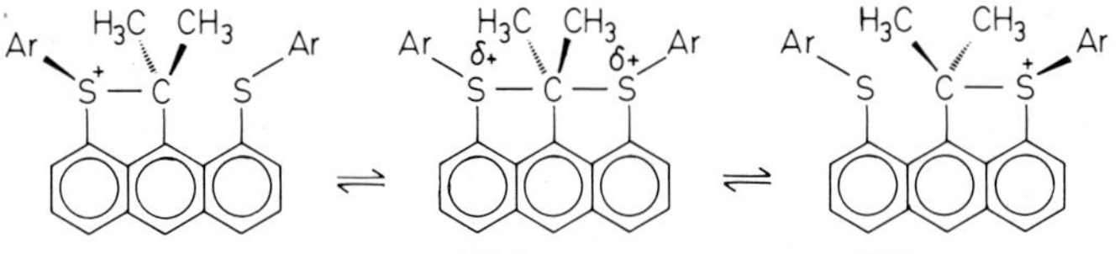

XIVa ⇌ XIVb ⇌ XIVc

The configuration of conjugated bonds in the reactive site XIV exactly corresponds to IX. Another example of an intramolecular methyl transfer is given by Eq. (7.3):

$$SO_2OCH_3 \ldots N(CH_3)_2 \longrightarrow SO_3^- \ldots N^+(CH_3)_3 \qquad (7.3)$$

which is analogous to Eq. (7.1). Owing to considerable length and flexibility of the chain, which links the reactive centers, the molecular structure in one of the sterically accessible conformations is close to that defined by the transition state of the S_N2 type reaction requirements so that intramolecular transfer successfully competes with intermolecular transfer [7].

A close connection of the above-described approach to the design of compounds, which show fast intramolecular rearrangements induced by migration within a molecule of bulky atomic groups, with the Pauling seminal transition state stabilization theory of enzymic reactions [8] may be traced. According to this theory advanced fourty years ago, binding enzyme to substrate leads to a strained configuration of this molecule, which becomes similar in its geometry to that of the activated complex of a catalyzed reaction. While in enzymic reactions a total or partial compensation for the energy losses due to the strain of a substrate is provided by the binding energy of the enzyme-substrate complex, smaller but appreciable stabilizing intramolecular interactions between the approaching functionalities contribute to the relief of strain in reactive conformations of the rearranged molecules [3, 4].

Since the intramolecular reactivity often achieves the highest limits of the enzyme reaction rates and even rivals them [9, 10], a special attention has been paid to studying its sources. A variety of useful rules and concepts, such as entropy and stereopopulation control, orbital steering, propinquity, and spatiotemporal hypotheses, have been evolved and their scope and limitations critically reviewed [11–13]. While differing from one another in their terms and emphases, they are common in reflecting in their essence a general principle of steric fitness of initial and transition state structures of fast intramolecular reactions.

7.3 Cyclization Reactions

Baldwin [14] presented a set of rules which permit prediction of a favorable stereochemical route (*exo* or *endo*) of the ring-forming reactions. They are based on the stereochemical requirements of transition states for ring closure processes.

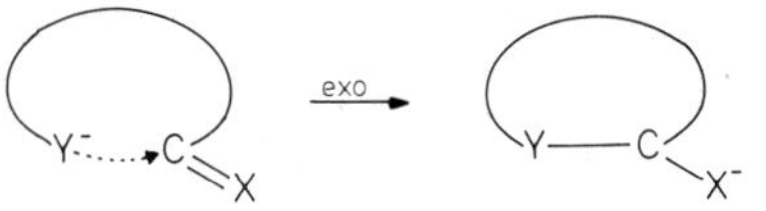

(7.4)

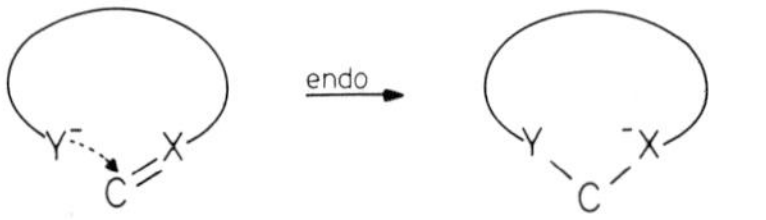

(7.5)

Equations (7.4) and (7.5), which feature these processes, represent, in effect, intramolecular additions to the double bond C=X carbon atoms (analogous schemes may be drawn for the single as well as for the triple bonds). Taking account of theoretical conclusions on the direction of nucleophilic attack on these bonds and assuming the angles of approach Y to the tetrahedral, trigonal, and digonal carbon atoms of the CX bonds to be, respectively, 180, 109, and 60°, Baldwin found correspondences between the structures of the expected transition states and the cycle sizes. The main statement is that the favored ring closures are those in which the length of the linking chain enables the terminal atoms to achieve the required trajectories. The rules are given in Table 7.1.

Baldwin's rules have gained a fairly wide acceptance among organic chemists. Regarding the cyclization reactions, in which carbon atoms at single or double

Table 7.1. Baldwin's rules for defining stereochemical control in ring closure reactions as in Eqs. (7.4) and (7.5). Numbers indicate the size of the forming ring

Type of carbon atom in the CX bond	Sterically favored reactions	Sterically unfavored reactions
Tetrahedral	From 3-*exo* to 7-*exo*	5-*endo*, 6-*endo*
Trigonal	From 3-*exo* to 7-*exo*; 6-*endo*, 7-*endo*	From 3-*endo* to 5-*endo*
Digonal	From 5-*exo* to 7-*exo*; From 3-*endo* to 7-*endo*	3-*exo*, 4-*exo*

bonds take part, these rules agree quite well with the calculations of optimal trajectories for a nucleophilic attack. On the other hand, in the case of the cyclization reactions involving triple bond, the assumed acute angle of 60° for nucleophilic attack is much smaller than the theoretical value of 127° (see Sect. 6.3). This can be explained as follows. In constructing the empirical rules of Table 7.1, the differences among stereochemistries of the nucleophilic, the electrophilic and the radical attacks were not taken into consideration, on the contrary, all factual material was examined as a whole without regard to individual cyclizations.

In actual fact, the angle of attack on the CX bond in addition reactions depends a great deal on the reaction type. Equally essential is the influence of reaction type on the character of rehybridization of the center being attacked occurring in the vicinity of the transition state structure. This may be illustrated by the data of nonempirical (3–21G) calculation [15] on reactions of the hydride ion (XVa), hydrogen atom (XVb) and proton (XVc) additions to the double-bond $C_{(2)}$ atom of propene:

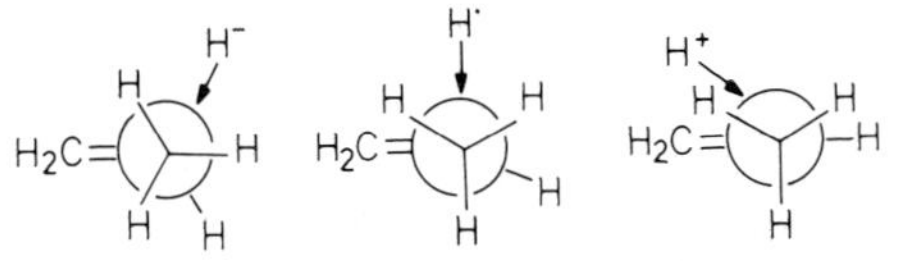

XYa, θ = 123° XYb, θ = 102° XYc, θ = 59°

Bearing in mind this circumstance, it must be emphasized that the rules of stereochemical control of cyclization reactions should be more rigorously adapted to the electronic nature of Y. One may state that they are well substantiated in regard to nucleophilic cyclizations involving intramolecular additions to single and double CX bonds. The situation is somewhat more complicated in the case of triple bond cyclizations. It has been shown recently by ab initio calculations [16, 17] that the obtuse angle for an attack of a nucleophile on the acetylenic triple bond may be not incompatible with the obsvered favored ring closures if the reaction involves a very early transition state with bent single bonds adjacent to the carbon–carbon triple bonds.

7.4 Topochemical Reactions

In the intramolecular reactions considered above, the drawing together and the favorable orientation of reactive centers were achieved by means of conformational potentialities of a molecule. In enzymic reactions, the fitting of a reaction site occurs as a result of specific interactions in the enzyme-substrate complex. A third possible mechanism of the fitting of a reaction site to the transition state demands is represented by freezing of interacting molecules or groups within a molecule in a crystalline matrix which holds them in a favorable starting position. Reactions in the crystalline phase proceed under minimal atomic and molecular displacement (topochemical principle) [18, 19]. If, however, the crystal stacking is such that the reacting molecules are oriented in accordance with stereochemical demands of a possible reaction, this reaction may develop at a rate that would exceed the reaction rate in solution.

A typical example is given by the following reaction of methyl transfer in Eq. (7.6), similar to intramolecular rearrangements of Eqs. (7.1), (7.3) considered earlier:

$$(CH_3)_2N-C_6H_4-SO_2OCH_3 \xrightarrow{\Delta} (CH_3)_3N^+-C_6H_4-SO_3^- \quad (7.6)$$

XVIa XVIb

In crystal, this reaction proceeds at a considerable rate even at room temperature, whereas in concentrated solutions under heating and in melt it develops rather slowly [20]. The reason for such behavior becomes clear from

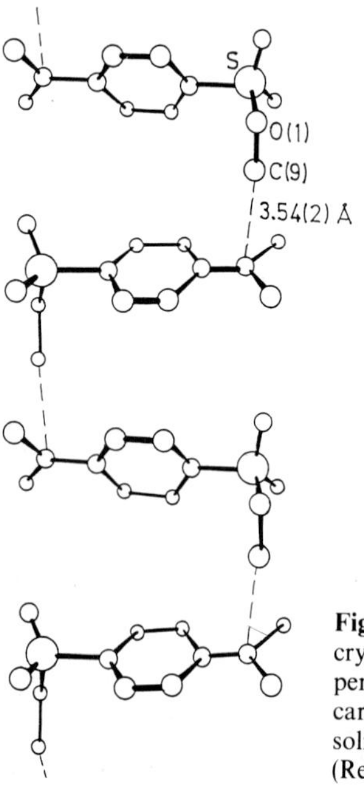

Fig. 7.2. A view of the stacking along one chain of molecules in crystals of methyl-*p*-dimethylaminobenzenesulfonate as seen perpendicular to the [101] plane. Distance indicated is that between the carbon atom of the methyl group which undergoes transfer in the solid state reaction and the nitrogen atom to which it moves. (Reproduced with permission from the American Chemical Society)

X-ray structural data [19, 20]. The molecules XVIa are packed in parallel stacks in which methyl sulfonyl and amino groups lie pairwise one over another in conformations which exactly correspond to steric demands of the transition structure of type III for cooperative methyl *O*, *N*-transfer.

Understanding of the nature of many other known—see the reviews of Refs. [18, 19, 21]—topochemical reactions and prediction of the new ones are based on the principle of correspondence of the crystal packing peculiarities to steric demands of the intrinsic mechanism of reactions.

References

1. Tenood L, Farooq S, Seibl J, Eshenmoser A (1970) Helv Chim Acta 53:2059
2. Minkin VI, Olekhnovich LP, Zhdanov YA (1977) Zh Vsesoyuzn Khim Obszh im DI Mendeleeva 22:273; (1981) Acc Chem Res 14:210
3. Minkin VI (1985) Sov Sci Rev Sect B Chemistry Reviews 7:51; (1989) Pure Appl Chem 61:661
4. Minkin VI, Olekhnovich LP, Zhdanov YA (1988) Molecular design of tautomeric compounds. Reidel, Dordrecht
5. Gillespie RJ (1972) Molecular geometry. Van Nostrand-Reinhold, New York
6. Martin JC, Basalay RJ (1973) J ACS 95:2572
7. King JF, McGarrity MJ (1982) Chem Commun:175
8. Pauling L (1946) Chem Eng News 24:1375
9. Lipscomb WN (1982) Acc Chem Res 15:232
10. D'Souza VT, Bender ML (1987) Acc Chem Res 20:146
11. Menger FM (1983) Tetrahedron 39:1013; (1985) Acc Chem Res 18:128
12. Bruice TC (1976) Annual Rev Biochem 45:331
13. Dorigo AE, Houk KN (1987) J ACS 109:3698
14. Baldwin JE (1976) Chem Commun:734, 738
15. Paddon-Row MN, Rondan NG, Houk KN (1982) J ACS 104:7162
16. Elliott RJ, Richards WG (1982) THEOCHEM 87:247
17. Elliott RJ (1985) THEOCHEM 121:79
18. Cohen MD (1975) Angew Chem Intern Ed Engl 14:386
19. Gavezzotti A, Simonetta M (1982) Chem Rev 82:1
20. Sukenik CN, Bonapace JAP, Mandel NS, Lau PY, Wood G, Bergman RG (1977) J ACS 99:851
21. Scheffer JR (1980) Acc Chem Res 13:283

Chapter 8

Radical Reactions

8.1 Specific Features of the Theoretical Analysis of Radical Reactions

The theory of the radical reactions has been developed not so thoroughly as that of the heterolytic reactions in which the closed electron shell of a reacting system is usually retained. The calculation of the systems with open electron shells is a rather harder task requiring the use of the UHF approximation and the inclusion of electron correlation corrections. For radical reactions, drawing together of the PES's of different electron states is characteristic and there is a high probability of changes in the multiplicity of an electron state in the course of reaction. The magnitudes of thermodynamic parameters of the radical processes depend crucially on the accuracy of the method of calculation applied, which has convincingly been demonstrated by detailed ab initio calculations on the radical substitution reactions:

$$F + CH_2{=}CH_2 \rightarrow CHF{=}CH_2 + H \tag{8.1}$$

$$F + CH_4 \rightarrow CH_3F + H \tag{8.2}$$

The data of Table 8.1 show that a satisfactory agreement with the experimental values of heats of reactions could be achieved only when the electron correlation had been taken into account by means of the not less than the third order Møller–Plesset perturbation theory. In this case, the HF solution has also to be sufficiently accurate involving the use of an extended AO basis set with polarization functions included.

The qualitative interpretation of the ease with which radical reactions occur, of their regio- and stereoselectivity also encounters additional difficulties. For example, a special role belongs in these reactions to exchange interactions between the unpaired electron of the radical center and the electrons of the bond attacked by it. Theoretical rationalization of the radical reactions is quite an important task in view of their exceptional significance for organic and biological chemistry.

Table 8.1. Calculated and experimental values of heats of Eqs. (8.1) and (8.2) in kcal/mol [1]

Method	$F + C_2H_4 \rightarrow CH_2H_3F + H$	$F + CH_4 \rightarrow CH_3F + H$
HF/3-21G	18.3	27.6
MP2/3-21G	−4.2	7.3
MP3/3-21G	2.3	12.8
MP4/3-21G	0.6	11.4
HF/6-31G*	10.3	17.2
MP2/6-31G*	−18.1	−8.5
MP3/6-31G*	−8.4	−0.6
MP4/6-31G*	−10.8	−1.3
$\Delta\sum 1/2h\nu$[a]	−4.5	−3.6
Experiment	-11.1 ± 2.0	-5.6 ± 2.0

[a] Zero point vibration energy corrections (ZPE). The calculated value of heat of a reaction is the sum of the ZPE and the differences between total energies of the reactants and the reaction products, given in the table

A sufficiently comprehensive classification of the radical reactions has been suggested in Ref. [2]. Regarding the structure of electron shells of the reactants, the radical reactions are subdivided into two classes. The first of these comprises the reactions in which the common electron shell is closed or has a triplet electron configuration (2n electrons), i.e., the reactions of recombination and those of radical pair and biradical formation. To the second class (free-radical reactions) the transformations belong in which one of the reactants has a closed shell and the other—one unpaired electron (total number of electrons is $2n + 1$).

8.2 Free-Radical Reactions

Notwithstanding the enormous diversity of free-radical processes, the following most common elementary reactions may be singled out: 1) cleavage of the atom X, usually a hydrogen atom, by a radical from the R—X bonds; 2) addition of the radical to the double bond and 3) reactions of radical (homolytic) substitution S_H2 proceeding without intermediates. Some other processes, for example, the widespread reactions of radical aromatic substitution and the type of Eq. (8.1) reactions occur as a combination of two elementary steps by the addition-elimination mechanism (AdE).

A qualitative examination of the free-radical processes was first carried out by Yamaguchi [3] who took account of the principal specificity of these reactions, namely, the presence of an unpaired electron. Using the approximation of the valence bond method with the electron exchange included, the correlation diagrams were considered for the doublet and quartet states of a three-electron model—one unpaired electron of the radical and two electrons of the R—X or π-bonds. That analysis led to the conclusion as to a basic difference between the behavior of the linear configuration of interacting centers typical of the cleavage of the hydrogen atom and that of the angular configuration that corresponds to the addition mechanism.

8.2.1 Bond-Cleavage and Addition Reactions

While the linear approach of a radical to the bond being attacked is allowed by the spin symmetry (there is no intersection between the doublet electron states of the reactants and the products), the angular approach is, on the contrary, forbidden (the terms intersect)—see Refs. [3, 4]. From this it follows that an attack upon the σ-bond R—X is, in reactions of the cleavage of a hydrogen atom by the radical, developing in a strictly collinear configuration I:

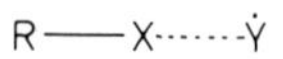

R——X·······Ẏ

I

H 1.093; 113.8°; C 1.597 H 0.783 T; 104.6°; H; H]‡

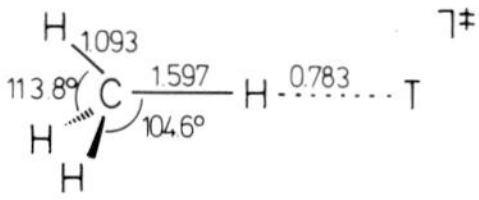

II

Indeed, the experimental and calculation data on the simplest reaction of the hydrogen atom abstraction $H + H_2 \rightarrow H$ corroborate the conclusions as to the preferability of the approach path of the type I [5].

The linear transition state structure II was revealed in the ab initio calculations [6], employing the IRC technique, on the reaction of the hydrogen atom abstraction from a methane molecule by the tritium atom:

$$T + CH_4 \rightarrow T - H + CH_3 \qquad (8.3)$$

In accordance with the theoretical rules, there occurs a collinear (C_1, S, R′) cleavage of R′ from a sulfur atom in the intramolecular cyclization of the radical III:

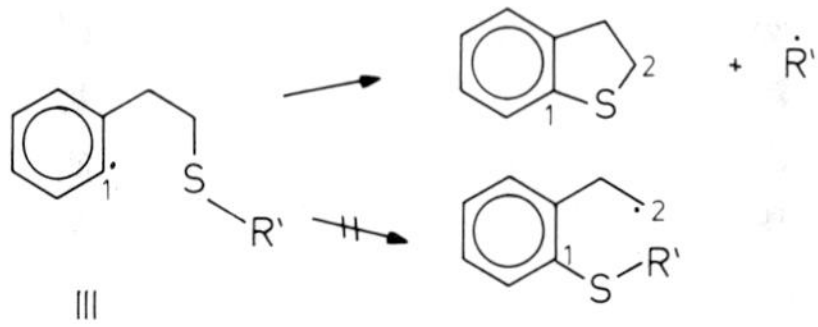

S, R', 1, 2, S, + Ṙ', 1, 2, S, R'

III

Another direction of the reaction, i.e., the formation of β-arylethyl radicals that would require angular approach of the radical center C_1 to the SC_2 bond does not materialize [7].

A nearly collinear abstraction of the hydrogen atom occurs, according to the ab initio calculations [8], also in the reaction $OH^{\cdot} + CH_4 \rightarrow OH_2 + CH_3^{\cdot}$. Making use of the data on the structure of the transition state, Dorigo and Houk have developed a technique founded on molecular mechanics for studying the intramolecular hydrogen abstractions. Relative activation energies are calculated there for intramolecular hydrogen abstractions by a variety of alkoxy radicals. The results show that there is no simple relationship, as was suggested by Menger [9], between calculated reaction rates and distances between the reacting atoms in the starting materials.

In the case of double bond addition reactions, when the attacking radical approaches an alkene in the π-bond plane (Sect. 6.2), the Yamaguchi rule implies that a terminal attack IV is more preferable than the central (triangular) approach V:

Two simultaneously operating demands in IV, those of orbital overlap and spin forbiddenness, should lead to an "early" transition state for radical addition reactions. Indeed, already the first semiempirical calculations [10] of the PES of the reaction $CH_3 + CH_2{=}CH_2$ showed the new bond C—C in the transition state to be quite long equalling 2.3 Å. The same value (2.352 Å) was obtained in the MERP calculations [11] of this reaction by the MINDO/3 method. The conclusion as to the "early", i.e., the reactant-like character of the transition state structure is consistent with the data of the fairly rigorous ab initio MP2/3–21G calculations [12] on transition state structures of the reactions of addition of hydrogen and fluorine to ethylene:

H ⌉‡ 1.845 107.4° 172.2° 1.346 167.4°
VI

F ⌉‡ 2.024 1.340 170.7° 177.1°
VII

The assumption of transition states of "early" type has been confirmed by high velocity, low activation energies (2–8 kcal/mol) and high exothermicity (15–20 kcal/mol) of the radical addition reactions. Furthermore, there is indirect evidence of a correlation between the properties of a transition state structure and those of the final radical product. Thus, the relative rates (and activation energies) of addition of CCl_3, CF_3 and some other electrophilic radicals correlate with the localization energies [13]. This contradiction was accounted for when analyzing redistribution of the electron and the spin density in reactions of addition of the methyl radical and the fluorine atom to ethylene [14].

According to RHF 4–31G calculations the odd electron is practically fully localized at the F atom, while changes affecting charges in ethylene are insignificant. However, the UHF calculations indicate considerable polarization of the charges in ethylene arising on account of the β-spin electron transfer from ethylene to fluorine and a decrease in electron density at the C_1—C_2 bond. The probability of the α-electron transfer is low.

This is important for understanding the peculiarity in the transition state structure: although the geometry and the full spin density point to an early reactant-like transition state, the substantial polarization of the α and β electrons as well as partial transfer of the β electron to the C—F bond and weakening of the double bond correlate with a late transition state.

Like other selection rules, the Yamaguchi rule is not absolute and by varying the structure of reactants its restrictions may be lifted. Salem and co-authors [7] have shown the conclusion as to preferability of the type I, IV approach to be fully valid only for the reaction between nonpolar reactants when covalent interactions prevail. However, in the case of zero overlap of the radical orbitals and one of the centers of the bond being attached also the triangular approach can become allowed. Such a situation may arise in reactions of radical addition to polarized alkenes that contain substituents at sp^2-carbon atoms.

Inclusion of the ionic structures in the three-electron model made it possible to take into account the effects of polarization upon the mechanism of free-radical reactions and to arrive at the following conclusion: addition of a free radical to a bond attacked by polar groups occurs in such a position for which the new bond that forms in the transition state possesses the most ionic character. If the reaction occurs under thermodynamic control, then the polarization effects stabilize the product thereby facilitating the development of reaction. Otherwise, products may form that are not thermodynamically the most stable. These qualitative conclusions have been supported [7] by ab initio calculations of the relative stability of reactants and products as well as the activation barriers for a series of the reactions:

$$H_2C{=}CHR + \dot{Y} \longrightarrow \begin{cases} H_2\dot{C}-CHRY \\ YH_2C-\dot{C}HR \end{cases} \qquad (8.4)$$

$Y = H, Cl, CH_3, CF_3$; $R = BH_2, CN, F, CH_3, NH_2$

For example, according to the above-formulated rule, the electronegative radical CF_3 must react regioselectively with fluoroethylene attaching to the unsubstituted group CH_2. Indeed, the calculated activation energy for this reaction is lower by 5 kcal/mol than in the case of addition to the substituted carbon atom, which is in good agreement with the experimental data on the yields of the isomeric products [13].

Using as an example the reactions of radical addition to a double and a triple bond, Schlegel and Sosa [15–18] have shown that the unwanted spin contamination must be removed with the aid of projection methods [19]. Thus, for the reactions $H + C_2H_4 \rightleftarrows C_2H_5$, $H + CH_2O \rightleftarrows CH_3O$ [15], $OH + C_2H_{2n} \rightleftarrows HOC_2H_{2n}$ [16], $H + C_2H_2 \rightleftarrows C_2H_3$ [17], and some others [18] the energy barriers computed by the Møller–Plesset perturbation theory based on the UHF reference determinant were found to be 5–20 kcal/mol too high because of serious spin contamination. When the largest spin contaminants are annihilated, the heights of energy barriers and the shapes of potential energy curves are greately improved. The agreement with the experimental activation energy is also made significantly better.

It should be noted that there exists relative insensitivity of radical addition transition structure geometries to computational methods [20]. As a result, the UHF/3–21G level of calculations proves sufficient for determining the qualitative features of transition state structures and relative activation energies for regioisomeric transition structures [20–21].

Noteworthy is also the fact that the heights of activation barriers are not always reliable characteristics for the determination of relative reaction rates. One may point by way of example to the study [22] of addition of methyl radicals to ethylene and acetylene. Although addition to ethylene is characterized by a lower activation energy, addition to acetylene proceeds at a higher rate due to the opposite and larger trend of preexponential factors.

8.2.2 Radical Substitution Reactions at the Tetrahedral Carbon Atom

Among numerous radical substitution reactions [23, 24] those of the S_H2 type at a saturated carbon atom have been studied in the most detailed fashion. The simplest and at the same time most important example is given by the substitution of a hydrogen atom of methane by tritium (or another hydrogen isotope):

$$T + CH_4 \rightarrow CH_3T + H \tag{8.5}$$

There is hardly another reaction of this type thus thoroughly investigated as this one. It has been studied experimentally not only in the region of ordinary thermal energies, but also in that of high vibrational excitations (hot hydrogen, deuterium, and tritium atoms) arising upon irradiation or during nuclear recoil. The substitution reaction of Eq. (8.5) competes with the reaction of hydrogen abstraction of Eq. (8.3) which has a lower activation energy (10–12 kcal/mol) than the former whose excitation occurs at a rather high threshold value of activation energy (35 kcal/mol).

For this value to be correctly reproduced, the use of the minimal or DZ basis set in the HF method is no longer sufficient and the electron correlation needs to be taken into account [25]. Table 8.2 lists data on the dependence of calculated activation energies of Eqs. (8.3) and (8.5) upon computational method. The energetically lowest transition state of the substitution reaction of Eq. (8.5) is the trigonal-bipyramidal structure VIII [25, 26]. Such a transition state leads, apparently, to inversion of configuration of the tetrahedral carbon atom in radical substitution reactions. No unambiguous experimental confirmation of this conclusion has been obtained so far, but there is indirect evidence in its favor [24].

Table 8.2. Data of the ab initio calculations [25] on activation barriers of hydrogen abstraction from Eq. (8.3) and radical substitution from Eq. (8.5)

Reaction	Transition state and reaction outcome	Activation energy, kcal/mol		
		Minimal basis set	DZ	DZ + CI
Eq. (8.3)	II, $CH_3 + TH$	36.1	34.2	17.0
	VIII, inversion of configuration	83.3	64.3	42.0
Eq. (8.5)	IX, retention of configuration	109.4	87.2	64.3
	X, retention of configuration	119.9	–	64.5

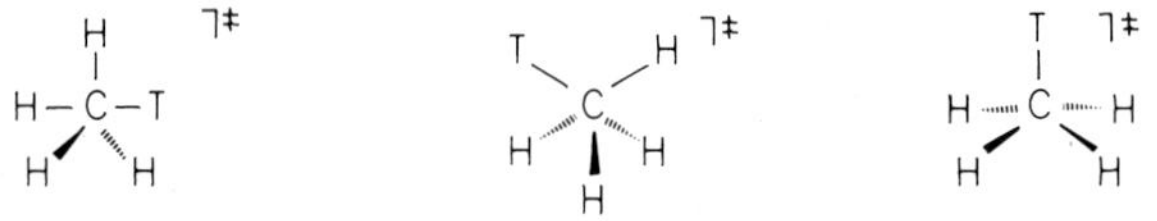

VIII, D_{3h} IX, C_s X, C_{4v}

The axial bonds in VIII are longer than the equatorial ones, as in the analogous D_{3h} structure of the transition state structure for the S_N2 reaction (see Sect. 5.1.1.2), though much less so than in the latter. The structures IX and X whose formation could retain the configuration of the central atom have much higher energy than VIII, as is apparent from the data of Table 8.2.

The data of this table and the results of some other ab initio calculations [6, 26] show that the axial abstraction of Eq. (8.3) is the most preferable channel for the reaction of methane with hydrogen isotopes. The substitution reaction of Eq. (8.5) must proceed exclusively from the vibrationally excited states, as has been verified experimentally.

8.3 Reactions with Formation of Biradicals

Compounds classified under biradicals are the ones in which the singlet-triplet splitting in the ground state is quite small (on the order of kT). In the orbital approximation, it means that two electrons occupy two degenerate or nearly degenerate weakly interacting orbitals [27–29]. Ordinarily, these orbitals are spatially separated, though this condition is not a necessary property of a biradical. Antiaromatic annulenes, conjugated hydrocarbons that cannot be described by classic structural formulas and carbenes with a small S–T gap may belong to biradicals. As a rule, the biradicals are highly reactive and relatively unstable compounds. This is particularly true of the organic 1,j-biradicals (numbers are given of the atoms in the common chain at which

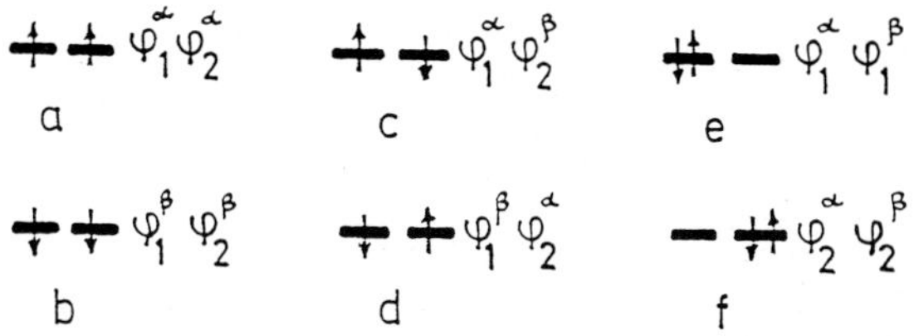

$\varphi_1^\alpha \varphi_2^\alpha$ a $\quad$ $\varphi_1^\alpha \varphi_2^\beta$ c $\quad$ $\varphi_1^\alpha \varphi_1^\beta$ e

$\varphi_1^\beta \varphi_2^\beta$ b $\quad$ $\varphi_1^\beta \varphi_2^\alpha$ d $\quad$ $\varphi_2^\alpha \varphi_2^\beta$ f

Fig. 8.1a–f. Possible electron configurations for the case of two electrons placed in two degenerate MO's γ_1 and γ_2 and the corresponding wave functions

the odd electrons are localized) whose formation is regarded as the crucial step of such reactions as the *cis-trans* isomerization of alkenes (1,2-biradicals), the pyrolysis of cyclopropanes (1,3-biradicals), the thermal dimerization of alkenes forbidden by the orbital symmetry conservation rules (1,4-biradicals), photochemical rearrangements and others [30].

Kinetical instability of biradicals emerging in those reactions creates considerable difficulties in their experimental study. Theoretical investigation of biradicals also requires special methods as the determination of a sufficiently exact wave function is in their case a much harder task than for the systems with closed electron shell. Figure 8.1 shows possible ways in which the electrons are located in the degenerate MO's φ_1, φ_2 of the biradical and the wave functions that correspond to these states. Two triplet (a, b) and two singlet (e, f) electron configurations are readily identifiable. The configurations c and d are not the eigenfunctions of the spin operator $\hat{S}^2$ [31], but their sum (third component of the triplet electron state) and difference (third singlet) satisfy this condition. A correct description of the third singlet cannot be made in terms of the RHF approximation, nor would this approximation be sufficient for the description of other singlet configurations. Moreover, the singlet and the triplet states, in the biradical's geometry which corresponds to the degeneracy of the MO's φ_1 and φ_2, are close in energy. This makes it necessary to apply more accurate calculational schemes, such as the configurational and multiconfigurational interaction methods (Sect. 2.2.4) as well as to take into account the probability of intersection of the PES's of various terms and to estimate the probabilities of nonadiabatic transitions (Sect. 1.6). The degeneracy of states and changes in their multiplicity necessitate an analysis of the stability of the pertinent Hartree–Fock solutions [4]. The problem may be illustrated by the reaction of thermal opening of the cyclopropane ring already considered in Sect. 1.4.

XI $\longrightarrow$ XIa $\qquad$ (8.6)

The highest bonding MO of cyclopropane is shown in Fig. 8.2a. As the cycle is being opened, the two-fold occupancy of this MO is disturbed and it splits

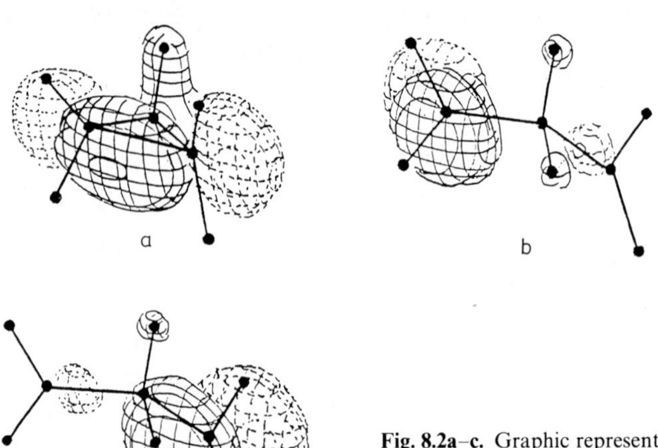

Fig. 8.2a–c. Graphic representation of the bonding MO of cyclopropane (**a**) and two degenerate nonbonding MO's of trimethylene (**b** and **c**) in the (FF) conformation, according to Ref. [4]

into two nonbonding orbitals of lower symmetry which correspond to the electrons with α and β spins. Thus, in the course of reaction there must arise triplet instability of the HF solution. This makes necessary a calculation in the UHF approximation using different orbitals for different spins. The orbitals of the 1,3-biradical XIa emerging in this calculation are also given in Fig. 8.2b, c. They are practically fully localized at the end atoms of carbon.

Clearly, the rupture of the bond in XI and the transition to the biradical form XIa proceeds continuously, rather than in a jumpwise fashion. The sharpest changes during this continuous process occur in the region of instability of the HF solutions. It is for this reason that the analysis of stability of the HF values is a convenient technique for determining the part of the configurational space in which the PES contains biradical structures. The appearance of triplet instability may be regarded as an indication of the transformation of structures with closed electron shell into those with two separated radical centers or biradicals.

Quantitatively the degree y of the biradical character is given (in %) by the equation suggested in Ref. [4].

$$y = \left(1 - \frac{2S_{ab}}{1 + S_{ab}^2}\right) \cdot 100\% \tag{8.7}$$

where $S_{ab} = \int \varphi_a \varphi_b d\tau$ is the overlap integral for the molecular orbitals φ_a and φ_b localized at the fragments a and b, similarly to the orbitals of the biradical XIa shown in Fig. 82b, c.

The biradical character is nil in the case of complete overlapping ($S_{ab} = 1$, nonradical case); for the fully biradical state $y = 100\%$ ($S_{ab} = 0$). The calculation of y allows quantitative evaluation of the degree of the biradical character which may usefully be compared with the contribution from an alternative dipolar structure. Thus, for an ozone molecule $y = 52\%$ indicating roughly equal contributions by the biradical structure XIIa and the ionic structure XIIb:

XIIa 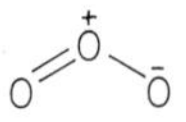 XIIb 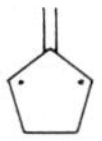 XIII

For trimethylene XIa $y = 80\%$. In the isoelectronic series (CH_2CH_2O, CH_2OO, CH_2NHO), the biradical character varies very broadly, with y being, respectively, 77%, 42%, and 2%.

Equation (8.6) is the prototype of an important family comprising the reactions of homolysis of the ordinary CC bond that lead to the formation of the 1,3-biradicals. If the bond in question is included in the sterically strained cycle, the gain in strain energy upon rupture of the bond may partially or even fully offset the loss in bonding energy. In this case, the thermodynamic stability of the biradical structure will be near to that of the isomeric covalent form. Thus, the cyclic derivative of trimethylene methane, 2-methylene cyclopentane-1,3-diyl, is more stable than the corresponding bicyclic isomer [30]. The emerging biradical may possess adequate kinetical stability, if the bond rupture involves forbidden crossing of the electronic states of the same symmetry (noncrossing rule) that correspond to the isomeric structures. A compound with covalent bonding transforms into the isomeric biradical via the transition-state structure in which the highest occupied MO of one isomer crosses the lowest vacant MO of the other. Stohrer and Hoffmann were the first to have pointed out the possibility of such isomerism (bond stretch isomerism), see Ref. [32]. A typical example is given by [2.2.2] propellane XIV in which the central CC bond is formed by the practically pure *p*-orbitals, thus providing for crossing of the states of the same symmetry when the bonds are stretched.

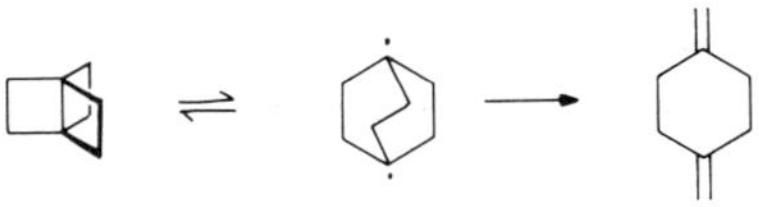

XIV XIVa XIVb

According to the ab initio CI calculations [33], the singlet biradical XIVa corresponds to the local minimum on the PES of C_8H_{12} ($r_{CC} = 2.51$ Å). Formation of the transition state at the XIV → XIVa step involves considerable stretching of the central bond ($r_{CC} = 2.35$ Å) and requires the energy of 29 kcal/mol relative to XIV. The biradical XIVa must extremely easily undergo the triply degenerate Cope rearrangement to give 1,4-dimethylene cyclohexane XIVb. This fact has been shown by the semiempirical INDO calculations [34] to lie at the root of many unsuccessful attempts to synthesize [2.2.2] propellane. The calculations have also demonstrated that in the triplet state the biradical XIVa is geometrically much nearer to XIV ($r_{CC} = 1.95$ Å), which prompted the idea to synthesize [2.2.2] propellane by irradiating XIVb in the presence of a triplet sensitizer with subsequent vertical relaxation of the formed triplet biradical XIVa to XIV. It is precisely in this fashion that [2.2.2] propellane XIV was first prepared (see Ref. [34]).

At the same time, according to the calculations [35], trimethylene cannot exist as a free-energy intermediate at accessible temperatures. The question whether substituted trimethylenes can correspond to intermediates remains, however, open. In greater detail this problem is discussed in Ref. [35] where ample literature is given about ab initio and semiempirical studies.

In order to predict transformations of biradicals, it is important to know the magnitude of the singlet-triplet gap, since the multiplicity of a biradical affects in an essential manner its reactivity. This may be illustrated by the well-known example of the reactions of 1,1-biradicals, namely, the carbenes considered below in the next section. In the case of 1,j-biradicals, it is a general rule that the stability of the singlet electronic state of a biradical is assisted by increased overlapping of the terminal obritals; the triplet state represents the ground state when this overlap is small [28, 36]. Ordinarily the S–T gap does not exceed 4–5 kcal/mol in the case of the 1,j-biradicals.

8.4 The Reactions of Carbenes

The effect of multiplicity of carbenes on their reactivity is most vividly marked in the following features rationalized by Skell et al. from experimental data [37–39]. First, the reaction of carbenes occurs in the singlet electron state at a much faster rate than in the triplet, with the absolute rates of typical reactions of addition to multiple bonds and of insertion into the C—H bonds exceeding, under normal conditions, the rate of intercombination conversion. Secondly, the singlet carbenes are characterized by one-step stereospecific addition to double bonds, as, for instance, in the cyclopropanation reaction, while the triplet carbenes react in a nonstereospecific way to form first an intermediate biradical through addition to one of the atoms of the double bond. The formation of a trimethylene radical, in the course of reaction of triplet methylene ($^3B_1$) with ethylene, has been confirmed by semiempirical [40, 41] and ab initio [42, 43] quantum chemical calculations.

These peculiarities of carbenes' reactivity, together with the fact that most experimentally accessible compounds of this series have singlet electron states, have aroused particular interest in the reactions of singlet carbenes which have been the subject of by far the greater number of the theoretical studies in this field. We are going to examine only the reactions of addition to the double CC bond as well as to the ordinary bonds, the study of which provided the facts for constructing the conceptual basis for the theory of mechanisms governing the reactions of carbenes and their analogs.

8.4.1 Addition to the Double Carbon–Carbon Bond

Judging from the electron configuration $^1(\sigma^2)$ of singlet carbenes, one is entitled to expect in them manifestations of both the electrophilic (vacant *p*-orbital) and the nucleophilic (electron pair of the σ-orbital) properties. In addition reactions to alkenes and other compounds with multiple bonds, the "philicity" of a carbene is determined by the width of the energy gap between frontier orbitals of the reacting molecules. The stabilizing two-electron interaction between these orbitals will be the stronger, the less is the width of the gap and the larger the overlap integral between the frontier orbitals in the region close to the transition state structure of the reaction. A case of relative positions of the frontier orbitals of reactants, the most typical for carbenes, is depicted in Fig. 8.3a, it is the case of prevalence of the electrophilic properties. Figure 8.3b displays destabilization, predicted by the ab initio calculations [44], of the *p*-orbital of carbene upon introduction of the strong π-donating substituents X, Y, which gives rise to the nucleophilic properties (the $\sigma - \pi^*$ interaction). Lastly a case of approximate equality of the $\sigma - \pi^*$ and $\pi - p$ interactions is possible (Fig. 8.3d) thus justifying the term of ambiphilic carbenes.

The philicity of a carbene directly depends on the structure of the transition state of an addition reaction. The rules of orbital symmetry conservation forbid the least-motion C_{2v}-symmetry reaction path [41]. For electrophilic carbenes, characterized by predominance of the $\pi - p$ interaction, preferable is the so-called π-approach (Fig. 8.3). In the case of nucleophilic carbenes, optimum conditions for the overlap between the σ_{CXY} and π^*-orbitals are provided by the asymmetrical σ-approach (Fig. 8.3b). By making use of certain assumptions, Rondan, Houk, and Moss [44, 45] calculated the overlap integrals S_{ij} between the corresponding frontier orbitals of carbene and alkene for the π- and the σ-approaches. Then, having computed the energies of those orbitals, they obtained the energies of stabilization of the composite system arising in two

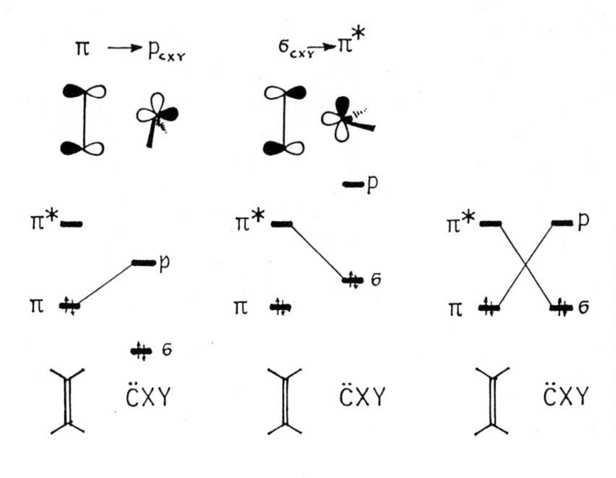

Fig. 8.3. Principal variants of relative positions of the energy levels of substituted carbenes CXY and ethylene depending on the substituents X, Y: **a** electrophilic carbenes; **b** nucleophilic carbenes; **c** ambiphilic carbenes. The order of energy level is adhered to for CCl_2 (**a**) and $C(NH_2)_2$ (**b**)

types of interaction between the frontier orbitals:

$$\Delta E \simeq S^2_{\text{HOMO-LUMO}}/(\varepsilon_{\text{LUMO}} - \varepsilon_{\text{HOMO}} - Q)$$

where $Q \simeq 5$ eV is the allowance made for the variation in the value of ε for the reactants in the course of their drawing together. The relation between the stabilization energies for the nucleophilic (σ) and the electrophilic (π) approach gives the theoretical value of the philicity:

$$\text{PI}_{\text{CXY}} = \Delta E_n/\Delta E_e \tag{8.8}$$

It is less than unity for electrophilic carbenes, more than unity for nucleophilic carbenes and is approximately equal to unity in the case of ambiphilic carbenes.

The calculated values correlate well with experiment: the experimentally obtained carbenes BrCCOOEt, CH_3CCl, CBr_2, PhCBr, PhCCl, PhCF, CCl_2, CFCl, and CF_2 have been assigned to the electrophilic species (with electrophilicity diminishing from the left to the right), the carbenes CH_3OCCl, CH_3OCF are ambiphilic and $(CH_3O)_2C$, $CH_3OCN(CH_3)_2$-nucleophilic. The philicity of a carbene varies also as a function of its conformation. Thus, according to the theoretical estimate [46] the *cis*-conformer of methoxyphenyl carbene possesses nucleophilic properties, while its *trans*-conformer is ambiphilic. In the review article [45], calculation data are given on differences among the orbital energies depending on substituents in alkene (with carbene unchanged). The addition of electrophilic carbenes proceeds more readily to electron-redundant alkenes, i.e., upon introduction of π-donors, while the nucleophilic carbenes exhibit enhanced reactivity when the π-bond of alkene is weakened, that is, upon introduction of electron-withdrawing substituents.

Clearly, the above-mentioned values of the philicity relate each time to a particular alkene, and the orders of philicity may be different with different alkenes [45]. Figure 8.3 visualizes this behavior: the stabilization or destabilization of the $\pi\pi^*$-levels of alkene, induced by substituents, may alter relative values of the contributions from the $\pi - p$ and $\sigma - \pi^*$ interactions. In practice, their values are often estimated without calculating ΔE confining oneself to finding the difference between the energies of frontier orbitals. Thus, the relative rate constants of addition of dichlorocarbene with substituted styrenes correlate well with the difference between the energies of the HOMO of styrene and the LUMO of carbene [47].

Direct calculations of the reaction pathways, by which carbenes are added to alkenes, clarify some important details of the intrinsic mechanism of such reactions. At the level of ab initio calculations, the reaction of singlet methylene with ethylene has been studied in the most detailed fashion [48]. Like the addition of silylene [49], this reaction proceeds without a barrier with the attacking particle moving under the plane of alkene ("internal" approach). "External" approach of methylene and other carbenes is energetically less

Table 8.3. Main geometry parameters of transition state structures of the carbenes-to-ethylene addition reactions, according to ab initio calculations

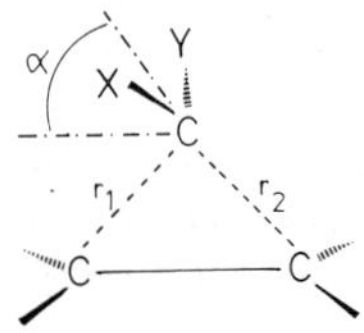

CXY	r_1, Å	r_2, Å	α	Basis set	Ref.
CCl_2	1.95	2.28	19.0	3–21G	[52]
CF_2	1.74	2.16	30.0	3–21G	[52]
FCOH	1.78	2.35	27.6	STO-3G	[44]
$C(OH)_2$	1.78	2.24	49.2	STO-3G	[53]
$C(NH_2)_2$	1.80	2.31	61.2	STO-3G	[53]

favored [50] leading to their insertion into the C—H bond [51]. Of particular theoretical interest are the data of calculations on the pathways of reactions of substituted carbenes on which transition state structures can be localized and identified. The "internal" route of approach of reactants is a hybrid of the π- and σ-approaches, and the structure of the transition states has for all reactions studied a similar form with $r_1 \neq r_2$. Results of some calculations are presented in Table 8.3.

A characteristic of transition states of the carbene-to-alkene addition reactions, particularly sensitive to the philicity of a carbene, is the angle of slope of the carbene plane relative to the double-bond plane. According to calculations [44, 45] one may hold that the carbenes for which in the transition states of addition to alkenes the angle $\alpha < 45°$ are electrophilic. The angle $\alpha > 50°$ is typical of nucleophilic carbenes, while the $45° < \alpha < 50°$ region relates to the ambiphilic carbenes. Ab initio [44, 52, 53] and semiempirical (MNDO) [54] calculations of pathways of addition reactions of various carbenes have verified this dependence.

Calculations of energy barriers and heats of the cycloaddition reaction have shown that the selectivity of carbenes (found at ambient temperature) is linearly related with their stability [44]. This conclusion in the spirit of the conventional activity-selectivity relationship does not, however, explain the experimental data referring to entropy control of the dichlorocarbene addition to alkenes [55], to the temperature-dependent selectivity of dihalocarbenes [56, 57] or to the zero or even negative activation energies for a number of cycloaddition reactions of carbenes [58, 59].

This last effect is habitually associated with the presence on the reaction coordinate of an intermediate complex stable to reactants. Such complexes corresponding to very early stages of the π-approach are indeed detected in ab initio calculations (basis sets of up to 6–21G level), however, with the effects of electron correlation included (MP2/3–21G), they no longer represent minima on the potential surface and, consequently, cannot exist as kinetically isolable particles [52, 60]. The authors of Ref. [52] had calculated the vibration frequencies and entropies of the carbene-alkene system at all its steps and advanced the following hypothesis. Since in the course of the reaction the entropy gradually decreases along the reaction coordinate and this change does not practically depend on the nature of alkene and carbene, the value of $\Delta G^{\neq}$, when the enthalpy rapidly falls for reactive carbenes, is controlled by the entropy term ($-T\Delta S^{\neq}$). For the reason that $\Delta G^{\neq}$, in consequence of the increase of that term (with ΔH falling along the reaction coordinate) attains its maximum, negative activation energies may be observed characteristic of the reactions of the most electrophilic carbenes with electron-redundant alkenes. The same authors [52] have proposed model potentials for ΔH and $-T\Delta S$ which reproduce quite well experimental data on reaction energetics without inclusion of the step of the intermediate complex formation.

8.4.2 Insertion into σ-Bonds

Reactions of addition of carbenes to single bonds associated with the rupture of these are referred to as the reactions of insertion into the bond. Being very characteristic of carbenes, they are the subject of numerous theoretical studies [61–67]. Calculations have shown that the insertion into the bonds H—H, C—H, Hal—Hal, C—C occurs without overcoming an energy barrier as a concerted reaction developing not along the least motion path. Two phases of insertion into the σ-bond may be singled out: 1) the electrophilic attack in which the vacant *p*-orbital of carbene interacts with the bonding orbital of the σ-bond being attacked to form a three-center bond; 2) the subsequent nucleophilic phase defined by the two-electron interaction of carbene's σ-orbital with the antibonding σ*-orbital. At this second step, stretching and breaking of the σ-bond takes place.

Insertion into the H—H bonds. Early semiempirical (EHMO) calculations predicted for the insertion of methylene into H_2, up to the step of the formation of a transition state, linear configuration >C···H—H characteristic of a mechanism of the abstraction-recombination type. However, more rigorous ab initio calculations [63, 64] indicated quite rapid formation of the three-center bonds at the initial step of the reaction. Methylene in the triplet ($^3B_1$) ground state, in contrast to its behavior in the singlet state, penetrates into the H—H bond after overcoming the activation barrier (Table 8.4) predicted by the ab initio calculations [68]. The transition state structure has a linear configuration with the distance H—H···CH_2 being 1.40 Å. Triplet carbenes have also a propensity to hydrogen abstraction reactions, such as $^3CH_2 + H_2 \rightarrow CH_3 + H$, with the energy barrier of this and like reactions amounting to *ca.* 25 kcal/mol [69].

Table 8.4. Ab initio calculation data on activation barriers of insertion reactions of carbenes and silylenes into the σ-bonds

Carbene, silylene	σ-bond	Reaction product	Calculation method	Activation barrier, kcal/mol	Refs.
$CH_2(^1A_1)$	H—H	–	DZ, CI	0	[64]
$CH_2(^3B_1)$	H—H	CH_4	DZ, CI	10.8	[68]
SiH_2	H—H	SiH_4	TZ + P, CI	6.3	[70]
CHF	H—H	CH_3F	MP4/6-31G*	15.2[a] (0.6)	[71]
CF_2	H—H	CH_2F_2	MP4/6-31G*	46.9	[71]
SiHF	H—H	SiH_3F	MP4/6-31G*	40.0	[71]
SiF_2	H—H	SiH_2F_2	MP4/6-31H*	65.0	[71]
	H—H	CH_2 / CH = CH	DZ, CEPA	40.8	[65]
$CH_2(^1A_1)$	H—CH_3	C_2H_6	MP3/6-31G*	0	[72]
$CH_2(^1A_1)$	H—CH_2CH_3	CH_3—CH_2—CH_3	MP3/6-31G	0.2	[66]
SiH_2	H—CH_3	SiH_3CH_3	MP3/6-31G*	27.5	[72]
$CH_2(^1A_1)$	H—SiH_3	SiH_3CH_3	MP3/6-31G*	0	[72]
SiH_2	H—SiH_3	Si_2H_6	MP3/6-31G*	0	[72]
$CH_2(^1A_1)$	CH_3—CH_3	$CH_3CH_2CH_3$	MP3/6-31G	46.0	[66]
$CH_2(^1A_1)$	CH_2 / CH_2——CH_2	H_2C—CH_2 / H_2C—CH_2	MP3/6-31G	2.2	[66]
$CH_2(^1A_1)$	H—NH_2	CH_3NH_2	MP4/6-31G*	12.5 (28.2)	[73]
SiH_2	H—NH_2	SiH_3NH_2	MP4/6-31G*	38.0 (25.0)	[74]
SiH_2	H—PH_2	SiH_3PH_2	MP4/6-31G*	20.0 (18.0)	[74]
$CH_2(^1A_1)$	H—OH	CH_3OH	MP4/6-31G*	0	[73]
SiH_2	H—OH	SiH_3OH	MP4/6-31G*	22.0 (13.0)	[74]
SiH_2	H—SH	SiH_3SH	MP4/6-31G*	14.0 (9.0)	[74]
$CH_2(^1A_1)$	H—F	CH_3F	MP4/6-31G*	0	[73]
SiH_2	H—F	SiH_3F	MP4/6-31G*	10.0 (7.0)	[74]
SiH_2	H—Cl	SiH_3Cl	MP4/6-31G*	8.0 (2.0)	[74]

[a] In parentheses, here and below, stabilization energy is given of intermolecular complex relative to separate reactants. In these cases, the activation barrier is counted starting from the energy of the complex. For $E_{stab.} > E_a$ the reaction proceeds without a barrier

To date no experimental evidence is available regarding the insertion of silylene into the H_2 molecule, however, according to the theoretical estimates in [75], the lower limit of the activation barrier amounts to 5 kcal/mol. Other theoretical estimates [70, 76] are consistent with this value (see Table 8.4). An analysis of structural alterations in the course of the reaction $^1SiH_2 + H_2 \rightarrow SiH_4$ has revealed that they are practically identical to those found in the reaction of methylene with hydrogen. The main reason for the appearance of an activation barrier lies in the difference between the orders of the singlet and the triplet states of silylene and in the sizeable magnitude of the S–T energy gap [77, 78].

The effect of substitution of hydrogen atoms in silylene on exothermicity and activation barriers of the reaction of insertion into the H—H bond has been studied on its fluoro- and difluoro-derivatives [71]. The activation barrier grows by about 20 and 30 kcal/mol upon introduction of, respectively, one and two atoms of fluorine. This result does not practically depend on the size of the basis set nor on whether the correlation energy is taken into account. Analysis of the structure of transition states has shown that they become more late with concurrent lessening of exothermicity of insertion reactions, which is consistent with the Hammond postulate.

A very high (40 kcal/mol) activation barrier is predicted in the ab initio calculations [65] for the insertion reaction of cyclic carbene of cyclopropenylidene into the H—H bond. This may be accounted for by the strong destablization of the "electrophilic" π^*-orbital of carbene and the obstacles associated with it to the development of reaction in the initial stage.

Insertion into the C—H and Si—H bonds. The insertion of singlet methylene into the C—H bonds occurs, according to experimental [79] and theoretical [66, 72] data, either without or with very low (0.2 kcal/mol) activation barrier, whereas for silylene a considerable (27 kcal/mol) barrier is predicted in agreement with experimental observations [80, 81]. The transition states have angular three-center bond structure.

Insertion of methylene and silylene into the Si—H bond does not require overcoming of any activation barrier. This experimental conclusion can be reproduced theoretically only on condition that the correlation corrections are very carefully accounted for [72].

Insertion into the C—C bond. According to ab initio calculations with an extended basis set of the 6–31G type and the correlation energy taken into account [66], the energy profile of insertion of methylene into the single C—C bond is, in the main, determined by steric factors. So the activation barrier for insertion into the C—C bond of ethane is very high (46 kcal/mol). At the same time, a similar reaction with cyclopropane must proceed with hardly any activation (a mere 2 kcal/mol). Upon insertion into a strained cycle C—C bond whose electron density exhibits a maximum off the bond line, strong electron interactions become possible at longer inter-nuclear distances, so the authors of Ref. [66] hold that such an insertion should proceed with no activation barrier. Table 8.4 contains calculated data on reactions of carbene insertion into some other bonds of the X—H type.

The above—considered reactions of singlet carbenes involving cycloaddition and bond insertion processes by no means represent all the reaction diversity they are capable of. At the same time, precisely these reactions exhibit all the principal steps of the most important intramolecular rearrangements of carbenes, such as the reactions of 1,2-shift, of cycle extension, etc. [82–84].

On account of their greater complexity, the reactivity of triplet carbenes is much less understood than that of singlet carbenes. Very little is still known about the philicity of triplet carbenes even though it has been postulated on the experimental basis that it may be explained through the stabilization of the 1,3-biradical intermediate by electron-donating or electron-withdrawing substituents [85–87].

A so far most detailed theoretical investigation into mechanisms governing the addition of triplet carbenes to ethylene has recently been reported in Ref. [88] where relevant literature is given. Using the 3–21G technique with the correlation included, the MERP's have been calculated of the addition of methylene in the $^3B_1$ state and of substituted methylenes in which the substituents stabilize the triplet state. Among the carbenes being considered are CHCN, which contains a π-withdrawing group, as well as CHBeH and CHLi, which possess two different δ-donating groups. According to calculations, the reaction becomes less favored as the electron-releasing character of substituent is enhanced and the reactivity of triplet carbenes decreases with the diminution of their electrophilic character—this behavior is much the same as that exhibited by singlet carbenes [44, 45].

References

1. Schlegel HB, Bhalla KC, Hase WL (1982) J Phys Chem 86:4883
2. Dneprovskii AS, Temnikova TI (1979) Theoretical foundations of organic chemistry (in Russian). Khimiya, Leningrad
3. Yamaguchi K, Fukutome H (1975) Progr Theor Phys 54:1599
4. Yamaguchi K (1982) Intern J Quant Chem 22:459
5. Murrell JN (1977) Structure and Bonding 32:93
6. Kato S, Fukui K (1976) J ACS 98:6395
7. Bonaccic-Koutecky V, Koutecky J, Salem L (1977) J ACS 99:842
8. Dorigo AE, Houk KN (1988) J Org Chem 53:1650
9. Menger FM (1985) Acc Chem Res 18:120
10. Basilevski MV (1967) Dokl Akad Nauk SSSR 172:87
11. Dewar MJS, Olivella S (1978) J ACS 100:5290
12. Schlegel HB (1982) J Phys Chem 86:4878
13. Tedder JM, Walton JC (1978) Adv Phys Org Chem 16:51
14. Clark DT, Scanlan IW, Walton JC (1978) Chem Phys Lett 55:102
15. Sosa, C, Schlegel HB (1986) Int J Quant Chem 29:1001; 30:155
16. Sosa C, Schlegel HB (1987) J ACS 109:4193
17. Sosa, C, Schlegel HB (1987) Int J Quant Chem Quant Chem Symp 21:267
18. Schelegel HB, Sosa C (1988) Chem Phys Lett 145:329
19. Schelegel HB (1986) J Chem Phys 84:4530
20. Houk KN, Paddon-Row MN, Spellmeyer DC, Rondan NG, Nagase S (1986) 51:2874
21. Delbecq F, Ilavsky D, Anh NT, Lefour JM (1985) J ACS 107:1623
22. Arnaud R, Barvone V, Olivella S, Sole A (1985) Chem Phys Lett 118:573
23. Wolfgang R (1969) Acc Chem Res 2:248; (1970) Acc Chem Res 3:48
24. Ingold KU, Roberts BP (1971) Free-radical substitution reactions. Wiley, New York
25. Morokuma K, Daies RE (1972) J ACS 94:1060
26. Fukui K, Kato S, Fujimoto H (1975) J ACS 97:1
27. Borden WT, Davidson ER (1981) Acc Chem Res 14:69
28. Salem L, Rowland C (1972) Angew Chem Intern Ed Engl II: 92
29. Parmon VN, Kokorin AI, Zhidomirov GM (1980) Stable biradicals. Nauka, Moskow (in Russian)

30. Berson JA (1980) in: Rearrangements in ground and excites states v 1:311, Academic Press, New York
31. Minkin VI, Simkin BYa, Minyaev RM (1979) Theory of the molecular structure. Vyschaya Schkola, Moskow (in Russian)
32. Stohrer WD, Hoffmann R (1972) J ACS 94:779
33. Newton MD, Schulman JM (1972) J ACS 94:4391
34. Cremaschi P, Gamba A, Simonetta M (1972) Theor Chim Acta 25:237
35. Doubleday Ch, McIver JW, Page M (1988) J Phys Chem 92:4367
36. Doubleday Ch, McIver JW, Page M (1982) J ACS 104:6533
37. Skell PS (1985) Tetrahedron 41:1427
38. Skell PS, Woodworth RC (1956) J ACS 78:4496
39. Skell PS, Garner AY (1956) J ACS 78:3409, 5430
40. Bodor N, Dewar MJS, Wasson JS (1972) J ACS 94:9095
41. Hoffmann R (1968) J ACS 90:1475
42. Goddard WA (1972) J ACS 94:793
43. Hay PJ, Hunt WJ, Goddard WA (1972) J ACS 94:638
44. Rondan NG, Houk KN, Moss RA (1980) J ACS 102:1770
45. Moss RA, (1980) Acc Chem Res 13:58
46. Moss RA, Shen S, Hadel LM, Kmiecik-Lawrynowicz G, Wlostowska J, Krogh-Jespersen K (1987) J ACS 109:4341
47. Kostikov RR, Drygilova EA, Golovkina EA, Komendantov AM, Molchanova P (1987) Zh Org Khim 23:2170
48. Zurawski B, Kutzelnigg W (1978) J ACS 100:2654
49. Anwari F, Gordon MS (1983) Isr J Chem 23:129
50. Apeloig Y, Karni M, Stang PJ, Fox DF (1983) J ACS 105:4781
51. Moreno M, Lluch JM, Oliva A, Bertran J (1985) J Chem Soc Perkin Trans II:131
52. Houk KN, Rondan NG, Mareda J (1985) 41:1555
53. Moreno M, Lluch JM, Oliva A, Bertran J (1986) J Chem Soc Perkin Trans II:186
54. Schoeller WW, Aktekin N (1982) J Chem Soc Chem Comm 20
55. Skell PS, Cholod MS (1969) J ACS 91:7131
56. Giese B, Meister J (1978) Angew Chem Int Ed Engl 17:595
57. Giese B, Lee WB (1980) Angew Chem Int Ed Engl 19:835
58. Moss RA, Lawrynowicz W, Turro NJ, Gould IR, Cha Y (1986) J ACS 108:7028
59. Wong PC, Griller D, Scaiano JL (1981) Chem Phys Lett 103:2423
60. Houk KN, Rondan NG, Mareda J (1984) J ACS 106:4291
61. Dobson RC, Hay DM, Hoffmann R (1971) J ACS 93:6188
62. Cremaschi P, Simonetta M (1974) J Chem Soc Faraday Trans II 70:1801
63. Kollmar H, Staemmler V (1979) Theor Chim Acta 51:207
64. Bauschlicher CW, Haber K, Schaefer HF, Bender CF (1977) J ACS 99:3610
65. Kollmar H (1978) J ACS 100:2660
66. Gordon MS, Boatz JA, Gano DR, Friederichs MG (1987) J ACS 109:1323
67. Jeziorek D, Zurawski B (1979) Int J Quant Chem 16:277
68. Bauschlicher CW, Bender CF, Schaefer HF (1976) J ACS 98:3072
69. Gordon MS (1984) J ACS 106:4054
70. Sax A, Olbrich G (1985) J ACS 107:4868
71. Sosa C, Schlegel HB (1984) J ACS 106:5847
72. Gordon MS, Gano DR (1984) J ACS 106:5421
73. Pople JA, Raghavachari K, Frisch MJ, Binkley JS, Schleyer PR (1983) J ACS 105:6389
74. Raghavachari K, Chandrasekhar J, Gordon MS, Dykema KJ (1984) J ACS 106:5853
75. John F, Purnell JN (1973) J Chem Soc Farad Trans II 69:1455
76. Gordon MS, Gano DR, Binkley JS, Frisch MJ (1986) J ACS 108:2191
77. Shaik SS (1981) J ACS 103:3692
78. Pross A, Shaik SS (1983) Acc Chem Res 16:363
79. Jones M, Moss RA (eds) (1972) Carbenes. v 2. Wiley, New York
80. Sawrey BA, O'Neal HE, Ring MA, Coffey D (1984) Int J Chem Kinet 16:31
81. Davidson IMT, Lawrence FT, Ostah NA (1980) J Chem Soc Chem Comm:659
82. Jones WH (1977) Acc Chem Res 10:353

83. Jones WM (1980) in: De Mayo P (ed) Rearrangements in ground and excited states. Academic Press, New York, p 95
84. Malzev AK, Korolev VA, Nefedov OM (1986) in: Kolotyrkin YM (ed) Physical chemistry. Current problems. Khimiya, Moscow (in Russian), p 144
85. Jones GW, Chang KT, Munjal R, Schechter H (1978) J ACS 100:2922
86. Giese B (1983) Angew Chem Int Ed Engl 22:753
87. Tomioka H, Ohno K, Izawa Y, Moss RA, Munjal RC (1984) Tetrahedron Lett 25:5415
88. Moreno M, Lluch JM, Oliva A, Bertran J (1988) THEOCHEM 164:17

CHAPTER 9

Electron and Proton Transfer Reactions

The electron and the proton are the lightest of the particles, taking part in a chemical reaction, whose electric charges are opposite in sign. These two factors predetermine the specific character of the electron and the proton transfer reactions, i.e., their extremely high velocities which, however, depend a good deal on the medium. Numerous works, both experimental and theoretical, have been devoted to studying those factors, to finding the relationships between kinetics and thermodynamics of such reactions. The present chapter will treat a different aspect of these processes which is associated with peculiar features characterizing the intrinsic mechanism of the elementary steps of electron and proton transfer and with the role they play in the overall stoichiometric mechanism of a number of organic and bioorganic reactions.

9.1 Electron Transfer Reactions

9.1.1 Single Electron Transfer Reactions in Organic Chemistry

Electron transfer as an elementary step is a necessary part of the stoichiometric mechanism of all redox and, as has recently been shown, some heterolytic reactions. The most important reactions are those of single-electron transfer (SET). The multi-electron transport in chemical and biochemical transformations consists, as a rule, of a series of consecutive single-electron transfer reactions, while the one-step two-electron transfers are relatively rare.

In organic chemistry, the reactions under consideration could first be registered in pure form by characteristic broadening of lines in the ESR spectra of radical-ions:

$$\text{C}_{10}\text{H}_8 + [\text{C}_{10}\text{H}_8]^{\cdot -} \rightleftharpoons [\text{C}_{10}\text{H}_8]^{\cdot -} + \text{C}_{10}\text{H}_8 \qquad (9.1)$$

$$\text{C}_6\text{H}_5^{\cdot -}\text{–}(CH_2)_n\text{–C}_6\text{H}_5 \rightleftharpoons \text{C}_6\text{H}_5\text{–}(CH_2)_n\text{–C}_6\text{H}_5^{\cdot -} \qquad (9.2)$$

Using commonly accepted terms for electron transfer reactions [2–4], the reaction of Eq. (9.1)—see Ref. [1]—corresponds to the case of the outer-sphere, while the reaction of Eq. (9.2) to that of the inner-sphere exchange [5]. The frequency of acts of electron transfer amounts in both processes to 10^7–$10^8\ s^{-1}$ at room temperature, the energy of activation is approximately 3 kcal/mol. Many other elementary reactions of single-electron transfer [2–4, 6–8] also proceed at very high rates (up to 10^{12}–$10^{15}\ s^{-1}$) and with low energy barriers ($\Delta G^{\neq} \ll k_B T$) see Ref. [9]. This fact, along with the formal possibility of devising a scheme of nearly every chemical reaction with the intermediates formed in a single-electron transfer incorporated, let alone experimental detection, in some cases, of radical or radical ion particles in the reaction mixture, may have tempted one to produce numerous hypothetical mechanisms of organic reactions with SET steps included. It should, however, be emphasized that precise experimental data confirming the electron transfer as an inherent step of the stoichiometric reaction mechanism and proving the vital rather than a subordinate role of the radical and ion-radical intermediates in the kinetical scheme have, up to-date, been obtained for relatively few organic reactions. Of course, among these few there are some important transformations, such as the radical nucleophilic substitution reactions in the series of nitro derivatives of alkanes [10] and aromatic (R = Aryl) compounds, i.e., the $S_{RN}1$ reactions [11, 12]:

$$
\begin{array}{ll}
(a) & RX + Y^- \longrightarrow RX^{\dot{-}} + Y^{\cdot} \\
(b) & RX^{\dot{-}} \longrightarrow R^{\cdot} + X^- \\
(c) & R^{\cdot} + Y^- \longrightarrow RY^{\dot{-}} \\
(d) & RY^{\dot{-}} + RX \longrightarrow RY + RX^{\dot{-}}
\end{array}
\tag{9.3}
$$

The reactions in Eq. (9.3) represent a typical chain process in which the steps (a) and (d), i.e., those of initiation and propagation of a chain, correspond to a single-electron transfer. In view of the fundamental novelty of the mechanism of radical nucleophilic substitution and great importance of reactions of this type for preparative organic chemistry, the separate steps constituting this mechanism (Eq. (9.3, a–d)] were thoroughly studied from the theoretical angle [13]. Before considering the principal results of that study, a brief introduction to the theory of electron transfer in molecular systems would be in order.

9.1.2 Elementary Act of Electron Transfer

The general theory of the elementary act of charge transfer in polar solvents has been well worked out and described in great detail in a number of monographs [2, 4, 14] and special reviews [15–19]. Here we offer some extracts from it that may give an idea of which energy and structural requirement must be satisfied by the system where the electron transfer takes place and how these characteristics affect the rate of a given reaction.

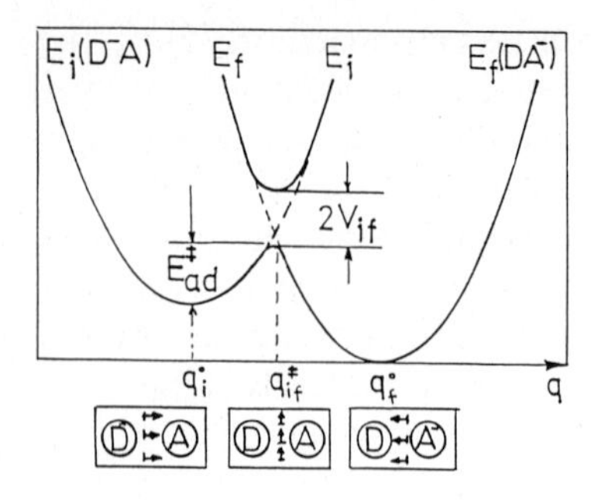

Fig. 9.1. PES of the adiabatic electron transfer reaction of Eq. (9.4) in polar solvent obtained through superposition of the terms of the initial E_i and the final E_f states (q is the reaction coordinate that includes the solvent reorganization). The process is shown schematically as a change in the polarization of the medium when passing from the equilibrium configuration (q_i^0) of the initial state to the equilibrium configuration (q_f^0) of the state with the transferred electron. The electron transfer occurs in the region $q_{if}^{\neq}$. When V_{if} is small, there exists a probability for the reaction trajectory to cross the transition state region without leading to product formation (nonadiabatic reaction)

Consider the reaction:

$$D^- + A \rightarrow D + A^- \tag{9.4}$$

in which a single-electron transfer occurs from the anion D^- (a donor) to the neutral particle A (an acceptor). The course of Eq. (9.4) may conveniently be described with the aid of a PES constructed through superposition of the terms (local PES's) of two extreme states—before $[E_i(q)]$ and after $[E_f(q)]$ of the electron transfer, as is shown in Fig. 9.1. The development of this reaction, namely, transition from one state into a new one, proceeds with the variation of the generalized reaction coordinate q which describes restructuring of D^- and A in the course of formation of the transition state of the electron transfer reaction as well as reorganization of the solvation shell of the reactants' system.

Both types of structural alterations are designed to take the system to that region of the configurational space ($q_{if}^{\neq}$ in Fig. 9.1) where the terms of the initial (D^-A) and the final (DA^-) state cross and these states become isoenergetical and isostructural. Precise coincidence of the energy levels and the geometrical characteristics of states of the system prior and subsequent to the act of electron transfer is a necessary condition for its realization. It is dictated by the demands of the conservation of energy and the impulse. As the time for electron transfer (10^{-15}–10^{-16} s) is less than the period of nuclear vibrations by 3–4 orders of magnitude, the internuclear distances and the velocities of nuclear motion do not change over the period of the electron transfer (the Frank–Condon principle). Thus, the kinetical energy and the impulse of the nuclei must in the electron transfer process stay unchanged, which is possible only in the region ($q_{if}^{\neq}$) where the change in the electron energy upon transition between two electron states is also zero.

Important for the choice of a physical mechanism of the electron transfer step is the degree of interaction between the electron states E_i and E_f in the area of their intersection q_{if}. In case the interaction is not spin-forbidden, the split of the energy level effected by it will be determined by the off-diagonal matrix element V_{if}, as in Eq. (1.42), whose magnitude is proportional to the overlap—see Eq. (4.3).

If the overlaping of the donor and acceptor orbitals in the $q_{if}^{\neq}$ region and the corresponding split of the terms are considerable, then the electron transfer occurs in a thermal reaction by the adiabatic mechanism as a motion of the system over the lower energy surface in the transition state zone denoted in Fig. 9.1 by a dashed line. The activation energy of the adiabatic electron transfer reaction $E_{ad}^{\neq}$ corresponds to the difference between the energy levels of the initial state and the saddle point (the top of the barrier).

Approximate estimates [4] lead to the conclusion that for a reaction to proceed in the adiabatic regime the interaction energy of the terms E_i and E_f in the $q_{if}^{\neq}$ region has to amount to ~ 0.5 kcal/mol. With a lesser split, the adiabatic character of reaction is disturbed and the transition from the surface $E_i(q)$ to $E_f(q)$ takes place through tunnelling under the barrier. Unlike the adiabatic mechanism, the nonadiabatic electron transfer can be realized with appreciable probability only on condition of repeated passage across the region $q_{if}^{\neq}$. The reaction rate constant contains a factor corresponding to the probability P_{if} of the nonadiabatic transition according to the Landau–Zener relationship of Eq. (1.42) which includes the quantity V_{if}.

For most electron transfer reactions in inorganic and organic systems, the adiabatic mechanism operates even when the centers of localization of the electron being exchanged are spaced quite a large distance apart. For example, for the complexes containing metal ions with variable valence, the inner-sphere electron exchange via π-conjugated bridges, such as:

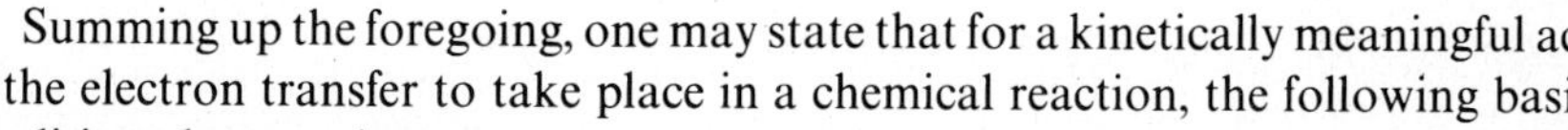

$(NH_3)_5\overset{2+}{Co}-N\langle\bigcirc\rangle-X-\langle\bigcirc\rangle N-\overset{3+}{Ru}(NH_3)_5$ Ia

$(NH_3)_5\overset{3+}{Co}-N\langle\bigcirc\rangle-X-\langle\bigcirc\rangle N-\overset{2+}{Ru}(NH_3)_5$ Ib

$X = CH_2, S, CH{=}CH, C{\equiv}C$

occurs adiabatically [$P_{if} \approx 1$ in Eq. (1.42)] even if the metal centers lie up to 20 Å apart [20].

Summing up the foregoing, one may state that for a kinetically meaningful act of the electron transfer to take place in a chemical reaction, the following basic conditions have to be met:

1) The configurational and conformational deformations of the initial system as well as the reorganization of the solvation shell, all of which bring the initial system into a region isoenergetic and isostructural to the system with the electron transferred, can be realized in thermal reactions at moderate activation energies $E_{ad}^{\neq}$.
2) In the transition state zone there must be present a considerable overlapping of the orbitals of interacting reactants partly or fully localized at the reactive centers of the intermediate complex being formed or at those inside one molecule.

In the intermolecular-type reactions, the fulfillment of the latter requirement often depends on the formation of an intermediate complex (DA) between the donor and the acceptor components. This complex is included as an intermediate in the reaction of the type of Eq. (9.4)—see Ref. [21, 22]. Its formation is thermodynamically advantageous when the ionization potential of the donor is smaller than the electronic affinity of the acceptor. Generally, endothermic reactions of electron transfer are possible too. In this case, the electron transfer step in the intermediate complex formed $[DA] \rightarrow [D^+A^-]$ determines the total reaction rate.

9.1.3 Theoretical Studies of the Mechanism of $S_{RN}1$ Reactions

So as to calculate electron transfer reaction rates in polar media,a number of effective approaches have been evolved based on model forms of the PES and on various models for evaluating the medium polarization and the energy of solvent reorganization [2, 17, 18, 20, 23–26]. Recently this analysis has been raised to the level of semiclassical trajectory calculations of the dynamics of above reactions [27]. At the same time, there are so far no sufficiently complete quantum mechanical calculations of the PES of the electron transfer reactions on account of complexities of the calculation involving extension of the basis sets and careful assessment of resolvation of the reactants in the development of the reaction process.

As an important example of the value of direct quantum mechanical calculations on the reactions that include electron transfer, let us consider some results of a theoretical analysis, done by Salem et al. [13] of the chain mechanism of the $S_{RN}1$ reaction of Eq. (9.3) whose idea was first put forward by Kornblum and Russell [10–12, 22]. The simplest compounds susceptible to this reaction are 2-chloro-2-nitropropane ($R = Me_2CNO_2$, $X = Cl$) and 2-nitropropyl anion ($Y = Me_2CNO_2$). In the example under consideration the methyl groups were replaced by hydrogen atoms and the overall reaction of Eq. (9.3) could be represented as a sequence of the corresponding steps:

$$
\begin{aligned}
&\text{a)}\ O_2NCH_2Cl + CH_2NO_2^- \rightarrow O_2NCH_2Cl^{\dot{-}} + CH_2NO_2;\\
&\text{b)}\ O_2NCH_2Cl^{\dot{-}} \rightarrow CH_2NO_2^{\cdot} + Cl^-;\\
&\text{c)}\ CH_2NO_2^{\cdot} + CH_2NO_2^- \rightarrow O_2NCH_2CH_2NO_2^{\dot{-}};\\
&\text{d)}\ O_2NCH_2CH_2NO_2^{\dot{-}} + O_2NCH_2Cl \rightarrow O_2NCH_2CH_2NO_2^{+} + O_2NCH_2Cl^{\dot{-}}.
\end{aligned}
\tag{9.5}
$$

Among not fully resolved questions concerning the mechanism of such reactions, we concentrate on two. First, what is the structural mechanism operative in the realization of the chain initiation step of Eq. (9.3a), i.e., with what orientation of the reactants optimum overlapping is achieved between the highest occupied MO of nitromethyl anion and the lowest vacant MO of chloronitromethane and how is accomplished the energy levelling between the initial reactants and the products of the electron transfer. Secondly, of interest is the question regarding stability of the intermediate radical anion generated at the step of Eq. (9.5) and the mechanim of its decomposition.

Since the elementary reactions in the stoichiometric mechanism of Eq. (9.5) include participation of anionic particles, the basis set in the ab initio calculations [13] contained polarization functions as a supplement to the standard set STO-3G and for a correct description of the acts of bond rupture [Eq. (9.5b)], several hundreds of mono- and double-excited electron configurations were taken into account. Allowance for the solvent (nine water molecules in the first coordination shell) was made in terms of the point charge model, as described in Sect. 3.3.2.

Calculations revealed two variants of the structural types in which the necessary conditions are satisfied for the electron transfer in the first step [Eq. (9.5a)]. In the first of these, nitromethyl anion II approaches in its stable planar form (starting from 3–3.5 Å) chloronitromethane in a "head-to-head" fashion. An even more positive overlapping of the frontier orbitals of the electron acceptor and electron donor (nitromethyl anion) is achieved if the latter undergoes certain structural transformations, namely, the rotation about the C—N bond and pyramidalization of the carbon atom (III):

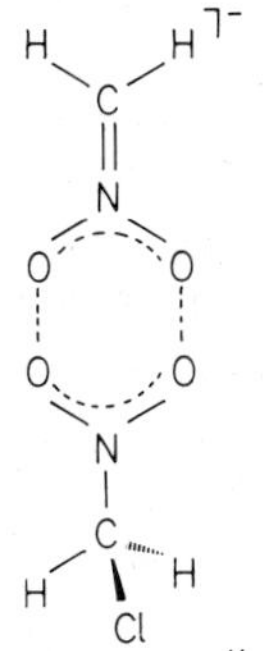

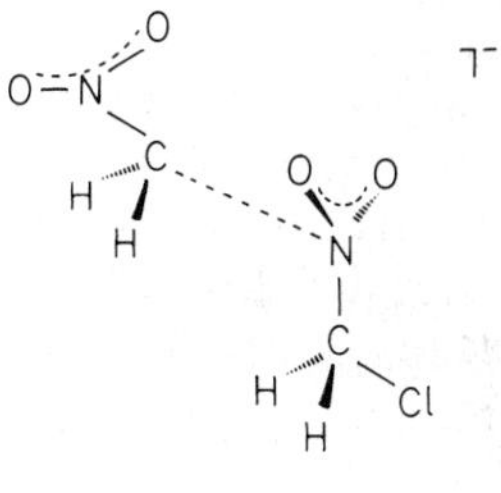

This rotation should also facilitate electron transfer by destabilizing the initial system and bringing its energy level closer to that of the products (Fig. 9.2, $r_{CCl} = 1.8$ Å). The calculations show, however, that a transition from the stable conformation of the nitromethyl anion to a structure having the nitro group turned in the dimer III would require too great an increase in energy (by 43 kcal/mol).

More realistic is another type of structural deformation making the system ready for the electron transfer, which is associated with the stretching of the bond C—Cl in II. As may be seen from Fig. 9.2, a stretching of this bond by 0.6 Å destabilizes the initial system by 25 kcal/mol while simultaneously

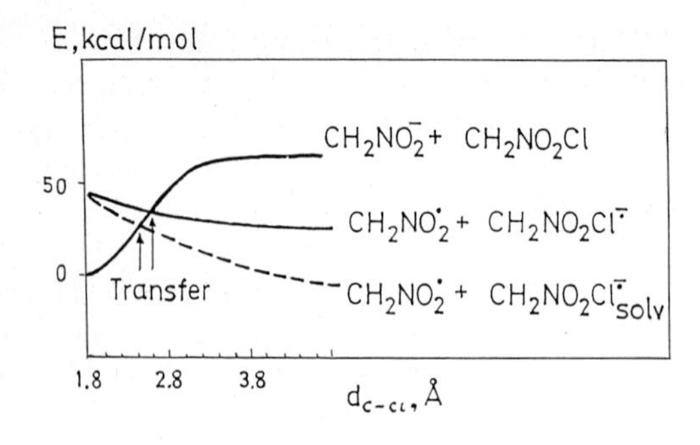

Fig. 9.2. Potential energy curves of the system of Eq. (9.5) at the stage of Eq. (9.5a) relative to the initial reactants as a function of the bond length C—Cl [13]. The dashed line represents calculation with the solvâtion energy taken into account. Arrows show the regions where the electron transfer takes place. (Reproduced with permission from the American Chemical Society)

stabilizing the products of the electron transfer providing thus for the achievement of the isoenergetic state required. The data in Fig. 9.2 suggest also that the inclusion of solvation efffects lowers the calculated activation energy of the thermal electron transfer.

The chloronitromethyl radical anion formed at the initiation stage is unstable (there is no minimum for it on the PES). The step of Eq. (9.5b) develops spontaneously with the stretching of the bond C—Cl modelling quite well the reaction coordinate. According to calculations [13], the decomposition of the radical anion is initiated also by a gain in the solvation energy (see Fig. 9.2). For the step of Eq. (9.5c) calculations have found no activation barrier and predicted considerable exothermicity of the reaction. The electron transfer at the step of Eq. (9.5d), as opposed to Eq. (9.5a), must be associated with a considerable exothermic effect since the calculated electron affinity of chlorinitramethane (+0.7 eV) is higher than in 1,2-dinitroethane. This reaction must be diffusion controlled.

Thus, the calculations [13] have produced evidence in favor of the Kornblum–Russell mechanism elucidating certain details of the nature and structural conditions for realization of its individual steps. Of course, so far only the simplest models have been considered, and the effect of the simplifying assumptions may in some cases be quite substantial. For example, experimental studies of radical-anions of various halogennitrobenzyl compounds IV using the impulse radiolysis technique [28] show that, unlike the chloronitromethyl radical considered earlier, the compounds IV do exist in solution for some finite time. Even so, their decomposition is, in accordance with the theoretical prediction, speeded up upon weakening of the C—X bond. The reaction proceeds by the scheme of intramolecular, inner-sphere electron transfer whose relatively low rate constant (10^7–10^0 s) can be accounted for by the small overlap of the π-orbital, which carries the unpaired electron in the initial form IV and is localized predominantly at the nitro group, and the σ^*-orbital C—X:

$$\underset{\text{IV}}{R-\underset{\displaystyle X}{\underset{|}{CH}}-C_6H_4-NO_2^{\dot{-}}} \longrightarrow R-\dot{C}H-C_6H_4-NO_2 + X^-$$

$$X = Cl, Br, TsO^-$$

Not only the nitroaromatic species, such as IV, but also some simpler compounds, which used to be considered as typical substrates of the S_N2 reactions, can be involved in multistep radical-forming nucleophilic substitutions. Evidence has been accumulating over the last years that the nucleophilic substitution with alkyl halides occurs, at least in some instances, by the single-electron transfer mechanism. It has been suggested [29, 30] that the SET and S_N2 mechanisms represent the extremes of a wide spectrum of mechanistic possibilities for substitution reactions. It has been deduced on qualitative theoretical grounds that the propensity of alkyl halide R—X to react with nucleophiles via an electron-transfer step depends crucially on the stability of the three-electron bond R—X in the initially formed radical-anion species. A more electronegative R will stabilize this bond and bring about a shift in the mechanism from the S_N2 to the SET type, which has then experimentally been shown to be a correct conclusion, see Ref. [30].

This brief theoretical analysis of the mechanism of the $S_{RN}1$ reaction of Eq. (9.3) shows how complicated the complete calculation may be of the reaction pathways associated with the electron transfer in polar solvents. But it also shows how one should, within the framework of realistic possibilities, approach the problems that have to do with the examination and identification of organic reactions of this type.

So as to ascertain a possibility of the electron transfer in a certain reaction phase, a quite useful criterion may be obtained through an analysis of the Hartree–Fock triplet (doublet in the case of radicals) stabilities of solutions (see Sect. 2.2.5). Since such an analysis is relatively simple, it should precede a more rigorous examination similar to that described in this section. Some examples of how an analysis of stability of the HF-solutions may be used in calculations on the reactions with a possible electron transfer are considered in Sect. 5.4, 6.1, and 8.3.

9.2 Proton Transfer Reactions

First assumptions as to the possibility of hydrogen changing its position in the process of transformation of one chemical particle into another one were voiced nearly 200 years ago [31], and the mechanism of this enormously important reaction has been attracting attention ever since. The reason for this interest is obvious seeing that the proton transfer reactions underlie the acid-base equilibria and determine the mechanisms of a broad range of biological phenomena such as the transport across membranes, enzymic catalysis, photosynthesis, the formation of the ATP acid, spontaneous mutations etc.

Similarly to the electron transfer reactions, a proton transfer is preceded by the formation of an intermediate associate which in this case is a hydrogen bond complex (H-complex) V:

$$A-H+B \rightleftarrows A-H\cdots B \rightleftarrows A\cdots H-B \rightleftarrows A+H-B \tag{9.6}$$

The proton transfer reaction proceeds as a transition from one H-complex (Va) to another (Vb) with the proton that moves along the hydrogen bond having to overcome a certain activation barrier. Let us consider some specific features of the intrinsic mechanism of this reaction.

9.2.1 Potential Energy Curves and Activation Barriers

The principal details of the proton transfer in Eq. (9.6) will become apparent after an analysis of the calculations [32–42] that determine the dependence of the form of potentials and activation barriers on the distance X.... X:

$$XH^+ + X \rightarrow X + HX^+$$
$$X = (H_2O)_n,\ NH_3,\ H_2S \quad n = 2, 3, 4, 5 \tag{9.7}$$

Figure 9.3 shows the PES sections of proton transfer in the ion $H_5O_2^+$ VI and its dihydrated form $H_9O_4^+$ VII:

VI VII

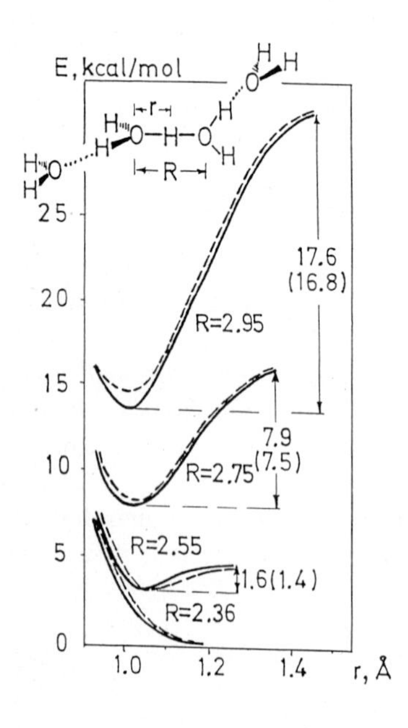

Fig. 9.3. Proton-transfer potentials for $(H_5O_2)^+$ (dashed curves) and $(H_9O_4)^+$ (solid curves). Energy barriers shown in parentheses refer to the dimer [32]. (Reproduced with permission from the American Chemical Society)

The equilibrium distance R = 2.36 Å corresponds to a single-well potential with symmetrical hydrogen bonding, with the proton occupying the symmetrical position in the bridge O—H—O. So as to provide for the proton transfer over this bridge from one oxygen atom to the other, the system must be vibrationally excited by stretching the equilibrium distance R. This extension of R in VI and VII gives rise to a second minimum. The energy barrier appearing in the double-well potential formed upon stretching of the hydrogen bridge is growing with the increase of R. The addition of the water molecules in VII does not practically alter the form of the potential curve of the proton transfer reaction of Eq. (9.7), which enables the model system $(H_5O_2)^+$ to be regarded as adequately representing the more complex chain oligomers.

An analogous mechanism is, according to calculations [36, 37, 39, 42, 43], operative in the nondegenerate proton transfer reaction in the heterodimer VIII:

VIII IX

These calculations were designed to study the effect of the basis sets and correlations on the magnitude of adiabatic barriers of proton transfer. At the equilibrium distance N···O equalling 2.709 Å (DZ + P basis set with about 3000 excited electron configurations included), the system VIII fits the single-well potential whose minimum corresponds to the structure $(NH_4)^+(OH_2)$ with $r_{N-H} = 1.035$ Å. The complex IX $(NH_3)(OH_3)^+$ cannot exist in the gas phase. A lengthening of the N···O distance gives rise to an additional minimum starting from R = 2.95 Å, which corresponds to this structure in the nonequilibrium configuration of the dimer VIII. Table 9.1 lists the magnitudes of the barriers of N → O and O → N proton transfers calculated in different approximations for various distances.

Unlike the H-complexes VI, VIII, for the systems NH_3—H···NH_3^+, H_2S—H···SH_2^+ and HO—H···OH^- double-well potentials are characteristic.

Table 9.1. Barriers of nondegenerate proton transfer in the system VIII IX [36, 37, 43]

	$E^{\neq}$, kcal/mol		
R(O···N), Å	4–31G	DZ + P	DZ + P + CI
		NH → O	
2.85	30.8	35.8	28.5
3.20	45.2	51.4	40.4
		OH → N	
2.95	2.2	4.3	0.5
3.20	12.2	16.0	8.5

Table 9.2. Activation barriers of the proton transfer reaction in $HOH\cdots OH^-$ and geometry characteristics of the ground and the transition states

Method of calculation [Ref.]	R_{∞} (equilib.)	R_{∞} (transit.)	$E^{\neq}$, kcal/mol
4–31 G [38]	2.469	2.433	0.2
6–31G** [38]	2.509	2.40	1.0
[54/31] + CI[a] [46]	2.465	2.422	0.1

[a] 50 000 single- and double-excited configurations

Calculations show that in the case of these H-complexes a transition from one asymmetrical structure to the other encounters quite low activation barriers. Experimental studies show that the proton is localized approximately midway between both oxygen atoms [44, 45]. It is particularly true of the anionic H-complex $(H_3O_2)^-$. The data given in Table 9.2 show that as the proton moves away from its equilibrium position at one of the oxygen atoms, there occurs not an increase, as in previous cases, but rather a reduction in the distance R_{∞}.

A methodologically important conclusion follows from the data in Tables 9.1 and 9.2: when using DZ basis sets of the 4–31G type, one may achieve the results close to those of the most rigorous calculations. This is explained by nonmonotonical dependence of calculation accuracy on the size of an AO basis set. An increase in the size can be compensated for by taking the electron correlation into account. The 4–31G basis set is widely used in calculations on H-complexes and dynamics of the proton transfer. Comprehensive lists of activation parameters of degenerate and nondegenerate proton transfer reactions may be found in Refs. [39, 40, 42, 47].

It should be kept in mind that the term "proton transfer" by no means signifies that the hydrogen migrating along the hydrogen bond line carries a single positive charge. In actual fact, the transfer reaction is associated with a redistribution of the electron density throughout the whole H-complex. Thus, for VI the increase in the positive charge at the hydrogen atom equals in the transition state only 0.03–0.04 e depending on R [32], and for VIII this increase amounts to a mere 0.024 e [36, 43].

It was quite a good while ago that Dogonadze et al. [48] and Marcus [49] extended the ideas of solvent intervention in electron transfers to proton-transfer processes in solutions. The central point in this theory has been to find out whether there is a correlation between the motions of the solvent molecules and the proton. A clearcut and exhaustive answer has not yet been given to this problem, although a partial solution may be found in Ref. [50] where the $(H_3O_2)^-$ and $(H_3O_2)^-$ $(H_2O)_2$ species were studied by ab initio 4–31G calculations. The participation of solvation parameters in the reaction coordinate points to a correlation between the motion of the water molecules and the proton transfer. The classic idea of considering the solvent as adjusting its position to the motion of the proton by means of a relaxation process may be changed to another view according to which the proton follows in its transfer process the solvent fluctuations. This assumption is borne out by the fact that when one permutes the solvation parameters of the 4–31G $(H_3O_2)^-(H_2O)_2$ minimum, a spontaneous proton transfer occurs. Consequently, the use of a continuum model to take into account the solvent effect on a proton transfer reaction is not adequate to describe the solvent molecules that intervene in the reaction coordinate [50]. Currently, the theory should concentrate on working out certain general rules that would regulate the use of the supermolecular and the continuum approaches to the description of the medium effects.

9.2.2 Stereochemistry

The data on energetics considered in the preceding Section relate to an optimum linear configuration V of the hydrogen bond bridge. This configuration can, on account of given steric conditions, far from always be implemented in real systems, such as biopolymers (proteins, carbohydrates, DNA), compounds with intramolecular hydrogen bonding etc. In this connection, the question arises as to how large admissible deviations from the linear three-center configuration of the hydrogen bond may be to allow the proton transfer to proceed with still relatively low energy barriers characteristic of Eqs. (9.6) and (9.7).

The theoretical analysis was based on calculations [36, 43] designed to determine the dependence of the magnitude of proton transfer barriers on the angular parameters that indicate deviations of the hydrogen bonds from linearity both for the heterodimer VIII (see VIIIa) and for the homodimers $(H_5O_2)^+$ VI and $(H_7N_2)^+$ X:

VIIIa

X

The calculation results with the AO basis set 4–31G at R = 2.95 Å, at which all potential curves are double-well, are presented in Table 9.3.

Table 9.3. The reaction barriers of proton transfer in the cationic dimers VI, VIII, IX, X at $R_{AB} = 2.95$ Å

α_1, α_2 (deg.)	VIII(NH→O)	IX(OH→N)	VI(OH→O)	X(NH→N)
0; 0	29.1	2.6	16.7	11.5
0; 20	30.8	3.3	18.6	12.9
0; 40	36.3	7.2	23.4	17.9
20; 20	32.5	5.5	19.8	14.1
40; 40	49.7	11.5	30.1	23.9
20; −20	33.4	4.5	20.5	15.0
40; −40	52.1	17.0	37.3	33.9

As seen from the data of Table 9.3, when the migrating proton is in the cone whose surface is determined by the deviation angles not greater than 20°, the increase in the transfer barrier never exceeds 3–5 kcal/mol. With greater distortions the barrier grows more sharply. This situation is similar to that considered earlier (Sect. 5.3.3.1) when an optimum angle of approach of the nucleophile to a carbon group was determined. The calculation results are consistent with the experimental data (see review under Ref. [51]) on geometry of hydrogen bonds for a wide series of crystalline structures of mono- and polysaccharides. While 10–20° deviations are ordinary, occasionally the angles α may become as large as 35°.

Thus, energetics of the proton transfer is quite sensitive to the effects of hydrogen bond geometry distortions (stretching and angular deformation). In Refs. [42, 47] this conclusion was shown to be correct for other systems also.

9.2.3 Proton Transfer in Systems with the Intramolecular Hydrogen Bonding

Transformations of this type play an important role in the mechanism of quite a few reactions, in particular, it is they which lie at the root of the intramolecular tautomerism [52]. It is therefore not surprising that the theoretical investigation of the intramolecular proton transfers is represented by a wide range of calculations of methodological importance. Particularly thoroughly was studied the reaction of the 1,5-proton shift in 3-oxy-2-propenal (*cis*-enol of malonaldehyde) XI which affords one of the most significant examples of the degenerate intrachelate tautomerism.

In the case of XI, the presence of two structures of C_s and C_{2v} symmetry may be assumed whose nature and interconversions will determine the PES profile (Fig. 9.4):

XIa, C_s ⇌ XIc, C_{2v} ⇌ XIb, C_s (9.8)

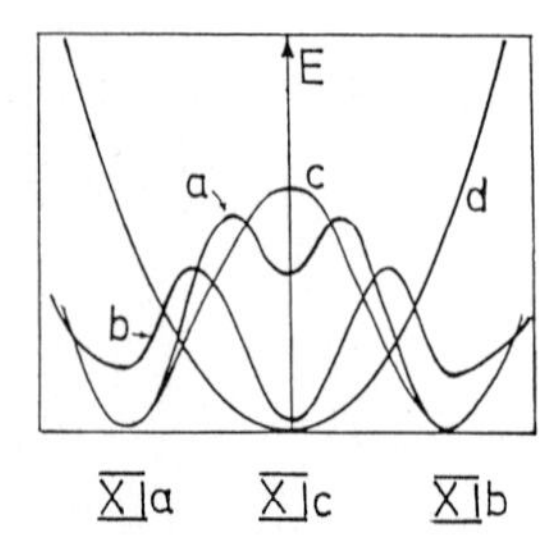

Fig. 9.4. Possible energy profiles of the intramolecular proton transfer reaction of Eq. (9.8) in malonaldehyde XI

The form of the PES, the nature of stationary points and the depth of minima all substantially depend on both the calculation scheme and the accuracy of optimization of geometry parameters. The conventional semi-empirical methods CNDO/2, MINDO/3 and MNDO prove inadequate for the description of Eq. (9.8). Thus, the CNDO/2 method compresses the reactive site, suggests non-planar structure of the ground state and reduces the activation barrier to a mere 1 kcal/mol [53]. An opposite result is given by the MINDO/3 method, which unjustifiably stretches the structure; the distance R_{∞} grows and, accordingly, the activation barrier, i.e., the difference between the energies of the C_{2v} and C_s forms, is increased to 25 kcal/mol [54]. These shortcomings of the most common semiempirical schemes and the need for an inexpensive method that could reproduce the H-bonding led to the development of special parametrization techniques CNDO/2H, MINDO/3H [55], and MNDO/H [56].

The nonempirical methods provide, in the main, consistent results. The size of the basis set and inclusion of correlation effects affect the quantitative estimates without altering the qualitative conclusion as to the energetic preferability of the C_s structure.

A need for a complete optimization of geometry may be illustrated by two nonempirical calculations done with nearly equivalent basis sets. The conclusion [57] that the C_{2v} form is more stable (the type-b PES in Fig. 9.4) was disproved through a more careful geometry optimization of the C_s structure [58]—the PES of type c in Fig. 9.4. An analysis of the force constants and the complete optimization of geometry [59] have made it possible to draw a quite definitive conclusion that in malonaldehyde there exists a symmetrical double-well potential with a transition state corresponding to an unstable structure of C_{2v} symmetry (potential curve c in the Fig. 9.4).

Table 9.4. Barriers to intramolecular proton transfer (energy differences between the C_s and C_{2v} structures) in malonaldehyde calculated in different appproximations

Method of calculation	$E^{\#}$ kcal/mol	Ref.
CNDO/2	1.0	53
MINDO/3	25.1	54
Ab initio STO-3G	6.7	58
4–31G	10.3	60
DZ	11.5	61
C, O(9S5P/4S2P), H(4S/2S)	11.5	59
4–31G + CI	9.8	60
DZ + CI	10.0	61
C, O(9S5P/4S2P), H(4S/2S) + CI	8.6	59
C, O(9S5P/4S2P), H(4S/2S) + CI[a]	8.1	59
C, O(9S5P/4S2S), H(4S/2S) + CI + ZPE[b]	5.0	59
6–31G[b] + MP	4.3	62

[a] Triple- and quadruple-excited configurations included;
[b] Zero point energy of vibrations included

The value of a theoretically assessed energy barrier of the proton transfer in XI is affected by both the completeness of the basis set and the degree to which the correlation effects are taken into account Table 9.4 lists some data on the magnitude of the barriers calculated in various approximations. Characteristic is here the nonmonotonical change of the energy difference between the structures C_s and C_{2v} with the growth of the basis set. Inclusion of the correlation lowers in all cases the magnitude of the barrier, albeit to a lesser degree than in the systems of the type VIII–X (Tables 9.1, 9.2).

The most accurate allowance for the correlation effects was made by Schaefer et al. [59]. An SCF calculation was supplemented by inclusion of the complete configurational interaction among all the single- and double-excited configurations (37528 for C_{2v} structure and 74740 for the C_s form). A lowering of the barrier was 2.9 kcal/mol and the allowance made for higher excitations added to this value 0.5 kcal/mol to give ultimately 8.1 kcal/mol.

In the case of the system XI, inclusion of the energy difference between zero vibrations of the C_{2v} and the C_s structure proved important: this difference amounting to about 3 kcal/mol is comparable to correlation corrections.

Thus, the apparently most accurate theoretical estimate of the barrier to proton transfer in a malonaldehyde molecule, determined as a difference between the energies of the structures XIa and XIc, is so far 4.3–5.0 kcal/mol. This value explains well fast ($k > 10^6\,s^{-1}$) tautomerization XIa ⇄ XIb observed in solution by the NMR method. Note, however, that calculations by means of a reaction surface Hamiltonian constructed for malonaldehyde [63] gave the barrier of 6.6 ± 0.5 kcal/mol.

The low value of the activation barrier accounts also for the ambiguity in interpretations of experimental results for XI by the methods of electron diffraction, IR spectroscopy, and even X-ray structural analysis. The most accurate experimental data on the structure of *cis*-enol of malonaldehyde were obtained by studying a microwave spectrum of a series of its isotopomers [64]. Consistently with theoretical predictions, the molecule has C_s symmetry. Experimental geometry parameters [64] are given below together with calculational data [62] on the ground (C_s) and the transition (C_{2v}) state.

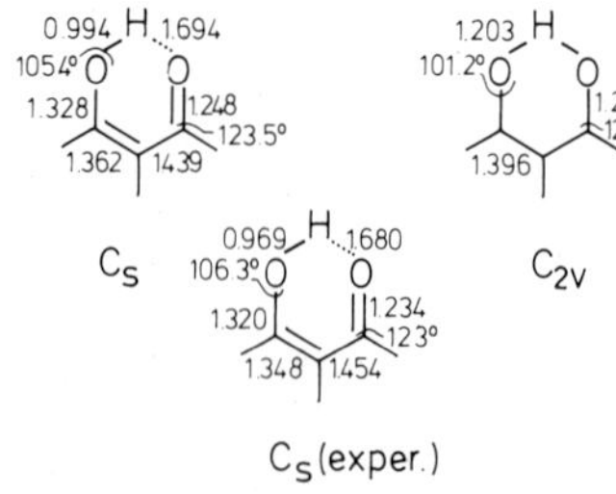

According to the calculations [62], the length of the hydrogen bond H···O is quite sensitive to correlation corrections. Thus, with the 6–31G** basis set, the calculated value of 1.880 Å is appreciably overestimated against experiment (1.68 Å). When correlation is taken into accout through the perturbation theory MP2, the agreement is essentially better (1.694 Å).

The hydrogen bond bridge in XI is sufficiently collinear and in the transition state its linearization is better still. These factors are quite compatible with the structural demands, considered earlier, with regard to the fast proton transfer reactions.

Such a compatibility cannot be achieved for the systems with intramolecular 1,3-proton transfer such as formic acid XII and formamidine XIII:

XIIa ⇌ XIIc ⇌ XIIb

XIIIa ⇌ XIIIc ⇌ XIIIb

Deviations from linearity of the bridge are too large in these molecules giving rise to quite high barriers to an intramolecular proton transfer. According to the symmetry rules (Sect. 1.2.4.3), the transition states XIIc and XIIIc are planar and have symmetry of C_{2v} type. For XIIc, a calculation [60] with the 4–31G AO basis set and the restricted configurational interaction 3 × 3 taken into account gives the energy of 63.1 kcal/mol relative to XIIa, b. Even when the electron correlation is more fully accounted for by using the CEPA scheme, this value, dropping as it does to 44.2 kcal/mol, still remains quite high.

An analogous structure of the transition state XIIIc was found in the calculations [66] using the 4–31G basis set and the IRC technique to analyze the proton transfer reaction path. This path is shown in Fig. 9.5 by a dashed line. The calculated barrier of activation is 59.2 kcal/mol. The imaginary

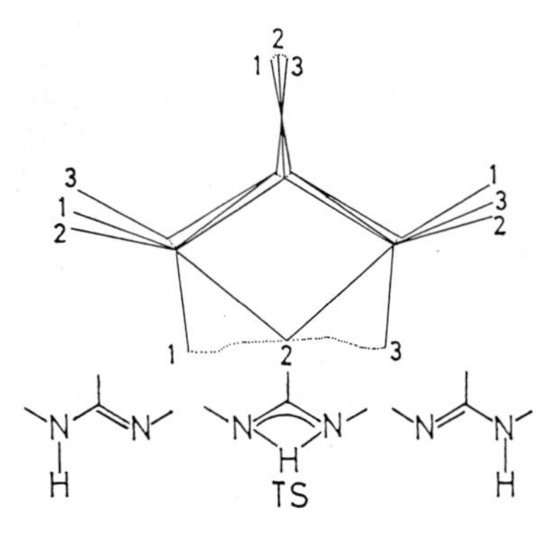

Fig. 9.5. MERP of the intramolecular proton transfer XIIIa ⇆ XIIIc ⇆ XIIIb and structural deformations of the reacting molecule according to ab initio calculations [66]. **1** XIIIa, **2** transition state XIIIc, **3** XIIIb. (Adapted from Ref. [66])

vibration frequency of the transition state structure XIIIc is quite high amounting to 2361 cm^{-1} indicating that the PES in the XIIIc zone is characterized by very steep sloping.

Similarly to the intermolecular proton transfer (Sect. 9.2.1), during the intramolecular transfer the charge at the migrating center is hardly changing, which is, in fact, characteristic of all XI–XIII structures. Its small magnitude ($\sim 0.2e$) indicates migration of hydrogen rather than a proton placing these reactions in the same category as the 1, j-sigmatropic shifts of hydrogen in hydrocarbons (Sect. 10.3).

Below a molecular diagram is given of the ground and the transition states for the intramolecular proton transfer reactions in the formamidine molecule XIII [66]:

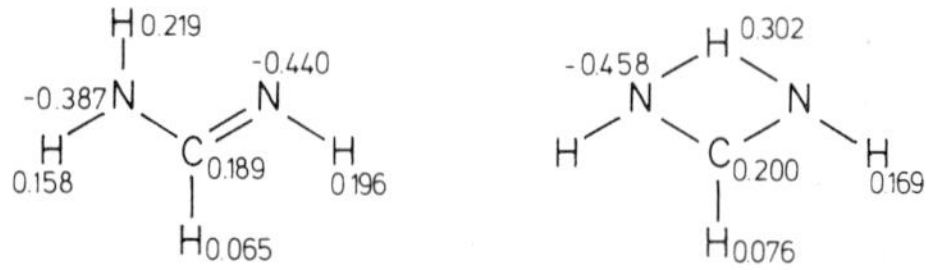

H0.219 -0.440 -0.387N N H C0.189 H 0.158 0.196 H0.065

-0.458 H 0.302 N N H C0.200 H 0.169 H0.076

High activation barriers for such reactions give rise to a particularly interesting situation in the case of nondegenerate systems, namely, a thermodynamically unstable tautomer can be isolated under conditions that would exclude all but the intramolecular interconversion channels. Such conditions occur in gas-phase reactions under deep vacuum, they are particularly effective in interstellar space.

Thus, the presence in space of gigantic clouds about 30 light-years in diameter of hydrogen isocyanate HNC XIV which is an unstable isomer of HCN XIVa ($\Delta E = 14.6$ kcal/mol according to the calculation in Ref. [67] using the DZ + P basis set with CI taken into account) is the straightforward consequence of the existence of an extremely high barrier to the intramolecular proton transfer reaction:

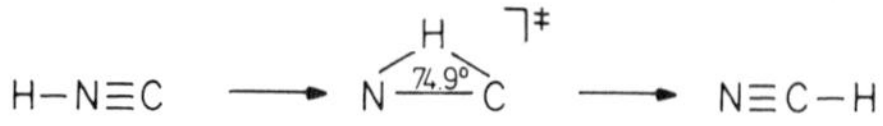

$$\underset{\text{XIV}}{\text{H}-\text{N}\equiv\text{C}} \longrightarrow \underset{\text{XV}}{\left[\text{N}\overset{\text{H}}{\underset{74.9^{\circ}}{\frown}}\text{C}\right]^{\ddagger}} \longrightarrow \underset{\text{XIVa}}{\text{N}\equiv\text{C}-\text{H}}$$

The transition state of the reaction, XV accurately located by the IRC scheme [68–70] has, according to Ref. [67], an energy higher by 49.4 kcal/mol than XIV. It is precisely due to the high magnitude of the activation barrier in rarefied HNC clouds (10^1–10^6 molecules/cm^3) that the exothermal reaction XIV–XIVa cannot take place[1]. The proton transfer reactions in compounds of the XII–XIV type proceed in the intermolecular manner (Sect. 9.2.5).

[1] The total amount of unstable hydrogen isocyanate detected in interstellar space is such that were it possible to initiate this reaction, its overall thermal effect would exceed all the energy obtained from the sun by our planet during its entire existence.

9.2.4 The Tunnelling Mechanism in Proton Transfer Reactions

Bell was the first to have pointed out the importance of tunnelling for the kinetics of the proton transfer [71, 72]. A contribution from the tunnelling to the overall mechanism of the intramolecular proton transfer XIa ⇄ XIb in malonaldehyde is confirmed by the presence of tunnel splitting of vibrational levels in the symmetrical double-well potential of this structure (Fig. 9.4). The magnitude of this splitting is, according to microwave spectrum studies [72], $23\,\mathrm{cm}^{-1}$, which corresponds to the tunnelling frequency of $6.3 \times 10^{11}\,\mathrm{s}^{-1}$. This value has been reproduced fairly well by the calculations [59]—see Table 9.4 in Sect. 9.2.3—using the Wentzel–Kramers–Brillouin approximation, according to which the magnitude of splitting is:

$$\Delta E = \frac{\hbar \nu_{\mathrm{cl}}}{\pi} e^{-\theta} \tag{9.9}$$

where ν_{cl} is the classic vibration frequency and, for one of the minima of the potential:

$$\theta = \frac{2\pi}{\hbar \nu_{\mathrm{im}}} (V_{\mathrm{eff}} - \sqrt{E_0 V_{\mathrm{eff}}} \tag{9.10}$$

where $E_0 = (1/2)\hbar\nu_{\mathrm{cl}}$ is the magnitude of the vibration quantum, ν_{im} is the imaginary frequency in the transition state and:

$$V_{\mathrm{eff}} = E^{\neq} + \sum_{k=1}^{F-1} \frac{1}{2}(h\nu_k^{\neq} - h\nu_k)$$

where ν_k and $\nu_k^{\neq}$ are the vibrational frequencies of the ground and the transition states, respectively, and $(F - 1)$ is the number of adiabatic vibrational degrees of freedom.

Having calculated with the DZ basis set the vibrational frequencies and the magnitude of the barrier with correlation corrections included (Table 9.4), Schaefer et al. [59] found the value for the tunnel splitting $\Delta E = 40\,\mathrm{cm}^{-1}$, which is about twice higher than the experimental value. One should, however, consider that the vibrational frequencies calculated with the DZ basis set are overestimated by about 10% (see Sect. 1.2.2.1) so that by reducing the value by this percentage a much better agreement with experiment may be obtained $\Delta E = E_{1\mathrm{u}} - E_{1\mathrm{g}} = 18\,\mathrm{cm}^{-1}$ (See Fig. 9.6).

The probability of proton tunnelling across the barrier of a double-well potential strongly depends on its symmetry [74, 75]. Figure 9.6 shows a symmetrical and an asymmetrical double-well potentials, their energy levels and eigenfunctions. In the case of the symmetrical PES profile (degenerate systems),

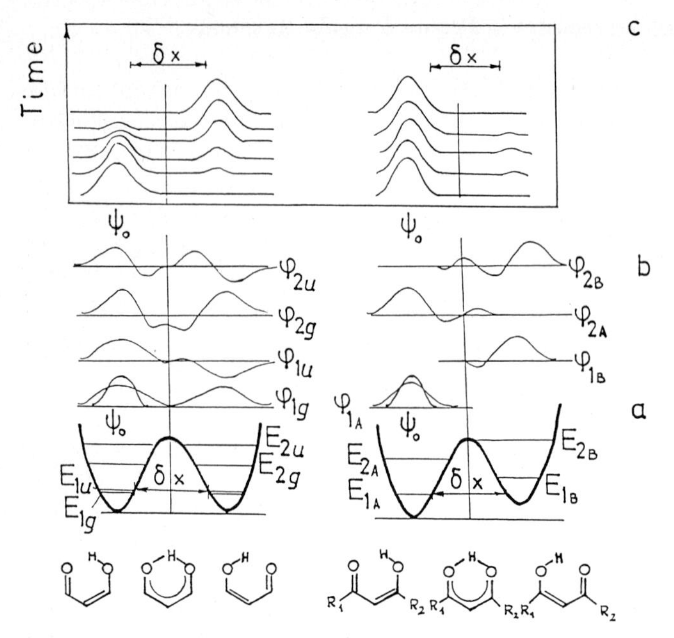

Fig. 9.6a–c. Schematic representation of the tunnelling mechanism in a symmetrical (malonaldehyde) and an asymmetrical (substituents $R_1 \neq R_2$) potential [74, 75]: **a** energy levels of the system; **b** wave functions of the corresponding states; **c** time dependence of the function of state ψ_0 (x, t) that describes the proton motion in a symmetrical and an asymmetrical double-well potential. (Adopted from Refs. [74, 75] with permission from the American Chemical Society)

the eigenfunctions are delocalized over two wells. The asymmetrical potential (nondegenerate systems) is marked by the localization of the eigenfunctions near the minimum and only upon approaching the top a probability arises of passing across the classic barrier and thereby of simultaneous presence of the proton in both minima of the potential. Figure 9.6 visualizes the dynamics of the proton motion for two types of the potential of the intramolecular hydrogen bond which is characterized by time dependences of the function of state of the system lying at the lowest vibrational level. In the case of the symmetrical potential, the function of state originally localized at one of the PES minima "oozes" slowly through the barrier and settles at the second minimum. With an asymmetrical potential such a possibility is completely ruled out.

A calculation of proton tunnelling frequencies in the following reactions shows the importance of symmerty for the effectiveness of tunnelling:

$$CH_3OH + {}^-OCH_3 \rightleftharpoons CH_3O^- + HOCH_3 \qquad (9.11)$$

$$CH_3OH_2{}^+ + HOCH_3 \rightleftharpoons CH_3OH + {}^+H_2OCH_3 \qquad (9.12)$$

$$H_2O \cdot H_2OH^+ + OH_2 \cdot OH_2 \rightleftharpoons H_2O \cdot H_2O + {}^+HOH_2OH_2 \qquad (9.13)$$

In the calculations of the PES in the first two reactions, the symmetry of the potential was upset by rotating the methyl group in the symmetrical H-complex XVI(s) giving thus rise to an asymmetrical H-complex XVI(A):

XVI (S) XVI (A)

As seen from Table 9.5, even a very insignificant alteration in the structure (the difference between the energies of the A and B forms is a mere 0.2 kcal/mol) disturbing the symmetry of the system results in a total inhibition of the tunnelling mechanism of the proton transfer. An analogous effect has been noted also for the system $(H_9O_4)^+$. The role of tunnelling is of importance only in those conformations of the oligomer VII which possess the symmetry elements associated with the central position of the migrating proton.

A similar effect of the symmetry of the double-well potential is predicted by calculations on the systems with an intramolecular hydrogen bonding. Thus, in the case of 3-oxy-2-methyl-2-propenal XVII, a proton tunnelling with the frequency of $7 \times 10^{10}\,s^{-1}$ is expected only for the conformation XVII(s) of the transition state, while a 90° rotation of the 2-methyl group totally prevents the operation of this transfer mechanism [75]:

XVII (S) XVII (A)

An interesting example of a combined experimental and theoretical study of the influence the symmetry has on the mechanism of proton transfer is provided by the system of naphthazarin XVIII. According to the data of the

Table 9.5. Dependence of proton tunnelling frequency on conformation of the system and potential symmetry in Eqs. (9.11)–(9.13), according to Ref. [75]

System	Potential symmetry	ΔE, kcal/mol	Tunnelling frequency, s^{-1}
$CH_3OHOCH_3^-$	S	9.5	4.0×10^{11}
	A		0
$CH_3OHHHOCH_3^+$	S	8.6	7.0×10^{11}
	A		0
$H_2OH_2OHOH_2OH_2^+$	S	7.6	1.0×10^{12}
	A		0

IR [76] and NMR [77] spectra, a rapid proton exchange XVIIIa ⇄ XVIIb takes place in the naphthazarin molecule both in solution and in the crystalline state. Such exchange is altogether absent in the monomethyl-substituted derivative where only one tautomeric form is realized. The introduction of the second methyl group into the unsubstituted nucleus restores symmetry and, accordingly, the rapid proton exchange resumes anew. Below a schematic representation is given of various proton migration pathways in the system. The route XVIII (a → c → b) describes a concerted proton exchange while the routes XVIII (a → c → b) and XVIII (a → d → b) refer to a stepwise exchange:

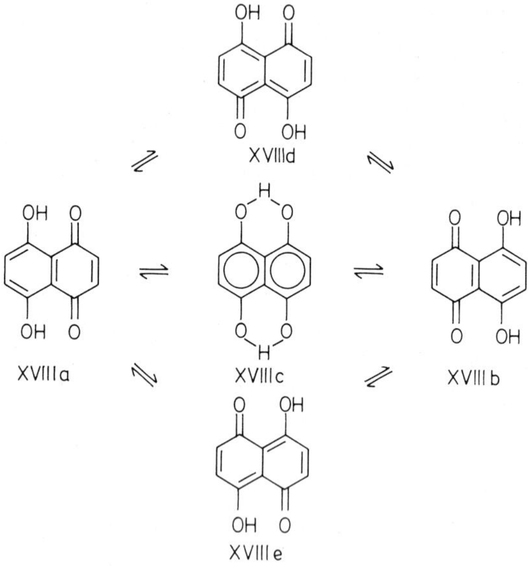

XVIIId

XVIIIa ⇌ XVIIIc ⇌ XVIIIb (9.14)

XVIIIe

According to the nonempirical STO-3G calculations [78], the forms XVIIIa and XVIIIb of naphthazarin, as well as its 2,6- and 2,7-dimethyl derivatives, correspond to the minima of the PES while the structures XVIIId and e are transition states lying higher by 2.5 kcal/mol than XVIIIa and XVIIIb. The structure XVIIIc of D_{2h} symmetry is located at the hill top lying higher by 28.2 kcal/mol than the minima.

The calculated tunnelling frequencies practically coincide for the concerted and the stepwise mechanism indicating their equal probability. In the case of naphthazarin and its symmetrical 2,6- and 2,7-dimethyl derivatives the tunnelling frequencies are 20–40 s^{-1}. For 2-methyl naphthazarin the difference in energy between the forms XVIIIa and XVIIIb is no more than 1.0 kcal/mol (the equilibrium constant $K = 0.23$ at room temperature). At the same time, the calculated exchange rate is $1.2 \times 10^{-9} s^{-1}$, which is equivalent to the tunnelling time of about 27 years.

9.2.5 Double Proton Migrations

A synchronous transfer of two protons, which in reaction (9.14) competes with a two-step process, is in some cases the predominant proton exchange mechanism. Such double proton migrations play an important role in many chemical and biochemical reactions in which the steric hindrances impeding proton transfer in a substrate molecule are removed thanks to the double proton exchange between substrate and enzyme [79]. The double proton transfers determine the mechanism of the bifunctional acid-base catalysis[80, 81]. The interest in the mechanism of double proton migrations in the H-bound complexes became especially keen after Löwdin [82] advanced in 1963 the hypothesis to the effect that it is precisely such processes in the DNA molecules that underlie the nature of spontaneous mutations.

The calculations [83], which for a long time were record-breaking as regards the number of the electrons taken into account and the overall size of the basis set, bore indeed witness to the energetic preferability of a synchronized shift of the protons H_1 and H_2 in the tautomerization of the pair guanine–cytosine XIX (G—C→G*—C*):

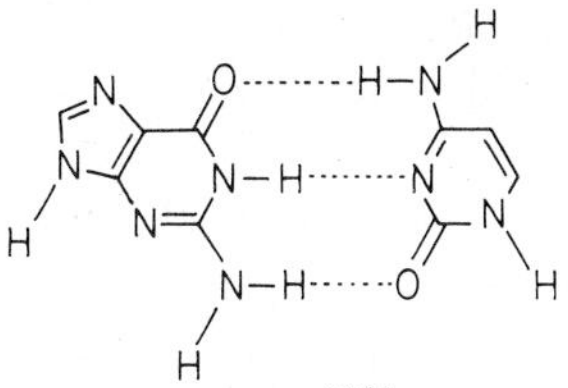

XIX

In view of unusual complexity of the system XIX, the calculations [83] were performed without geometry optimization, i.e., without taking into account the geometry relaxation of the system caused by the motion of protons over the hydrogen bond bridges. Meanwhile, detailed theoretical studies of the double proton migrations in degenerate dimers, for example, in the dimers of the formic acid XX (R = H), have pointed to a quite substantial effect the geometry relaxation has on the magnitude of the activation barrier of Eq. (9.15):

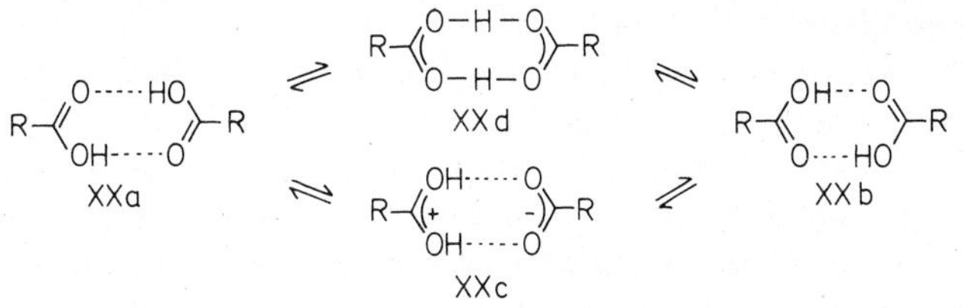

XXa ⇌ XXd ⇌ XXb; XXa ⇌ XXc ⇌ XXb (9.15)

Also the structure and symmetry of the transition state of Eq. (9.15) is a question deserving of serious attention. One may, by analogy with

Table 9.6. Activation energies and expected mechanism of Eq. (9.15) according to the data of quantum chemical calculations

Method	Geometry optimization	E_{act} kcal/mol	Reaction path	Ref.
CNDO/2	No optimization	167	asynchronous	[84]
CNDO/2	Partial optim.	18	synchronous	[85]
PRDDO	No optimization	44	asynchronous	[86]
PRDDO	Complete optim.	23	synchronous	[86]
ab initio (DZ)	No optimization	44	asynchronous	[83]
ab initio STO-3G	No optimization	48	asynchronous	[87]
ab initio STO-3G	Complete optim.	15	synchronous	[87]
ab initio 4–21G	Complete optim.	12	synchronous	[87]

intramolecular proton transfers [Eq. (9.14)], think of several possible reaction channels for a double proton exchange. The route XX (a → d → b) runs via the structure XXd of D_{2h} symmetry, which may correspond to both a minimum and a transition state on the PES. In the latter case, one may assume concerted, though not necessarily synchronous, shift of two protons. The structure XXc on the route (a → c → b) of reaction XX may equally correspond to either the minimum or the transition state. The existence of a minimum corresponding to the structure XXc would indicate stepwise consecutive single proton migration.

The conclusions as to the reaction mechanism and the magnitude of the barriers of the double proton exchange depend directly on the completeness of geometry optimization (Table 9.6).

The most accurate data of calculation listed in Table 9.6 are in excellent agreement with the experimental barriers of activation of 14–18 kcal/mol for Eq. (9.15) in the gas phase and in solution [88, 89]. On the other hand, the activation barrier of this reaction determined by the NMR method for the dimer of the *p*-toluic acid (XX, R = $C_6H_4CH_3$) equals a mere 1.1 kcal/mol [90]. The authors explain such a sharp lowering of the barrier in the crystal by defects in the crystalline packing which impede conformational possibilities of R and raise the symmetry of the potential. As a consequence, the role of tunnelling is enhanced, which, according to Ref. [90], is insignificant for Eq. [9.15] reactions in solution where they proceed by the classic overbarrier route.

A somewhat different opinion on the reason for the the above-discussed disparity between activation barriers was expressed in Ref. [91] to the effect that "a dimer may be compressed in crystal packing in order to maintain its structure on balance of inter- and intramolecular forces. Thus, the equilibrium structure of the dimer in the crystal may differ from that in the isolated state". Calculations by a molecular mechanics method gave some substance to this assumption: the O···O distance in a dimer was shown to be compressed approximately 0.03–0.04 Å in the crystal in comparison with an isolated dimer. As a result, the calculated activation barrier is reduced though not so drastically.

Probably both factors pointed out in Refs. [90, 91] should have to be taken into account when making comparisons between experimental results for the crystalline phase and calculation data for isolated dimers.

The geometries of a stable dimer of the formic acid and of the transition state XXd with D_{2h} symmetry calculated by an ab initio method for the synchronous double proton transfer [87] are as follows.

With the geometry relaxation taken into account, the calculated barrier of Eq. (9.15) (using the basis set STO-3G) is lowered from 48 kcal/mol to 14 kcal/mol. Characteristic is the reduction of the nonvalent distance O···O from 2.73 Å in the starting dimer to 2.30 Å in the transition state. This trend toward "compression" of the reactive site and, as a consequence, lowering in the activation barrier was noted earlier in connection with intramolecular proton transfer in malonaldehyde.

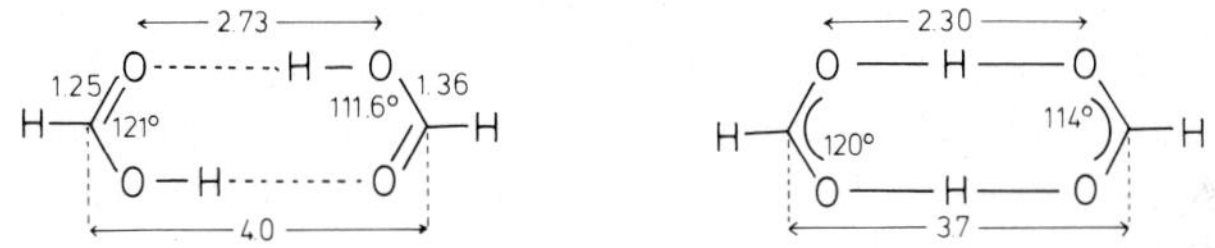

XXa (R=H) XXd (R=H)

There have already been several theoretical investigations, using mainly nonempirical methods, into intermolecular concerted dihydrogen transfer concerned, for example, with hydrogenation of ethylene by ethane [92] or with reactions between hydroxymethylene and ethylene [93], acetic acid and methanol [94, 95], methanol and formaldehyde [96].

Nondegenerate double proton migrations between a solute molecule and the molecules of the solvent (water) may determine the mechanism of intermolecular proton transfers that give rise to a new tautomeric form. Thus, using the relaxation method of temperature jump, it was found [97] that the tautomerization of 6-methoxy-2-pyridone proceeds as a nondissociative concerted process of bifunctional interaction with one molecule of water:

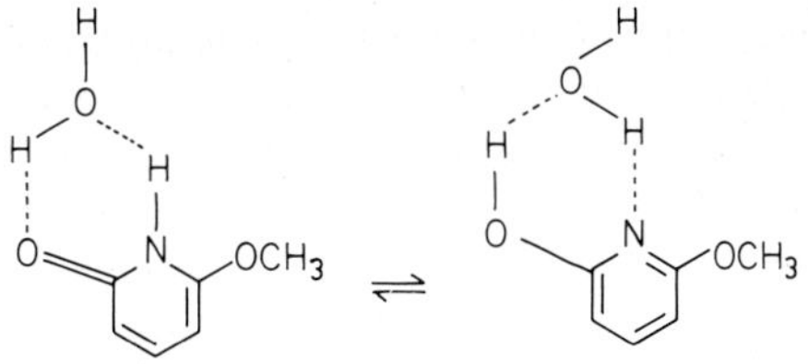

Such processes have been studied theoretically by means of ab initio 3–21G and 6–31G calculations for hydrated complexes of formamide XXIa, its imidole tautomer XXIb and formamidine XXIIa [98]:

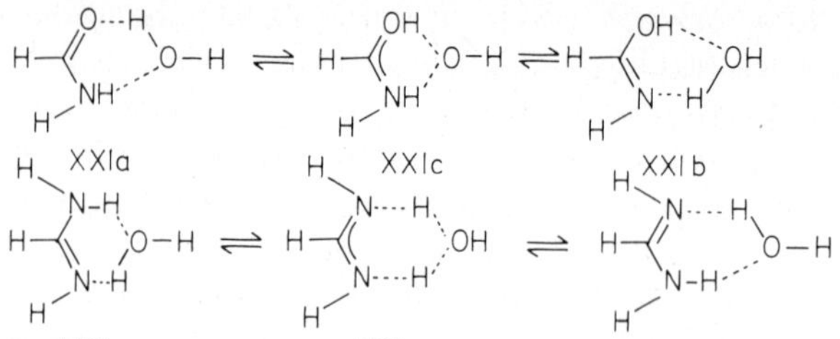

XXIa ⇌ XXIc ⇌ XXIb

XXIIa ⇌ XXIIc ⇌ XXIIb

Inclusion of an additional water molecule leading to a six-membered cycle, as in the malonaldehyde XI, provides more favorable steric conditions in XXI, XXII for proton transfer. Thus, for formamidine hydrate the calculated barrier of tautomerization XXIa ⇄ XXIIb, when intermolecular mechanism is operative, (17 kcal/mol with the 3–21G and 21 kcal/mol with the 6–31G basis set) is three times lower than in the case of the intramolecular reaction XIIIa ⇄ XIIIb (see Sect. 9.2).

The change in the activation barrier is even more drastic in the case of the proton transfer HNC → HCN when four water molecules have been added [98]. Stereochemistry of the 1,2-sigmatropic hydrogen shift requires precisely that number of water molecules for an energetically favored transition state to be formed. Note that the question has not been resolved yet as to the probabilities of formation of the to such a degree ordered chains of water molecules in solution and, consequently, as to the compatibility of the conclusions drawn from the model description of the medium with the statistical nature of the liquid phase.

The double proton migrations can occur also in intramolecular reactions, for example, in diotropic systems [52]. These comprise the compounds whose transition into a tautomeric form requires the transfer of two migrants (such as the protons), and, in terms of the classic structural notions, this transfer must be concerted since a single transfer would result in the formation of a nonclassic structure of the zwitterionic or biradical type. A simplest example of such systems is provided by the molecule of the oxalic acid for which, depending on the nature of the stationary point on the PES that corresponds to a structure with one transferred proton, two in principle different reaction mechanisms may be realized:

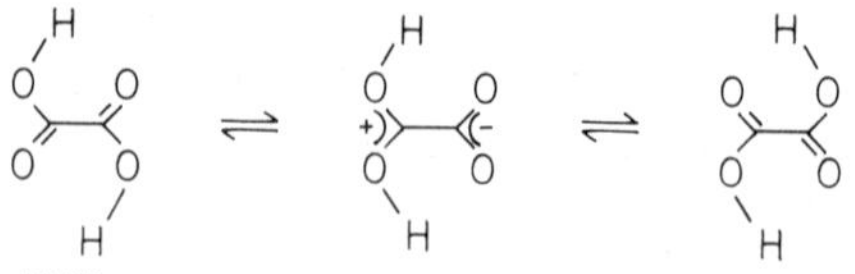

XXIIIa ⇌ XXIIIc ⇌ XXIIIb

In case this structure corresponds to a minimum on the PES, the reaction XXIIIa ⇄ XXIIIb develops stepwise including the zwitterionic intermediate

XXIIIc. But if the structure XXIIIc represents a transition state, the reaction proceeds as a concerted double proton transfer.

MINDO/3 calculations [54, 100] have shown that in the case of 1,4-diotropic proton migrations, such as XXIIIa ⇄ XXIIIb, XXIVa ⇄ XXIVb, XXVa ⇄ XXVb, the products of the single migration XXIIIc, XXIVc, XXVc correspond, contrary to expectations, to the minima on the PES. Thus, these reactions are characterized by a stepwise mechanism, which can be accounted for by unfavorable stereochemistry of the reactive site and by the reaction centers being spaced fairly wide apart.

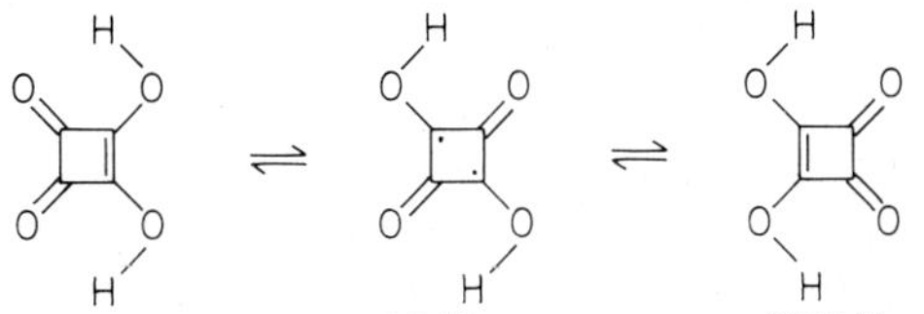

XXIVa ⇌ XXIVc ⇌ XXIVb

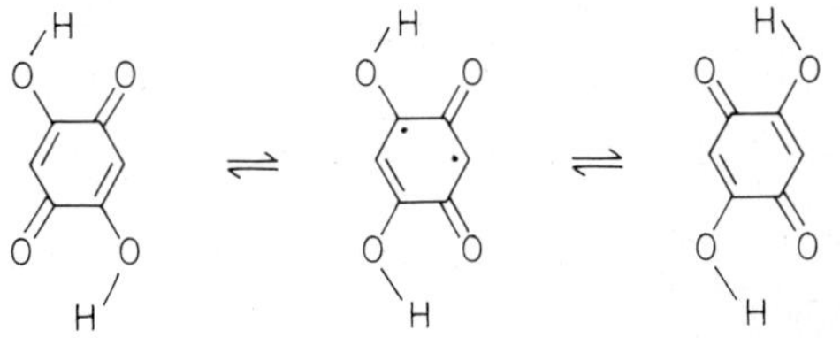

XXVa ⇌ XXVc ⇌ XXVb

Later this conclusion was borne out by findings in Refs. [101, 102]. Dewar and Merz [101] using the AM1 method studied the automerization of azophenine by a double proton shift. The results suggest, in accordance with Refs. [54, 100], that the reaction takes place neither in a synchronous manner nor via a symmetrical transition state, but in two steps, each involving transfer of a single proton. The predicted activation energy for the single proton transfer is 50.1 kcal/mol, which was the value of activation energy of automerization of tetraphenylazophenine found in Ref. [94]. Dewar and Merz argue that the automerization reaction proceeds by vibrationally assisted tunnelling [101, 103]. According to the MINDO/3 calculations, the mechanism of the thermal rearrangement of *trans*-4a, 10a-dihydrobenzocyclooctene into 7, 8-dihydrobenzocyclooctene consists of two consecutive 1,5-hydrogen shifts.

At the same time, in the diotropic systems with a stereochemistry more favorable for a proton transfer, there are no minima on the PES that would

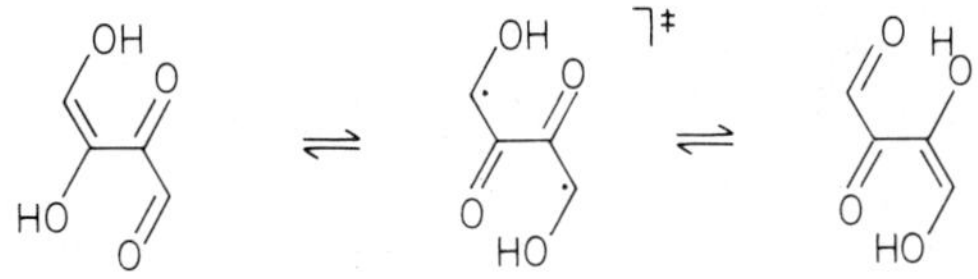

XXVIa ⇌ XXVIc‡ ⇌ XXVIb

correspond to b-type structures. As a consequence, in the 1,5- and 1,6-diotropic systems the reaction mechanism of double proton transfers is concerted [100], for example:

References

1. Ward RL, Weissman SI (1957) J ACS 79:2086
2. Cannon RD (1980) Electron transfer reaction. Butterworths, London
3. Chanon M (1982) Bull Soc Chim: 197
4. Sutin N (1975) in: Eichgorn GB (ed) Inorganic biochemistry. Elsevier, Amsterdam v. 2, Chap 19
5. Voevodski VV, Solodovnikov SP, Chibrikin VM (1959) Dokl Akad Nauk SSSR 129:1082
6. Szwarc M (1968) Carbanions, living polymers and electron transfer processes. Wiley, New York
7. Okhlobistin OY (1974) Electron transfer in organic reactions. Rostov Univ, Rostov-on-Don (in Russian)
8. Chanon M (1985) Bull Soc Chim: 209
9. Barbara PF, Jarzeba W (1988) Acc Chem Res 21:195
10. Kornblum N (1975) Angew Chem Intern Ed Engl 14:734
11. Bunnett JF (1978) Acc Chem Res 11:413
12. Beletskaya IP, Drozd VN (1979) Uspekhi Khim (Russ Chem Rev) 48:793
13. Bigot B, Roux D, Salem L (1981) J ACS 103:5271
14. Taube H (1970) Electron transfer reactions of complex ions in solution. Academic Press New York
15. Marcus RA (1968) J Phys Chem 72:891: (1964) Ann Rev Phys Chem 15:155
16. Rather MA (1978) Int J Quant Chem 14:675
17. Dogonadze RR, Kuznetsov AM (1974) Zh Vsesoyuzn Khim Obsch im DI Mendeleeva 19:242
18. Sutin N (1982) Acc Chem Res 15:275
19. McLendon G (1988) Acc Chem Res 21:160
20. Larsson S (1982) Chem Phys Lett 90:136; (1981) J ACS 103:4034
21. Taube H (1984) Science 226:1028
22. Todres ZV (1978) Uspekhi Khim (Russ Chem Rev) 47:260
23. Ovchinnikov AA, Ovchinnikova MYa (1969) Zh Experiment Theor Phys 56:1278
24. Helman AB (1983) Chem Phys 79:235
25. Marcus R, Sutin N (1985) Biochim Biophys Acta 811:265
26. Eberson LE (1987) Electronic transfer reaction in organic chemistry. Springer-Verlag, Berlin. (Reactivity and structure, v 25)
27. Warshel A (1982) J Phys Chem 86:2218
28. Bays JP, Blumer ST, Baral-Tosh S, Behar D, Neta P (1983) J ACS 105:320
29. Pross A, Shaik SS (1983) Acc Chem Res 16:363
30. Pross A (1985) Adv Phys Org Chem 21:99
31. Grotthus CJT (1806) Ann Chim (Paris) 58:54
32. Scheiner S (1981) J ACS 103:315
33. Scheiner S, Harding LB (1981) J ACS 103:2169
34. Scheiner S, Harding LB (1981) Chem Phys Lett 79:39
35. Scheiner S (1982) J Phys Chem 86:376
36. Scheiner S (1982) J Chem Phys 77:4039
37. Scheiner S, Harding LB (1983) J Phys Chem 87:1145
38. Scheiner S, Szczesniak MM, Birgham LD (1983) Int J Quant Chem 23:739
39. Cao HZ, Allavena M, Tapia O, Evleth EM (1985) J Phys Chem 89:1581
40. Scheiner S, Redfarn P (1986) J Phys Chem 90:2969
41. Scheiner S, Bigham LD (1985) J Chem Phys 82:3316
42. Scheiner S (1985) Acc Chem Res 18:174
43. Scheiner S (1983) Int J Quant Chem 23:753
44. Abu-Dari K, Freyberg DP, Raymond KN (1979) Inorg Chem 18:2427
45. Bino A, Gibson D (1982) J ACS 104:4383
46. Roos BO, Kraemer WP, Diercksen GHF (1976) Theor Chim Acta
47. Hillenbrand EA, Scheiner S (1986) J ACS 108:7178

48. Dogonadze RR, Kuznetsov AM, Levich VG (1967) Electrokhimiya 3:648
49. Marcus RA (1968) J Phys Chem 72:891
50. Andres JL, Duran M, Lledos A, Bertran J (1986) Chem Phys Lett 124:177
51. Cecarelli C, Jeffrey GA, Taylor R (1981) J Mol Struct 70:255
52. Minkin VI, Olekhnovich LP, Zhdanov YA (1988) Molecular design of tautomeric compounds. Reidel Dordrecht
53. Kato S, Kato H, Fukui K (1977) J ACS 99:684
54. Simkin BY, Golyanski BV, Minkin VI (1981) Zh Org Khim 17:3
55. Noor SM, Hopfinger AJ (1982) Int J Quant Chem 22:1189
56. Burstein KY, Isaev AN (1984) Theor Chim Acta 64:397
57. Isaacson AD, Morokuma K (1975) J ACS 97:4453
58. Del Bene JE, Kochenour WL (1976) J ACS 98:2041
59. Bicerano J, Schaefer HF, Miller WH (1983) J ACS 105:2550
60. Bouma WJ, Vincent MA, Radom L (1978) Int J Quant Chem 14:767
61. Karlström G, Jönsson B, Roos B, Wennerström H (1976) J ACS 98:6851
62. Frish MJ, Scheiner AC, Schaefer HF (1985) J Chem Phys 82:4194
63. Carrington T, Miller WH (1986) J Chem Phys 84:4364
64. Baughcum SL, Smith Z, Wilson EB, Duerst RW (1984) J ACS 106:2260
65. Rodwell WR, Bouma WJ, Radom L (1980) Int J Quant Chem 18:107
66. Yamashita K, Kaminoyama M, Yamabe T, Fukui K (1981) Theor Chim Acta 60:303
67. Pearson PK, Schaefer HF, Wahlgren U (1975) J Chem Phys 62:350
68. Ishida K, Morokuma K, Komornicki A (1977) J Chem Phys 66:2153
69. Müller K (1980) Angew Chem Intern Ed Engl 19:1
70. Müller K, Brown LD (1979) Theor Chim Acta 53:75
71. Bell RP (1973) The proton in chemistry. Chapman & Hall, London
72. Bell RP (1980) Tunnel effect in chemistry. Chapman & Hall, London
73. Baughcum SL, Duerst RW, Rowe WF, Smith Z, Wilson EB (1981) J ACS 103:6296
74. Brickmann J, Zimmermann H (1969) J Chem Phys 50:1608
75. de la Vega JR (1982) Acc Chem Res 15:185
76. Bratan S, Stohbush F (1980) J Mol Struct 61:409
77. Lyerla JR, Yannoni CS, Fyfe CA (1982) Acc Chem Res 15:208
78. de la Vega JR (1982) J ACS 104:3295
79. Jenks WP (1969) Catalysis in chemistry and enzymology. McGraw-Hill New York
80. Litvinenko LM, Oleynic NM (1981) Organic catalysts and homogeneous catalysis (in Russian). Naukova Dumka Kiev
81. Jenks WP (1976) Acc Chem Res 9:425; 13:161
82. Lövdin PO (1965) Adv Quant Chem 2:213
83. Clementi E, Mehle J, Neissen W (1971) J Chem Phys 54:508
84. Morita H, Nagakura S (1972) Theor Chim Acta 27:325
85. Socalski WA, Romanovski H, Jaworski A (1977) Adv Mol Relax Interact Process 11:29
86. Scheiner S, Kern CW (1979) J ACS 101:4081
87. Graf F, Meyer R, Ha TK, Ernst RR (1981) J Chem Phys 75:2914
88. Morita H, Nagakura S (1972) J Mol Spectr 42:539
89. Costain CC, Srivatava GP (1964) J Chem Phys 41:1620
90. Meier BH, Graf F, Ernst BR (1982) J Chem Phys 76:767
91. Nakamura R, Machida K, Hayashi S (1986) THEOCHEM 146:101
92. Feller DF, Schmidt MW, Ruedenberg K (1982) J ACS 104:960
93. Ahmed SN, McKee ML, Shevlin PB (1985) J ACS 107:1320
94. Limbach HH, Hennig J, Gerritzen D, Rumpel H (1982) J Chem Soc Farad Trans 74:229
95. Gerritzen D, Limbach HH (1984) J ACS 106:869
96. McKee ML, Shevlin PB, Rzepa HS (1986) J ACS 108:5793
97. Bensaud O, Dubois JE (1977) Compt Rend Acad Sci Ser C 285:503
98. Zeilinski TJ, Poirier RA, Peterson MR, Csizmadia IG (1983) J Comput Chem 4:419
99. Bertran J, Lledos A (1983) THEOCHEM 123:211
100. Simkin BY, Golyanski BV, Minkin VI (1981) Zh Org Khim 17:1793
101. Dewar MJS, Merz KM (1985) THEOCHEM 124:183
102. Baumann H, Cometta-Morini C, Oth JFM (1986) THEOCHEM 138:229
103. Dewar MJS, Merz KM, Stewart JJP (1985) J Chem Soc Chem Commun:166

CHAPTER 10

Pericyclic Reactions

According to Woodward and Hoffmann ([1] p. 182), the pericyclic reactions comprise all concerted intermolecular and intramolecular (electrocyclic, sigmatropic) cycloaddition reactions. The rules of selection of preferable structures for transition states of these reactions based on the principles of orbital approach have found wide acceptance and may serve as an example of an effective qualitative theory. One should not, however, forget that the formulation of the rules [1] rests on analysis of the general topology rather than on specific geometry of alternative structures of transition states. As will be shown below direct calculations of the PES of pericyclic reactions often introduce quite substantial corrections into conventional notions regarding the coordinate and the structure of a transition state of the pericyclic reaction. On many occasions, only such calculations enable us to answer the question as to whether a reaction is indeed concerted (in other words, proceeding without formation of intermediates) and, if so, whether the bond-making and bond-breaking processes are synchronized.

A shortcoming of the orbital approach which underlies the formalism of Woodward–Hoffmann and similar schemes (see Chap. 4) is that they fail to take into account the electron repulsion. For this reason, the selection rules for pericyclic reactions do not depend on the multiplicity of the state and the configurational interaction cannot be taken into account in a sufficiently rigorous manner to be able to allow for electron correlation. This point is particularly inconvenient when analyzing symmetry-forbidden reactions.

In the following, we are going to consider calculation data on the PES and MERP of some representative pericyclic reactions. The emphasis will be placed on the specific features of the intrinsic mechanism which do not follow from the above-examined qualitative notions on the steric course and the one-step character of pericyclic reactions. As a matter of fact, the data from rigorous quantum chemical calculations have assisted in forming novel ideas concerning the detailed mechanism of these reactions thus modifying quite substantially the assertion that the pericyclic transformations represented the "reactions without a mechanism".

10.1 Reactions of Cycloaddition

10.1.1 [2 + 2]-Cycloaddition

A simplest example of a cycloaddition reaction is provided by the dimerization of ethylene with the formation of cyclobutane, i.e., [2 + 2]-cycloaddition. According to the Woodward–Hoffmann rules, the supra-antara route I of the reactants' approach is thermally allowed.The parallel supra-supra approach II is unfavored in the electron ground state but must materialize in an excited state.

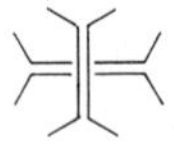

I  II

However, the data of direct quantum chemical calculations on the PES of this reaction indicate strong steric repulsion in the case of the supra-antara approach I, which makes this mechanism energetically unfavored [2, 3]. But the route II of the parallel approach with the synchronous formation of two bonds C—C is equally energetically unrealizable. The MINDO/3 calculations with precise localization of the transition state by minimization of the gradient norm lead to the structure III. A three-center C_1—C_2—C_3 interaction, predictable from the perturbation theory [5], takes place in this structure. The form of the transition vector III reflects the character of the carbon atom shifts which determine the reaction path where the processes of breaking and making of the C—C bonds are sharply asynchronous. The ab initio calculations [3, 6] lead to the same conclusion, they indicate a transition state with the structure IV in which two CC bonds lying in parallel planes are spaced 2.237 Å apart.

The calculated activation energies of 58.8 [4] and 62 kcal/mol [6] are in good agreement with the experimental estimate of 62.6 kcal/mol.

On the PES of the reaction, obtained by the semiempirical MINDO/3 method, there are no minima of the biradical type, which is an evidence for the concerted character of the process [4]:

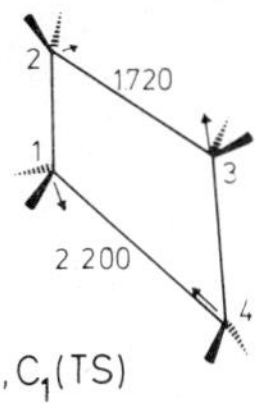

2 1720 1 3 2 200 4

III, C_1(TS)

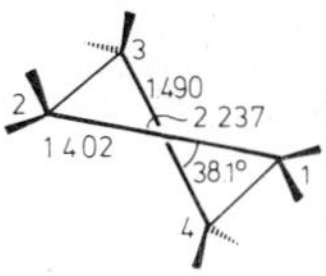

3 1.490 2 2.237 1.402 38.1° 1 4

IV, C_2(TS)

On the other hand, the ab initio calculations [7–10] with the STO-3G, 4–31G basis sets and CI point to the presence of biradical-type intermediates V, i.e., *gauche*- and *trans*(VI)-tetramethylenes.

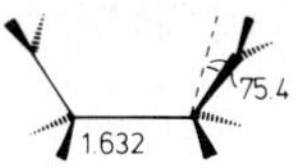

The PES's of the [2 + 2]-cycloaddition were also computed for the following reactions: formaldehyde + formaldehyde and ethylene + singlet oxygen [10]. The results of these computations indicate that for a [2 + 2]-cycloaddition only step-wise reaction paths exist involving diradicaloid transition states and intermediate *gauche*- and *trans*-forms.

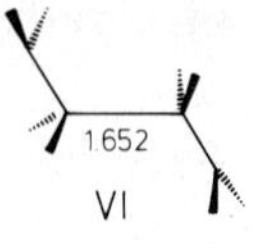

In the electron-excited state, when there are, according to the Woodward–Hoffmann rules, favorable conditions for the synchronous supra–supra approach, exact inclusion of the electron repulsion and the correlation effects bears witness to a more complex mechanism:

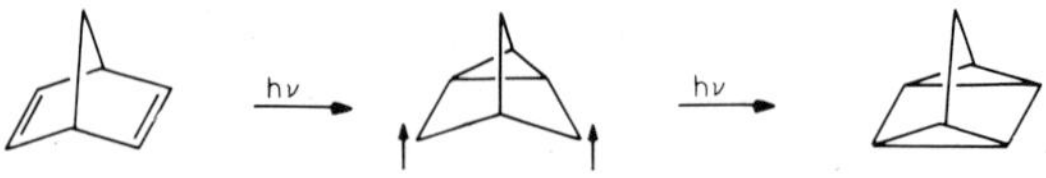

Ab initio calculations [11] have shown that the photochemical cyclization of norbornadiene VIII to quadricyclan IX will occur most readily in the triplet-excited state, which could not be predicted while staying within the framework of orbital concepts. The processes of making of the bonds C_1C_3 and C_2C_4 in the reaction VII → IX are asynchronous, as may be seen from the structure VIII which corresponds to the most stable form of the lowest triplet state common to VII and IX. The length of the bond C_1C_3 in VIII calculated in the MP2/6–31G approximation is 1.5 Å, while the distance C_2C_4 amounts to 2.365 Å.

The results of calculations of the mechanisms of the polar [2 + 2-]cycloaddition make it possible to explain as well as to predict the effect the substituents may have on the activation barrier, the character of the PES and the nature of the transition state [4]. There are several qualitative predictions as to the mechanisms of polar cycloaddition [12] which assume the formation of bipolar intermediates of the chain X or the cyclic XI type.

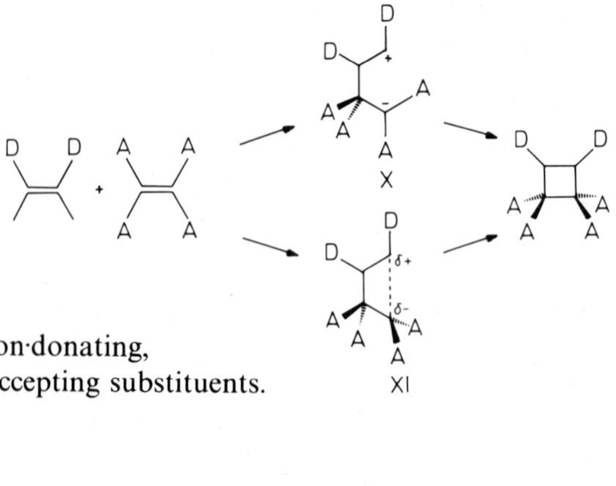

D are the electron-donating,
A the electron-accepting substituents.

According to Fukui [5], the reaction starts from three-center interaction between the carbon atoms of the donor molecule and the most electrophilic carbon atom of the acceptor molecule.

Duran and Bertran [4] studied by means of the MINDO/3 method possible mechanisms of the reaction of 1, 1-dicyanoethylene with oxyethylene which may lead to two different products, i.e., 1-oxy-3, 3-dicyanocyclobutane XII and 1-oxy-2, 2-dicyanocyclobutane XIII:

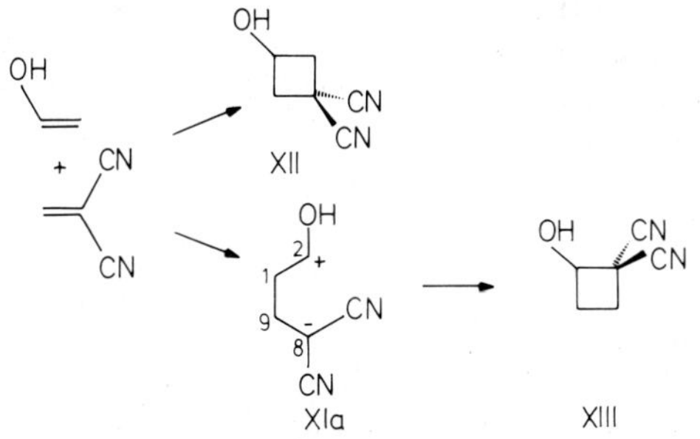

The formation of the less favored product XII proceeds in a manner similar to the dimerization of two molecules of ethylene. There are no intermediates on the reaction PES, a type III three-center interaction arises also in the transition state. First a bond is formed between the unsubstituted carbon atoms. Even though the reaction is concerted, the process of the formation of two new carbon-carbon bonds is asynchronous. The calculated activation barrier is 59.7 kcal/mol.

On the other hand, the more stable product XIII is formed by a different mechanism. Calculations predict the reaction to be of two-step type, the concerted character of the process is disturbed. On the PES of the reaction, the intermediate XIa of the chain type (XIV) appears with the distance $R_{1-9} =$ .74 Å and $R_{2-8} = 3.32$ Å. At the stage of the formation of XIa from starting olefines a bond between the atoms C_2 and C_8 has already been established, while covalent interaction between the atoms C_2 and C_8 is still absent. The energy level of the intermediate XIa is higher by 25.6 kcal/mol than that of the isolated molecules. The PES in the vicinity of XIa proved from the side of the starting molecules so smooth as to make impossible localization of a transition state.

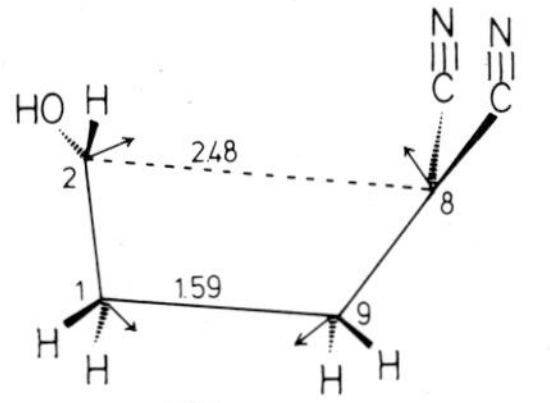

The transition state has at the stage of passing from the intermediate XIa to 1-oxy-2, 2-dicyanocyclobutane XIII the structure XIV:

In this case, there is no type III three-center interaction, the activation barrier equals 44.1 kcal/mol. The charge transfer in the transition state predicted by calculations is rather high amounting to 0.53 e.

Thus, according to data of quantum chemical calculations, the reaction mechanism depends essentially on the degree of the charge transfer. In reactions of the polar [2 + 2]-cycloaddition, which lead to the most stable product, a two-step mechanism is operative that includes an intermediate of chain structure with a well-pronounced 1,4-dipole character.

10.1.2 [4 + 2]-Cycloaddition

The [4 + 2]-cycloaddition reaction (Diels–Alder reaction) is the most effectual method for the synthesis of six-membered carbon cycles. Since this reaction had been discovered in 1929 by Diels and Alder, numerous works have been devoted to studying its mechanism. However, in spite of quite extensive experimental and theoretical investigations (detailed bibliographical information may be found in the reviews of Refs. [13–16], the mechanism of this reaction is still open to discussion with two variants being usually argued about. According to the first of these, which has gained wider currency among organic chemists, the [4 + 2]-cycloadditions represent concerted synchronous reactions with symmetrical transition states of the XV type. The second variant is a two-step reaction in the course of which a stable intermediate emerges of the biradical or zwitter-ionic type XVI. In this case, it is usually assumed that the formation of the intermediate which rapidly transforms into the final product is the limiting step.

Calculations on the PES of the reaction of [4 + 2]-cycloaddition were made in different approximations for a simplest reaction of this type, namely, the thermally allowed suprafacial addition of 1,3-butadiene and ethylene with the formation of cyclohexene XVII. Both semiempirical [17–21] and ab initio [10,22–28] calculational methods were employed. The most complete calculations were carried out by Dewar by means of the MINDO/3 and AM1 methods with the configurational interaction taken into account in the UHF

XV

XVIa

XVIb

XVII

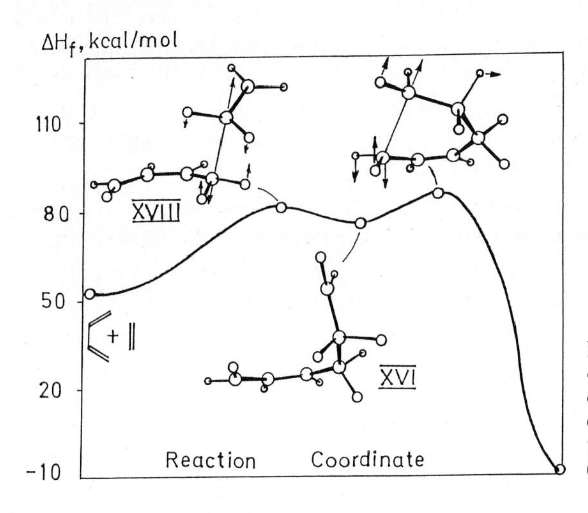

Fig. 10.1. Schematic representation of the MERP of the [4 + 2] cycloaddition reaction of 1,3-butadiene and ethylene, according to the data of the UMINDO/3 calculations [18] (Adapted from Ref. [18])

approximation (UMINDO/3 and UAM1)—see Refs. [18, 20]. The scheme of the mechanism that followed from the calculations and was subsequently corroborated by additional MNDO calculations and found to be consistent with recent ab initio (MC—SCF—4–31G) calculations [29] turned out rather unforeseen. In the case of a gas-phase reaction, a non-concerted mechanism is expected with sharply asynchronized processes of the formation of two carbon–carbon bonds with the limiting step being not the formation of the biradical XVI but rather its closure into the product of cycloaddition. The MERP scheme of the [4 + 2]-cycloaddition and the calculated structures of XVI and two transition states are shown in Fig. 10.1.

The stationary points on the PES were identified from an analysis of the Hess matrix. The activation barriers at the stages of the formation of the biradical XVI and its transformation into cyclo-hexene XVII amount to (with certain corrections taken into account) 28.0 and 32.0 kcal/mol, respectively. The cyclic structure XV has the energy of 46 kcal/mol relative to the original reactants with two negative force constants found by the calculations. This implies that the structure XV is not a transition state. An identical two-step mechanism with a biradical intermediate XX and the limiting step at the stage where the intermediate passes into the cyclic structure was determined by MNDO and MINDO/3 calculations of the reaction of cyclodecomposition of 3,6-dihydropyridazine XV into 1,3-butadiene and nitrogen [31]:

XIX → [ ]‡ → XX → [ ]‡ → + N_2

Table 10.1. Optimized bond lengths (Å) of stationary points on the PES of reaction of [4 + 2]-cycloaddition of 1,3-butadiene to ethylene according to MINDO/3 [18] and ab initio [28] calculations (latter in parentheses)

Bond	XVII	XVI	XVIII (TS)	XV
1–2	1.35	1.40 (1.43)[a]	1.44 (1.48)[a]	1.42 (1.38)[b]
2–3	1.50	1.37 (1.33)	1.35 (1.31)	1.38 (1.39)
3–4	1.52	4.05 (4.0)	5.29 (5.0)	2.19 (2.24)
4–5	1.52	1.46 (1.47)	1.35 (1.42)	1.37 (1.39)
5–6	1.52	1.51 (1.65)	2.06 (1.91)	2.19 (2.24)
6–1	1.50	1.48 (1.47)	1.38 (1.44)	1.38 (1.38)

[a]CAS2 STO-3G calculations; [b]CAS1 STO-4–31G calculations

The calculated enthalpy and entropy of activation agree quite well with the experimental data.

Thus, according to the MINDO/3 calculations, the Diels–Alder reaction is not a concerted process. Table 10.1 lists the geometry characteristics, calculated by this method, which correspond to the most important stationary points on the PES of the reaction of [4 + 2]-cycloaddition of 1,3-butadiene to ethylene, compared with the results of the ab initio calculations [28].

It should be noted that certain doubts have been thrown upon the conclusions as to the nonconcerted character of [4 + 2]-cycloaddition reactions. It was, for example, pointed out [32, 33] that the semi-empirical methods based on the NDO approximation (Sect. 2.4) cannot account for the role of the four-electron destabilization in Eq. (4.10), which for the nonsymmetrical structures of the XIV, XVIII type is greater than in the case of the symmetrical structures XV.

According to this opinion, it is to be expected that, provided the calculational schemes were employed that accurately accounted for overlap integrals, a concerted mechanism would be preferable. Indeed the results of the ab initio calculations [22–28] using the STO-3G and 4–31G basis sets, with the configurational interaction (CI 3×3) included, have led to a conclusion that both synchronous and non-synchronous paths exist, but the former is favored by 2 kcal/mol over the latter with the symmetrical transition state XV.

Table 10.2 lists calculated energy parameters of the reaction of ethylene with 1,3-butadiene.

An important conclusion throwing some light on the reasons for the discrepancy between the semiempirical and the ab initio calculations of the mechanism of [4 + 2]-cycloaddition reactions was recently arrived at in Ref. [26] where the effect the correlation corrections have on the PES of this reaction was studied by means of the Møller–Plesset perturbation theory (see Sect. 2.2.4). The authors used the minimal STO-3G basis set and restricted themselves to the calculation of the section of the PES assuming the sum of the lengths of the forming δ-bonds to be equal to 4.4 Å. Figure 10.2 shows

Table 10.2. Heats of reaction ΔH_{298} and heats of activation of direct and reverse reaction of [4 + 2]-cycloaddition of 1,3-butadiene to ethylene (kcal/mol)

Method	$\Delta H^{\neq}$ (direct	$\Delta H^{\neq}$ (reverse)	ΔH°_{298}	Mechanism	Ref.
MINDO/3	28.4	102.6	−61.2	biradical	[17, 18]
UMINDO/3	25.4	70.7	−61.2	biradical	[18]
MINDO/3 + CI	33.9	82.7	−58.2	biradical	[18]
STO-3G	38.6	74.4	−36	concerted	[25]
7s 3p	20.7	110.7	−90	concerted	[23]
UHF/3–31G	26.9	54.7	−27.8	concerted	[25]
4–31G + 3 × 3 CI	42.5	79.7	−37.2	concerted	[24]
exp.	27.5	65.9	−40.5		[34]

variation in the profile of the PES with successive increase in correlation corrections.

Inclusion of the correlation effects (MP3) changes drastically the profile of the PES obtained by an ab initio scheme, bringing it closer to the results of the MINDO/3 calculations. Clearly, the effectiveness of the semiempirical methods, such as MINDO/3, MNDO, and AM1, for the description of the MERP of cycloaddition reactions depends on the character of parametrization of these methods which would make it possible to include a certain part of correlation effects into empirical parameters.

The conclusion as to the nonconcerted charactcr of thc [4 + 2]-cycloaddition reactions and sharp asynchronization of the process of formation of the δ-bonds in the course of the reaction has been borne out by a careful kinetical analysis, performed by Dewar, of reactions of typical dienophile maleic anhydride with a number of furane derivatives as well as by critical examination of other experimental data on the Diels–Alder reactions [29, 35]. A similar conclusion has also been drawn for the [4 + 1]-cycloaddition reaction [31], for some other

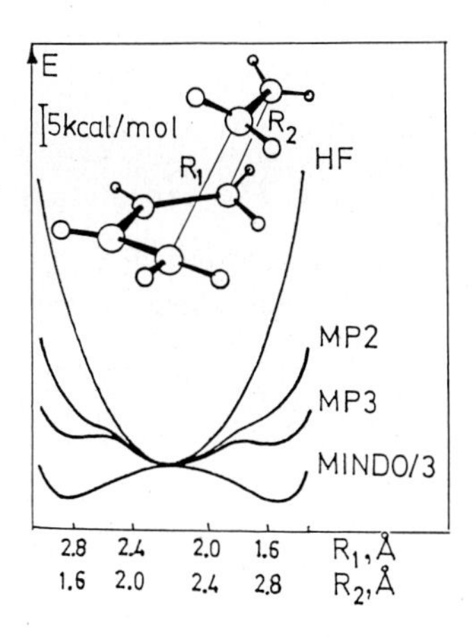

Fig. 10.2. Energy profile of the cycloaddition reaction of butadiene and ethylene corresponding to the section surface for $R_1 + R_2 = 4.4$ Å (Adapted from Ref. [26])

pericyclic transformations and a number of reactions resulting in the making and breaking of two and more bonds (multibond reactions).

These results led Dewar to formulate a new rule: the synchronous multibond reactions are usually forbidden [30]. Even though this rule cannot be deduced from strict theoretical premises as, for example, the rule governing the selection of transition state structures by their symmetry properties (see Sect. 1.3.3.2), it has been verified, according to an analysis in Ref. [30], for all the multi-bond reactions for which a reliable examination of the PES's has been carried out. This fact requires a certain reassessment of the customary view in regard to the pericyclic transformations as necessarily synchronous reactions in those cases when they are allowed by the rules of orbital symmetry conservation.

It should be noted in conclusion that the above-considered (Sect. 9.2.5) double proton migrations may also be regarded as typical multibond reactions. The rule cited above is actually quite applicable to these reactions. At the same time, given a favorable stereochemistry, a concerted synchronous mechanism may be realized for some systems of this type. Analysis of the factors that may provide for the possibility of such exceptions in the case of this and some other multibond reactions would be of particular interest for understanding the nature of cooperative processes.

10.2 Electrocyclic Reactions

The electrocyclic reactions are intramolecular $(\pi + \sigma)$ cycloadditions, a most important example of which is provided by the cyclization of 1,3-butadiene XXI to cyclobutene XXIII:

XXI → XXII → XXIII

The usual notion of the coordinate of electrocyclic reactions is associated with the rotation of the end groups about double bonds. The conrotatory motion is thermally allowed for the reaction XXI–XXIII. Semiempirical [2, 36–42] and ab initio [43–45] calculations of the critical regions of the PES of this reaction and of the still simpler cyclization of the allyl cation to the cyclopropyl cation have greatly refined the overall picture of the intrinsic mechanism and revealed some important distinguishing features common to all electrocyclic transformations.

At an invariable distance R between the methylene links in XXI, their conrotatory motion involves very great energy expenditure (over 100 kcal/mol) and for the reverse reaction this value exceeds 400 kcal/mol. Thus, rotation of the methylene links does not determine the reaction coordinate at the initial stage. Based on earlier calculations [36, 37, 39, 40, 43] it was assumed that the changes of the parameters R and θ are practically independent of each other all along the reaction path. At the initial stage of the opening of cyclobutene XXIII to butadiene XXI, the reaction coordinate is the quantity R, i.e., the stretching of the bond CC.

When R is changing from the equilibrium value of 1.534 Å for XXIII to the value of about 2.270 Å, the methylene groups do not rotate. Rotation starts only when R attains 2.07 Å (MINDO, Ref. [39]) or, according to ab initio calculation in Ref. [43], at 0.270 Å, with R not changing any more once the methylene groups are involved in rotation. It is precisely at this stage of reaction that the difference emerges between the energetics of the conrotatory and the disrotatory channels (see Fig. 2.3). As soon as the rotation of the methylene groups stops after having produced the magnitude of the angle θ demanded by the structure XXI, the stretching of R resumes up to the equilibrium value of 2.819 Å in butadiene XXI. A nonplanar transition state of C_2 symmetry XXII with the dihedral angle $\varphi = 54°$ appears in the zone where the reaction coordinate passes from the stretching of the CC bond to the rotation of the methylene groups.

However, a more rigorous construction of the reaction path for XXI ⇄ XXIII as the steepest descent line from the transition state point has shown that the rotation for the methylene groups starts at the early stage of the reaction path concurrently with the change in R [38, 44], which is clearly seen in Fig. 10.3, and this change of the quantities R, θ_1 and θ_2 ends only at the point which corresponds to the final state of the system.

The success of the Woodward–Hoffmann rules in accounting for the preferability of the thermal conrotatory process can be explained by the fact

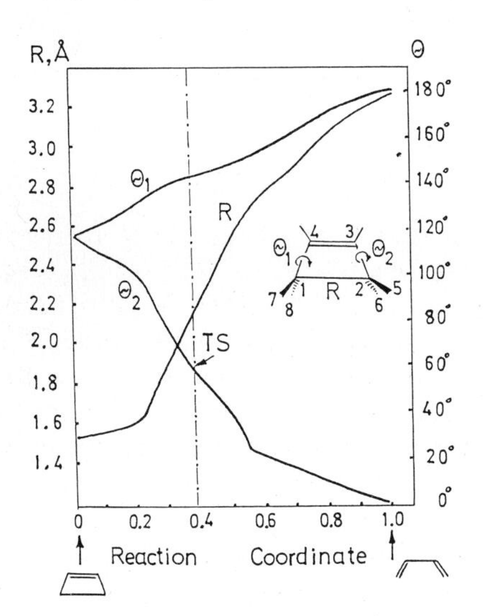

Fig. 10.3. Variation of principal geometry parameters for the conrotatory conversion of cyclobutene into butadiene (Adapted from Ref. [38])

that the change in R, θ_1 and θ_2 within the limits mentioned does not affect relative position and the symmetry of the frontier orbitals and the orbitals close to these. Therefore, the conclusions drawn from the analysis of the equilibrium structures prove valid also for that portion of the PES where the allowed conrotatory and the forbidden disrotatory paths can actually be distinguished.

For a correct description of the latter, the effects of electron correlation must be exactly accounted for, as was noted earlier. This is necessary in the first place because of crossing of the orbitals when reaction moves along the disrotatory route giving rise to the 1,4-biradical structure of the transition state XXIIa [37]. Although a planar carbon skeleton is retained in XXIIa, its structure clearly indicates an asynchronous character of rotations of the methylene links on the MERP:

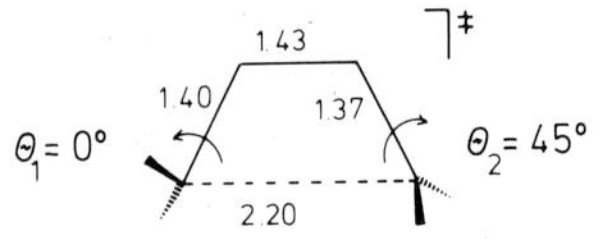

The difference between the energy of the disrotatory transition state XXIIa forbidden by the orbital symmetry conservation rules and that of the allowed conrotatory transition state XXII does not exceed 15 kcal/mol, as seen from Table 10.3.

According to the ab initio calculations [49, 50][1], the following mechanism of photoreaction is operative. 1) Excitation $S_0 \rightarrow S_1$; 2) Nonadiabatic transition

Table 10.3. Activation barriers for conrotatory and disrotatory isomerization of cyclobutene to *cis*-butadiene (kcal/mol)

Method	Conrot.	Disrot.	Ref.
MINDO/3	49	55.6	[39]
MNDO	49.8	—	[38]
AM 1	36.0	—	[41]
ab initio (1971)	58.9	72.3	[43]
ab initio (1984)	42.4	58.4	[44]
ab initio (1988)	36.6	—	[48]
Exper.	34.5[a]	47.9	[46]
	32.9[a]		[47]

[a] See caption to Fig. 2.4

[1] Symmetry conditions were imposed on the MERP being calculated: C_2 for conrotatory channel and C_s for disrotatory channel

$S_1 \rightsquigarrow S_2$ (on the PES of the S_1 state no path capable of reaction was found). 3) Nonadiabatic transition $S_2 \rightsquigarrow S_0$ from the minimum on the PES of the S_2 state with the energy gap between the PES's of the excited S_2 state and the S_0 state in the zone of the conrotatory channel of the photoreaction being much smaller than in that of the disrotatory channel, which implies greater effectiveness of the disrotatory photochemical cyclization. 4) Structural relaxation to cyclobutene on the PES of the ground state.

As has been shown by an initio calculations [45] and experimental studies [51–53], the electronic factors also determine stereoselectivity of reactions of opening of substituted cyclobutenes. Donor substituents at C3, such as OR, display considerable preference for outward rotation in the electrocyclizations XXIV → XXV. Such an effect is observed for the compound with R = t-Bu. In spite of the large steric effect, *tert*-butyl rotates inward due to the electronic effect of the small hydroxy group.

XXIV → XXV

Electrocyclic reactions of thermo- and photocyclization of 1,3,5-hexatriene to 1,3-cyclohexene have recently been investigated by the semiempirical MINDO/3 method [54, 55]. For the transition state on the disrotatory thermocyclization route, allowed by the Woodward–Hoffmann rules, a point on the PES was located and identified which corresponds to the structure with nonplanar carbon skeleton but C_s symmetry. For the first time an analogous structure of TS was studied in Ref. [56]. The barrier of the direct reaction calculated in Refs. [54, 55] was 34 kcal/mol, while the experimental value was 29.6 kcal/mol [57]. A study of the reaction path of the forbidden conrotatory thermocyclization has revealed disturbance of C_2 symmetry or rather total lack of the symmetry elements. When moving along this path with the length R of the forming carbon–carbon bond being the reaction coordinate, first the rotation of one methylene group occurs and only in the region of $2.1\,\text{Å} \leqslant R \leqslant 2.5\,\text{Å}$ the second group starts rotating. Then the bond length R acts once again as the reaction coordinte. The calculated reaction barrier for the forbidden path is higher by 15 kcal/mol as compared to the allowed disrotatory channel.

For the photochemical path of cyclization of 1,3,5-hexatriene to 1,3-cyclohexadiene a nonsymmetrical route capable of reaction was found on the PES of the excited S_1 state which nevertheless led to the topomer of 1,3-cyclohexadiene predicted by the orbital symmetry conservation rules. The photoreaction follows, according to calculations, this route: 1) excitation of 1,3,5-hexatriene to the S_1 state; 2) overcoming of the 5 kcal/mol activation barrier on the nonsymmetrical route of conrotatory motion; 3) transition, nonadiabatic or due to internal conversion, to the PES of the ground state; 4) relaxation on the PES of the S_0 state to the final product, namely, 1,3-cyclohexadiene.

An analogous mechanism must govern also the reverse photoreaction. According to the calculations [58], the photocyclization of 1-(3-phenyl) butadiene-1,3 to 1,8a-dihydronaphthalene proceeds by the same route.

Analogous to the opening of 1,3-cyclohexadiene is the electrocyclic reaction of ring opening of 2H-pyrans XXVI → XXVII:

$R^1 = H, NH_2, OH, NO_2, CHO$
$R^2 = H, OH$
$R^3 = H, OH, NH_2$

Calculations [59, 60] of the MERP of a series of these reactions carried out by means of the MINDO/3 method, have shown that the length of the bond C_{spiro}—O in the transition states whose structure has been identified by calculating the matrices of force constants equals 1.95 Å practically being independent of substituents and annelation.

A similar scheme of reaction is also characteristic of the transformation of bicyclobutane XXVIII into 1,3-butadiene which is a double electrocyclic process:

The MINDO/3 calculations [61] reveal on the PES of this reaction a biradical intermediate XXIX. According to the rule formulated in the preceding section, the breaks of two σ-bonds in XXVIII are completely asynchronous. The calculated value for the activation barrier of 40.3 kcal/mol for the route XXVIII → XXIX → XXI is in excellent agreement with the experimental value of 40.6 kcal/mol [62].

However, recently conducted ab initio calculations (MP2/6–31G* + CI)—Ref. [63]—have shown that the thermal reaction XXVIII → XI proceeds concertedly (in one step) via genuine nonsymmetrical transition state corresponding to conrotatory motion of two methylene groups (the breaking C_1C_4 and C_2C_3 bonds are of different length in the transition state equalling 2.313 and 1.618 Å, respectively). The calculated activation barrier of 43.6 kcal/mol turned out to be exactly equal to the experimental value. Unfortunately, in Ref. [63] no use was made of the channel of opening of dicyclobutane XXVIII via the biradical intermediate structure XXIX. Even though additional theoretical investigations into ring opening of heterosubstituted bicyclobutanes [64] apparently speak well for a concerted mechanism, the question as to the preferability of either the concerted or the stepwise mechanism will without a detailed comparison between the two still remain open.

10.3 Sigmatropic Rearrangements

The simplest type of sigmatropic rearrangements, the 1,j-rearrangement, is associated with migration of a single atom or a monovalent atomic group from a position neighboring on the π-system to one of the positions within it. Consider as an example the degenerate 1,3-rearrangement of the hydrogen atom in a propene molecule.

XXX XXXI, C_{2v} XXXII, C_s XXXa

The orbital symmetry conservation rules predict preferability of an antara-surface migration of hydrogen by the scheme XXX ⇄ XXXI ⇄ XXXa. Non-empirical calculations [65, 66] of this rearrangement have revealed two possible reaction paths and two transition states, XXXI with the symmetry of C_{2v} type corresponding to a thermally allowed antara-surface shift and XXXII with the symmetry C_s corresponding to a forbidden supra-surface shift. It should be noted that, contrary to the prediction, XXXII has a lower energy than XXXI. A calculation of the barriers for XXXI and XXXII with the DZ basis set and the electron correlation taken into account by the CEPA scheme has yielded 95.6 and 92.9 kcal/mol, respectively [66]. Another technique of inclusion of the electron correlation (CI 3×3) using the 4–31G basis set has given a still more strongly marked difference of 114.9 and 93.1 kcal/mol [65]. The values of activation barriers calculated for the forbidden supra-surface reaction path are in excellent accord with the experimental value of activation enthalpy of the reaction XXX ⇄ XXXa which equals 87.7 kcal/mol [67].

It is to be noted that a correct relationship between the energies of the structures XXXI and XXXI can be achieved only on condition that the electron correlation is accounted for. This is necessary in view of the fact that the structure XXXII represents a trimethylene biradical with a conformation close to the (EE) type (see Sect. 1.3). In the zone of the XXXII structure, the PES is considerably flattened containing, probably, several additional local minima that correspond to the radical pair $C_3H_5 \cdots H$. The presence of biradical states and, consequently, the nonconcerted character are the typical features of pericyclic reactions developing along the route forbidden by the orbital symmetry conservation rules.

A simple, physically clear-cut criterion of concertedness of 1, j-sigmatropic rearrangements is provided by the comparison of the activation barriers with the energy of rupture of the migrating σ-bond [68]. For propene XXX the energy of the C—H_1 bond (90 kcal/mol) is lower than the activation energy of the reaction XXX $\leftrightarrows$ XXXa which is an evidence for its unconcerted character. By contrast in the case of the 1,5-sigmatropic rearrangement of *cis*-1,3-pentadiene XXXIII the activation energy amounts to a mere 35–37 kcal/mol [69] indicating that the reaction is likely to proceed in a concerted manner. The ab initio calculations [70, 71] are definitive in pointing to preferability of the C_s structure XXXIV of the transition state for the suprasurface migration of the methyl hydrogen atom:

H
XXXIV, C_s
CH_3
CH_3
XXXIII
XXXIIIa
H
XXXV, C_{2v}

An alternative planar C_{2v} structure XXXV of the transition state is by 19.3 kcal/mol less favored than XXXIV (6–31G*/MP2 basis set). It has two negative force constants or two imaginary frequencies at the UHF/3–21G level. The second imaginary frequency corresponds to a distortion of the C_{2v} structure XXXV to the C_s structure XXXIV [71].

The kinetic isotopic effect calculated for this reaction by ab initio [71, 72] and MINDO/23 [73] methods is about 2.50–2.53, which is in poor agreement with the experimental result of 4.97 [69]. Dewar et al. [73] suggested that the discrepancy might be due to tunnelling from vibrationally excited states. However, Jensen and Houk [71] hold that an improvement may be expected with a more accurate calculation of unharmonic frequencies.

The mechanism of the 1, j-sigmatropic rearrangements becomes a lot more complex when the migrating group has its own π-system. Figure 10.4 presents some calculation results for the PES of positional topomerization and isomerization of nitrosocyclopropene [74]:

The calculated PES of 1,3-sigmatropic shifts in XXXVI (Fig. 10.4, upper plane) resembles the PES of ozone topomerization (Fig. 1.2) but is more complicated containing, apart from the minimum of the fully symmetrical

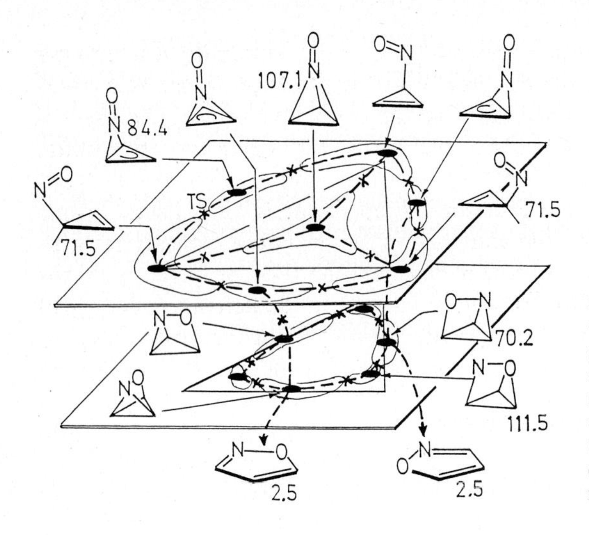

Fig. 10.4. Schematic C_3H_4NO PES *Crosses* denote the transition states, *dark circles*—the minima, *dashed lines*—the paths of topomerization and isomerization reactions. Values near the structures are the heats of formation (kcal/mol) calculated by the MINDO/3 method [67])

XXXVI a ⇌ XXXVI b ΔH_f = 71.5 kcal/mol ⇌ XXXVI c

XXXVII, C_{3v}

ΔH_f = 107.1 kcal/mol

XXXVIII, C_s

ΔH_f = 84.4 kcal/mol

η^3-isomer XXXVII, an additional minimum of the biradicaloid η^2-structure XXXVIII.

As is evident from Fig. 10.4, two routes of isomerization of nitroso cyclopropene XXXVI are possible. The first of these is associated with the $\eta^1 \to \eta^2$ transition and includes formation of the intermediate XXXVIII. The initial η^1 nitroso cyclopropene XXXVI and the η^2-isomer XXXVIII are separated by a rather low activation barrier of 14.2 kcal/mol. Another route of isomerization runs via the intermediate η^3-structure XXXVII. The transition state of the $\eta^1 \rightarrow \eta^3$ rearrangement is, however, characterized by rather high energy of 40.6 kcal/mol relative to XXXVI. The calculations [74] have shown that energetically unfavored are also the gas-phase migrations of the nitroso group along the three-membered ring proceeding by the dissociative mechanism via the ion states $C_3H_3^+ + NO^-$ $C_3H_3^- + NO^+$ or else via the radical pair $C_3H_3^{\cdot} + NO^{\cdot}$. In the case of solutions, the solvation energy may introduce certain corrections, for example, the mechanism of positional topomerization of η^1 azido cyclopropene has to do with the dissociation of the C—N bond and the formation of the ion pair $C_3H_3^+ \cdot N_3^-$ with its subsequent randomization [75].

10.4 Haptotropic Rearrangements

The above-considered reaction of topomerization of η^1-nitroso cyclopropene belongs to the general type of 1, j-sigmatropic rearrangements of cyclic compounds associated with the formation of intermediate half-sandwich structures, located on the reaction coordinate, with an elevated hapto (η)-number. Hoffman [76, 77] coined the term haptotropic rearrangements to denote such pericyclic reactions in which the character of coordination of the migrating center changes relative to the polyene system. Qualitative analysis of the mechanism of such rearrangements is made easier when the use is made of: the electron count rules for determining the stability (possibility of assignment of the PES minima) of organic π-complexes of various structural types [78, 79], the principle of isolobal analogy (Sect. 4.4) and skeletal electron count rules [80–83] in the case of organometallic compounds. Consider as an example the topomerization reaction of derivatives of the Dewar thiophene XXXIX:

XXXIXa XXXIXb XL XXXIXc XXXIXd

The activation energy found by the dynamic NMR method for the perfluorotetramethyl derivative amounts to 22 kcal/mol (for a review on migrational rearrangements associated with shifts of the σ-bond along the periphery of 3–8-membered cycles see Ref. [84]). If the reaction mechanism includes the formation of a transition state or an intermediate of the pyramidal type XL (η^4-structure), the rearrangement may be assigned to the haptotropic type. However, an anlaysis of the orbital interactions that stabilize nonclassic pyramidal structures formed through addition to cyclic polyene of a nontransition atom or a σ-bond has led to a conclusion that such structures are stable only on condition that the eight-electron rule is satisfied [78, 79, 85]. Valence orbitals of the apex and cyclic polyene in type XL pyramidal structures form only three bonding and one nonbonding MO, therefore the total number of the π-electrons of the basal fragment and the valence electrons of the apical group must not exceed eight. In XL this number is ten, so the rearrangement reaction XXXIXa ⇄ XXXIXb ⇄ ··· may be expected to develop as a series of supra-surface 1,3-shifts C—S along the circumference of the four-member ring. Indeed, the MINDO/3 calculations [85] of the PES of this reaction have shown that the structure XL corresponds to the top of a rather steep hill while the transition state has a C_s structure XLI (Fig. 10.5a).

Upon the removal from the molecule XXXIX of two electrons, the pyramidal form of the type XL becomes, in accordance with the eight-electron rule, stable. This is confirmed by MINDO/3 and ab initio (STO-3G basis set) calculations

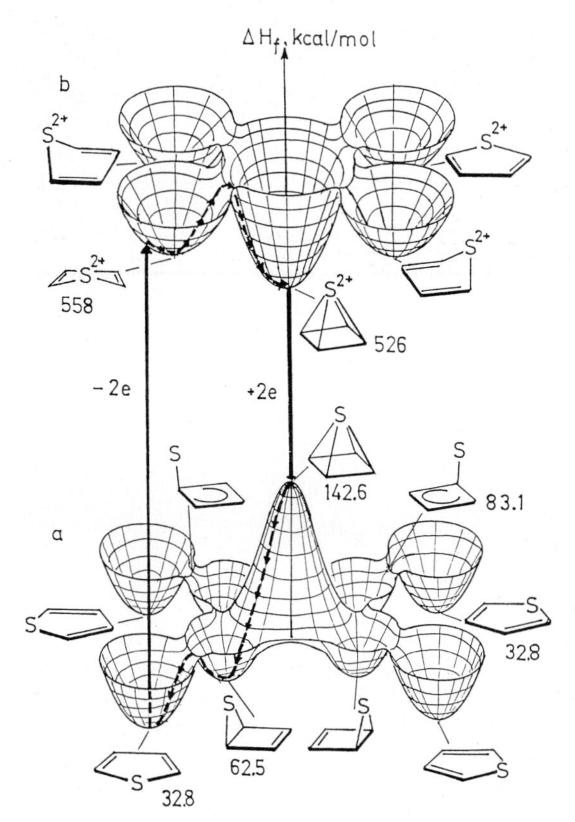

Fig. 10.5a. Schematic PES of the XXXIXa ⇄ XXXIXb ⇄ XXXIXc ⇄ XXXIXd reaction calculated by the MINDO/3 method [79]. Also shown is the channel of the electrocyclic conversion of Dewar thiophene into thiophene; **b** same as in **a** for the topomerization of the thiophene dication [79, 80]. Pyramidal structures of the XL type correspond to different types of the stationary points on two PES's

[86, 87] on the PES of the topomerization of thiophene dication (Fig. 10.5b) which may, accordingly, be regarded as a degenerate haptotropic rearrangement.

XLI XLIIa XLIIb

The most general class of haptotropic rearrangements are the shifts of organometallic groups of the ML_n, MC_pL_m type, and other groups isolobal to these, over the plane of polynuclear cyclic polyenes LXIIa ⇄ LXIIb. Among these rearrangements, reversible reactions have recently been found [81, 88] with the activation energies well within the scale of the tautomeric reactions.

The general principle of a theoretical analysis of haptotropic rearrangements [77] consists in the singling out of the principal orbital interactions that determine the binding of ML_n to the cyclic π-system and studying the changes in this binding along the route of a reaction of the XLIIa ⇄ XLIIb type. The most important interactions are those between the highest occupied MO's of the π-system and the *e*-orbitals of the organometallic fragment depicted in Fig. 4.5. In case the overlap between those orbitals is retained throughout the whole route of the haptotropic rearrangement, it may be expected that it would require the overcoming of a moderate activation barrier. Otherwise the reaction would have to surmount a high barrier or follow some other route, for instance, the intermolecular one.

Consider as an example a reaction of the degenerate rearrangement of the π-complex of chromium naphthalene-tricarbonyl XLIII. It would be natural to assume as an intermediate structure or a transition state the structure XLIIIa in which the group $Cr(CO)_3$ is located between two nuclei on the least-motion path.

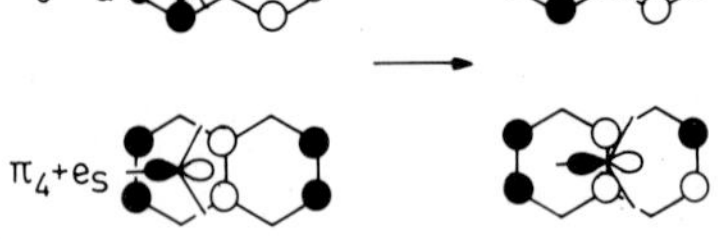

$\pi_5 + e_a$

$\pi_4 + e_s$

XLIII XLIIIa

It is not hard to see that in the expected transition state structure XLIIIa no interaction takes place between the π-orbitals of naphthalene and the lower-lying free orbitals of $Cr(CO)_3$. Indeed, the haptotropic rearrangement reaction in the naphthalene system requires a quite high activation barrier of over 30 kcal/mol [77, 78]. A direct calculation of the PES done by the EHMO method [77] has shown that the least-motion path is not advantageous and the displacement of the $Cr(CO)_3$ group over the naphthalene cycle proceeds with the formation of an intermediate XLIV in which the necessary $(\pi - e)$ overlap is kept intact. By contrast, in the pentalene complexes XLV, when the migrant is shifted to the center along the least-motion path, the overlapping is not lost and a low activation energy may be expected for the haptotropic rearrangement XLVa ⇄ XLVb. Non the less the calculations of the PES have shown that in this case, too, the most favored is not the least-motion path but a less symmetrical route with a transition state of C_s symmetry in which the organometallic migrant, when located over the central bond, is shifted away from this bond's center.

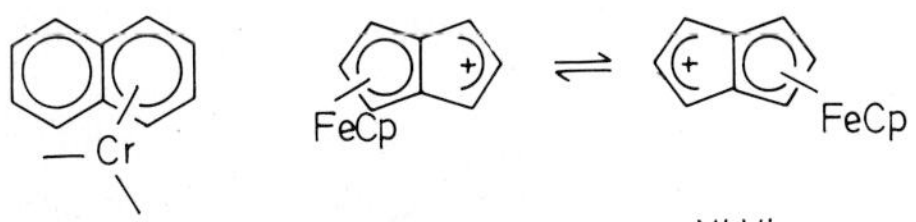

Cr FeCp ⇌ FeCp

XLIV XLVa XLVb

The above-considered examples of theoretical studies of haptotropic reactions have emphasized once again that a comprehensive analysis of the intrinsic mechanism of these reactions must be based on the admittedly fruitful semiquantitative orbital approach as well as on the quantitative studies of the PES.

10.5 Ion-Radical Pericyclic Reactions

Interest in reactions of this type has been aroused by the detection of the strongest catalytic effects associated with the inclusion of ion-radical intermediates in the processes of electrocyclic isomerizations and cycloadditions.

Thus, the isomerization of benzonorcaradiene XLVI to benzocycloheptatriene XLVII that results from two successive reactions of the 1,3-sigmatropic shift and the electrocyclic disrotatory opening of the six-membered ring and is, in a neutral molecule, quite a slow process can in the anion-radical be accelerated by 15 orders of magnitude! [89].

Although the derivatives of cyclobutadiene are more favored thermodynamically than the isomers or derivatives of tetrahedran, the XLIII → L transition is forbidden by the orbital symmetry conservation rules [1, 90]. For this reason, the transformation of tetra-*tert* butyltetrahedran into tetra-*tert* butylcyclobutadiene ($R = t\text{-}C_4H_9$) can occur only upon heating to 135°C and higher. The calculations [91–93] show that the reaction proceeds in several steps that include the formation of intermediate isomeric (according to the type of bond stretching) biradicals XLIX (see Sect. 8.3):

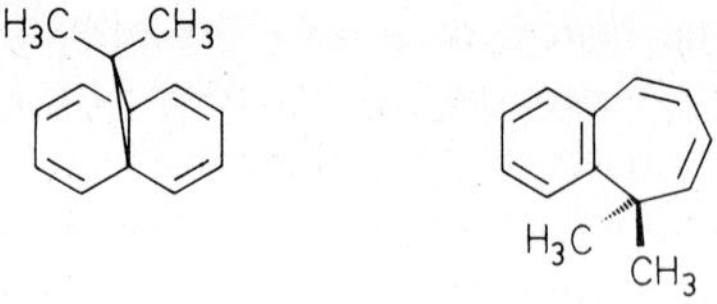

XLVI XLVII

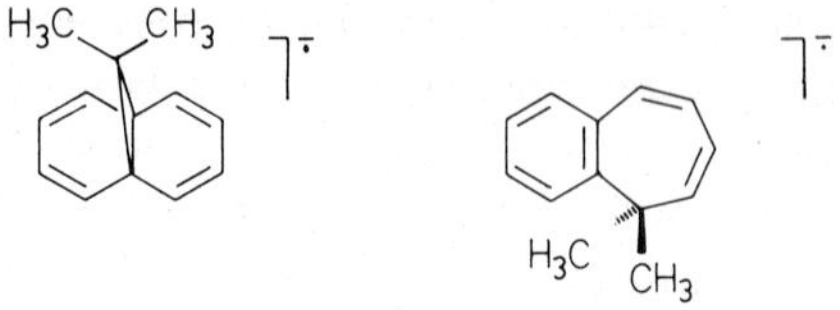

XLVIa XLVIIa

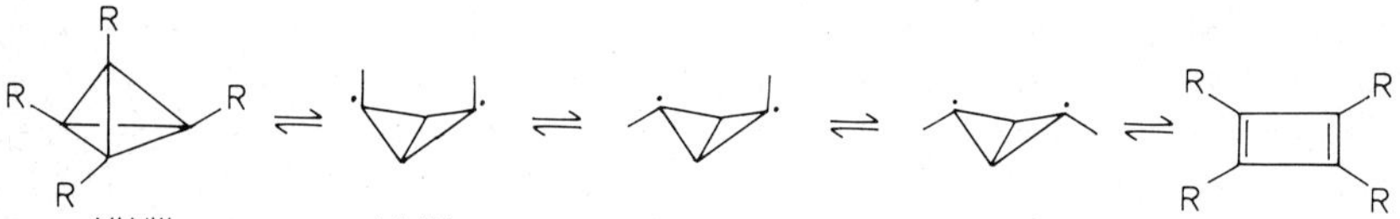

XLVIII XLIXa XLIXb XLIXc L

As opposed to it, the cation-radical XLVIII (R = t-C_4H_9) that is readily formed from XLVIII upon one-electron oxidation rearranges barrierlessly into the cation-radical La (R = t-C_4H_9), Ref. [90].

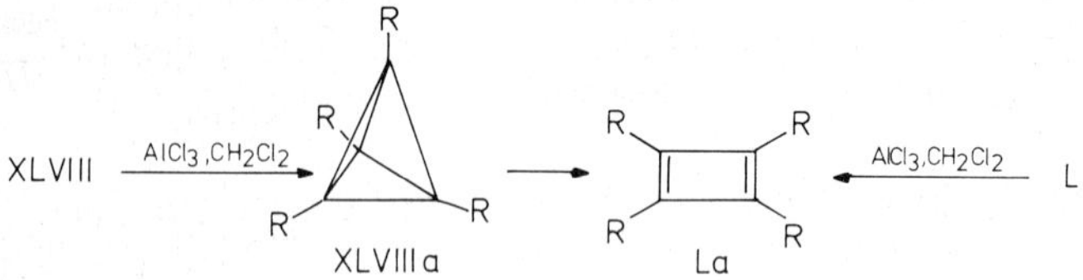

XLVIIIa La

In practical terms, particularly important are the cycloaddition reactions catalyzed through the formation of cation-radicals. A theoretical analysis of the PES of these reactions [94] has helped to evolve some general notions concerning the mechanism of ionradical pericyclic transformations.

Consider a cation-radical reaction of cycloadditiion [4 + 1] and [3 + 2]:

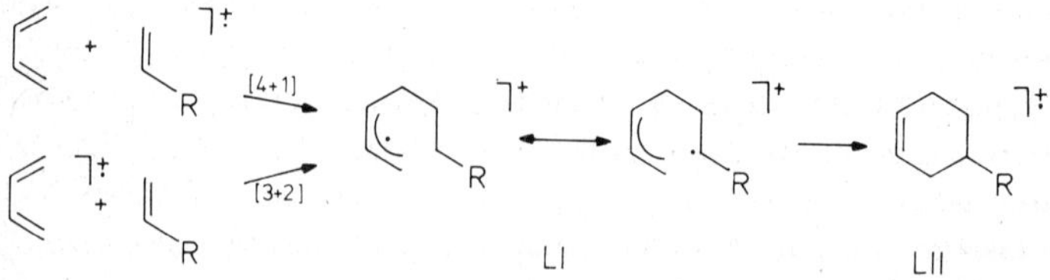

LI LII

Arguments based on the orbital approach and direct calculations by the MINDO/3 and MNDO methods indicate that the cation-radical route does not provide any thermodynamic advantage over the reaction of neutral molecules (Sect. 10.1.2) not leading to any greater stability of the cyclic cation-radical LII in comparison with the starting reactants. On the contrary, the calculated heat of the [4 + 2] dimerization of neutral butadiene ($R = CH{=}CH_2$) equal to -44.0 kcal/mol exceeds the corresponding value for the cation-radical reaction of -42.5 kcal/mol. The reason for the acceleration of reaction lies in a sharp lowering (down to complete disappearance) of the energy barrier at the stage of formation of the intermediate cation-radical LI, which is in contrast to the analogous formation of the biradical XVI (nonsynchronous process) or the final product (synchronous process) in the reaction of neutral molecules. That lowering is brought about by strong stabilization of the cation-radical LI with respect to the starting reactants. Indeed, whereas the formation of the cation-radicals of butadiene or ethylene requires removal of an electron from the strongly bonding higher occupied MO's of the initial molecules, the orbital of the unpaired electron is in the cation-radical LI nonbonding. Therefore the starting reactants are in the cation-radical cycloaddition strongly destabilized in comparison with those in a neutral reaction and the intermediate products XV and LI have heats of formation that are close in value. MINDO/3 calculations on the reaction of [4 + 1]-cycloaddition of butadiene to its cation-radical ($R = CH{=}CH_2$) have yielded the following values of energy barriers for the formation steps LI and LI $\rightarrow$ LII: 8.0 and 9.3 kcal/mol, while in the case of a reaction of neutral molecules the total activation energy amounted in the same calculation [95] to 36 kcal/mol. For the cation-radical reaction of [4 + 1]-cycloaddition of butadiene and ethylene, the calculated energy barrier was at the LI formation step 16 kcal/mol but it was 28 kcal/mol for the formation of the biradical XVI from neutral molecules [18, 35].

Such a thermodynamically favorable condition for the formation of intermediate cation-radicals undoubtedly requires complete asynchronization of the processes of making the two new σ-bonds in LII which are even more strongly pronounced than in the case of the neutral molecules. Indeed, the calculations [95] have shown the symmetrical routes for the formation of transition states in the cation-radical reactions of [4 + 1]- and [3 + 2]-cycloaddition to be extremely unfavored.

The scheme considered is fully valid also in the case of cation-radical [2 + 1]-cycloadditions. These reactions, like the corresponding thermal reactions of the [2 + 2]-cycloaddition of neutral molecules, are forbidden by the orbital symmetry conservation rules. The same calculations [95] have shown that the addition of the cation-radical of ethylene to a neutral ethylene molecule proceeds in an unconcerted and nonsynchronous fashion. Unlike the [2 + 2]-cyclodimerization of ethylene (Sect. 10.1.1), the [2 + 1]-cycloaddition involves the formation of an intermediate LIII with the energy barrier calculated for this highly exothermal step being extremely low (1.3 kcal/mol by the MNDO method). A barrier lower still (1.0 kcal/mol) is expected for the step of transformation of LIII into the cation-radical of cyclobutane LIV in which the forming bond is considerably lengthened as contrasted to other CC bonds (data of a calculation for R = H):

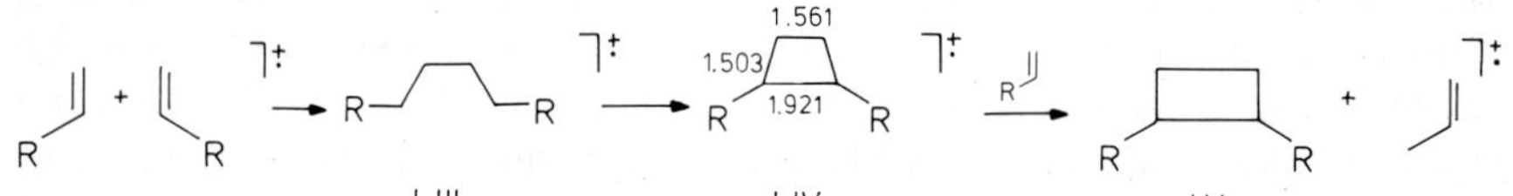

LIII LIV LV

Transition to the final product, the cyclobutane LV, can be greatly facilitated by the use of electron-donating substituents R. In this case, the electron transfer from substituted ethylene to the orbital of the cation-radical LIV occupied by one electron switches on the chain mechanism represented on the above scheme, similar to that for the reactions of the $S_{RN}1$ type (Sect. 9.1). Such mechanisms are a characteristic feature for the class of reactions, catalyzed by electron transfer, in which the sequence 1) ionization (electron capture), 2) reaction in the ion-radical, 3) electron capture (ionization) – is a very convenient route for effecting the needed transformation [96].

It is to be noted that in the general case the mechanism of the ion-radical pericyclic reaction may differ from that of the corresponding reaction of neutral molecules in that it may include additional framework rearrangements of initial or intermediate ion-radicals. For example, considerable lowering of the activation energy for the conrotatory electrocyclic transformation of the cyclobutene cation-radical LVI into the cation-radical of butadiene LVII that is observed in mass-spectrometrical and radiolytic experiments cannot be explained by assuming a concerted one-step mechanism for the transformation LVI → LVII, similar to that for XXI → XXIII (Sect. 10.2), see Ref. [97].

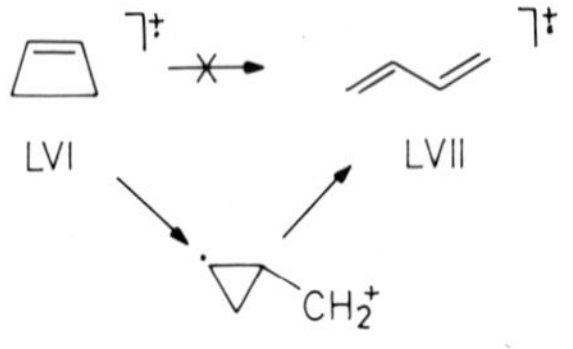

LVI LVII LVIII

The activation energies for the reaction LVI → LVII calculated by the MINDO/3 and MNDO methods amount, respectively, to 34 and 31 kcal/mol, while the experimental value for this transformation lies within 7–18 kcal/mol and for neutral cyclobutene it is 33 kcal/mol [47]. The calculations [97] have shown that a more advantageous reaction path is represented by the two-step mechanism LVI → LVIII → LVII. The MINDO/3 and ab initio (6–31G basis set) methods predict, in good agreement with each other, the energy barrier of the rearrangement of LVI into the cyclopropylcarbinyl cation-radical to be 20–21 kcal/mol and a mere 2.6 kcal/mol for the step LVIII → LVII.

The analysis, done in Chaps. 5–10, of data on the mechanisms for some of the principal classes of organic reactions has shown what indisputably important information the quantum chemical calculations may afford concerning the details of reaction coordinates and the structure of transition states. Fundamental theoretical study of reaction mechanisms by the methods of quantum chemistry is still in its infancy. It may, however, be expected that this study will in the near future take important steps forward thanks to realistic prospects opening up of more rigorous calculations on quite large molecules both in the gas-phase and in solution.

References

1. Woodward RB, Hoffmann R (1969) The conservation of orbital symmetry. Academic Press, New York
2. Jug K (1980) Theor Chim Acta 54:263
3. Burke LA, Leroy G (1979) Bull Soc Chim Belg 88:379
4. Duran M, Bertran J (1982) J Chem Soc Perkin Trans II:681
5. Fujimoto H, Inagaki S, Fukui K (1975) J ACS 97:6108; (1976) ibid 98:2670
6. Bernardi F, Bottoni A, Robb MA, Schlegel HB, Tonachini G (1985) J ACS 107:2260
7. Segal GA (1974) J ACS 96:7892
8. Bernardi F, Olivucci M, Robb MA, Tonachini G (1986) J ACS 108:1408
9. Bernardi F, Olivucci M, McDouall JJW, Robb MA (1987) J ACS 109:544
10. Bernardi F, Bottoni A, Olivucci M, McDouall JJW, Robb MA, Tonachini G (1988) THEOCHEM 165:341
11. Raghavachari K, Haddon RC, Roth HD (1983) J ACS 105:3110
12. Epiotis ND (1973) J ACS 95:1191
13. Sauer J, Sustmann R (1980) Angew Chem Intern Ed Engl 19:779
14. Konovalov AI (1983) Uspekhi Khim (Russ Chem Rev) 52:1852
15. Bauld NL, Bellville DJ, Lorenz KT, Pabon RA, Reynolds DW, Wirth DD, Chiou HS, Marsh BK (1987) Acc Chem Res 20:371
16. Basilevsky MV, Ryaboy VM (1987) in: Veselov MG (ed) Current problems of quantum chemistry. The quantum chemical methods. The theory of intermolecular interaction and solid state. Khimia, Moscow (in Russian)
17. Dewar MJS, Griffin AC, Kirschner S (1974) J ACS 96:6225
18. Dewar MJS, Olivella S, Rzepa HS (1978) J ACS 100:5650
19. Oliva A, Fernandez-Alonso JT, Bertran J (1978) Tetrahedron 34:2029
20. Dewar MJS, Olivella S, Stewart JP (1986) J ACS 108:5771
21. Basilevsky MV, Shamov AG (1977) J ACS 99:1369
22. Burke LA, Leroy G, Sana M (1975) Theor Chim Acta 40:313
23. Burke LA, Leroy G (1977) Theor Chim Acta 44:219
24. Townshend RE, Ramunni G, Segal G, Hehre WJ, Salem L (1976) J ACS 98:2190
25. Brown FK, Houk KN (1984) Tetrahedron Lett 25:4609
26. Ortega M, Oliva A, Lluch JM, Bertran J (1983) Chem Phys Lett 102:317
27. Houk KN, Lin YT, Brown FK (1986) J ACS 108:554
28. Bernardi F, Bottoni A, Field MJ, Guest MF, Hiller IH, Robb MA, Venturini A (1988) J ACS 110:3050
29. Dewar MJS, Pierini AB (1984) J ACS 106:203
30. Dewar MJS (1984) J ACS 106:209
31. Dewar MJS, Chantranupong L (1983) J ACS 105:7152
32. Basilevsky MV, Shamov AG, Tikhomirov VA (1977) J ACS 99:1369
33. Caramella P, Houk KN, Domel-Smith LN (1977) J ACS 99:4514
34. Barnard JA, Parrot JK (1976) J Chem Soc Farad Trans II 1:2404
35. Dewar MJS (1978) in: Further perspectives in organic chemistry Elsevier, Amsterdam
36. McIver JW, Komornicki A (1972) J ACS 94:2625
37. Dewar MJS (1975) Chem Brit 11:95

38. Jensen A (1983) Theor Chim Acta 63:269
39. Dewar MJS, Kirschner S (1974) J ACS 96:6809
40. Basilevsky MV, Shamov AG (1981) Chem Phys 60:347
41. Dewar MJS, Zoebisch EG, Healy EF, Stewart JJP (1985) J ACS 107:3902
42. Kikuchi O (1974) Bull Chem Soc Japan 47:1551
43. Hsu K, Buenker RJ, Peyerimhoff SD (1971) J ACS 93:2117
44. Breulet J, Schaefer III HF (1984) J ACS 106:1221
45. Rondan NG, Houk KN (1985) J ACS 107:2099; Rudolf K, Spellmeyer DS, Houk KN (1987) J Org Chem 52:3708
46. Cooper W, Walters WD (1958) J ACS 80:4220
47. Garr RW Jr, Walters WD (1965) J Phys Chem 69:1073
48. Spellmeyer DC, Houk KN (1988) J ACS 110:3412
49. Van der Lugt WTA, Oosterhoff LJ (1969) J ACS 91:6042
50. Grimbert D, Segal G, Devaquet A (1975) J ACS 97:6629
51. Kirmse W, Rondan NG, Houk KN (1984) J ACS 106:1871
52. Dobier WE Jr, Korohiak H, Burton DJ, Heinze PL, Bailey AR, Shaw GS, Hansen SW (1987) J ACS 109:219
53. Houk KN, Spellmeyer DC, Jefford CW, Rimbault CG, Wang Y, Miller RD (1988) J Org Chem 53:2127
54. Pichko VA, Simkin BYa, Minkin VI (1987) Dokl Akad Nauk SSSR 292:910
55. Pichko VA, Simkin BYa, Minkin VI (1989) THEOCHEM (in press)
56. Komornicki A, McIver JW (1974) J ACS 96:5798
57. Lewis KE, Steiner H (1964) J Chem Soc :3080; Marvell EN, Caple G, Senatz B (1965) Tetrahedron Lett :385
58. Simkin BY, Pichko VA, Minkin VI (1988) Zh Org Khim 24:1569
59. Simkin BY, Makarov SP, Minkin VI (1982) Khim Heterocycl Soed N8:1028
60. Simkin BY, Makarov SP, Minkin VI (1984) Khim Heterocycl Soed N6:747
61. Dewar MJS, Kirschner S (1975) J ACS 97:2931
62. Frey HM, Stevens IDR (1969) Trans Farad Soc 61:90
63. Shevlin P, McKee ML (1988) J ACS 110:1666
64. Budzelaar PHM, Kraka E, Cremer D, Schleyer PR (1986) J ACS 108:561
65. Bouma WJ, Vincent MA, Radom L (1978Int J Quant Chem 14:767
66. Rodwell WR, Bouma WJ, Radom L (1980) Int J Quant Chem 18:107
67. Rosenstock HM, Draxl K, Steiner BW, Herron JT (1977) J Phys Chem Ref Data (Suppl I) 6:1
68. Gajewski JJ, Conrad ND (1978) J ACS 100:6268
69. Roth WR, König J (1966) Liebigs Ann Chem 699:24
70. Hess BA, Schaad LJ (1983) J ACS 105:7185
71. Jensen F, Houk KN (1987) J ACS 109:3139
72. Dewar MJS, Healy EF, Ruiz JM (1988) J ACS 110:266
73. Dewar MJS, Merz KM, Stewart JJP (1985) J Chem Soc Chem Commun :166
74. Minyaev RM, Yudilevich IA, Minkin VI (1986) Zh Org Khim 22:19
75. Kessler H, Feigel M (1982) Acc Chem Res 15:2
76. Anh NT, Flian M, Hoffmann R (1978) J ACS 100:110
77. Albright TA, Hofmann P, Hoffmann R, Lillya CP, Dobosh PA (1983) J ACS 105:3396
78. Minkin VI, Minyaev RM, Zhdanov YA (1987) Nonclassical structures of organic compounds. Mir, Moscow
79. Minkin VI, Minyaev RM (1982) Uspekhi Khim (Russ Chem Rev) 51:586
80. Wade K (1975) Chem Brit 11:177
81. Mingos DMP (1984) Acc Chem Res 17:311
82. Rudoph R (1976) Acc Chem Res 9:446
83. Teo BK (1985) Inorg Chem 24:4209
84. Childs RF (1982) Tetrahedron 38:567
85. Minkin VI, Minyaev RM (1979) Zh Org Khim 15:1569
86. Minyaev RM, Minkin VI (1982) Zh Org Khim 18:2009
87. Minkin VI, Minyaev RM, Orlova GV (1984) THEOCHEM 110:241
88. Ustynuk YA, (1982) Vestn MGU (ser 2 Khimia) 23:605
89. Gerson F, Huber W, Müllen K (1978) Angew Chem 90:216
90. Salem L (1982) Electrons in chemical reactions: first principles. Wiley-Interscience, New York

91. Glukhovtsev MN, Simkin BYa, Minkin VI (1985) Uspekhi Khim (Russ Chem Rev) 54:86
92. Bally T, Masamune S (1980) Tetrahedron 36:343
93. Kollmar H, Carrion F, Dewar MJS, Bingham RC (1981) J ACS 103:5292
94. Bock H, Roth B, Maier G (1984) Chem Ber 117:172
95. Bauld NL, Bellville DJ, Pabon R, Chelsky R, Green G (1983) J ACS 105:2378
96. Chanon M (1982) Bull Soc Chim France 2:197
97. Bellville DJ, Chelsky R, Bauld NL (1982) J Comput Chem 3:548

List of Abbreviations

ab initio	nonempirical quantum mechanical method of calculation
AM1	Austin model 1
AO	atomic orbital
CEPA	coupled electron-pair approximation
CI	configuration interaction
CNDO	complete neglect of differential overlapping
DZ	double-zeta basis
EHMO	extended Huckel method
GTO	Gaussian-type orbital
HOMO	highest occupied molecular orbital
INDO	intermediate neglect of differential overlapping
IRC	intrinsic reaction coordinate
LCAO	linear combination of atomic orbitals
LUMO	lowest unoccupied molecular orbital
MC	Monte Carlo method
MCI	multiconfiguration interaction
MD	molecular dynamic method
MEP	molecular electrostatic potential
MERP	minimal energy reaction path
MINDO	modified INDO
MNDO	modified neglect of diatomic overlapping
MO	molecular orbital
MPN ($N = 2 \div 4$)	Moller–Plesset perturbation theory of order $2 \div 4$
NDDO	neglect of diatomic differential overlapping
PES	potential energy surface
PLM	principle of least-motion
PRDDO	partial retention of diatomic differential overlapping
RHF	restricted Hartree–Fock method
SCF	self-consistent field
SET	single-electron transfer
STO	Slater-type orbital
UHF	unrestricted Hartree–Fock method
ZDO	zero differential overlap

Subject Index